Artificial Intelligence Using Federated Learning

Federated machine learning is a novel approach to combining distributed machine learning, cryptography, security, and incentive mechanism design. It allows organizations to keep sensitive and private data on users or customers decentralized and secure, helping them comply with stringent data protection regulations like GDPR and CCPA.

Artificial Intelligence Using Federated Learning: Fundamentals, Challenges, and Applications enables training AI models on a large number of decentralized devices or servers, making it a scalable and efficient solution. It also allows organizations to create more versatile AI models by training them on data from diverse sources or domains. This approach can unlock innovative use cases in fields like healthcare, finance, and IoT, where data privacy is paramount.

The book is designed for researchers working in intelligent federated learning and its related applications, as well as technology development, and is also of interest to academicians, data scientists, industrial professionals, researchers, and students.

Intelligent Manufacturing and Industrial Engineering

Series Editor: Ahmed A. Elngar, Beni-Suef University, Egypt.
Mohamed Elhoseny, Mansoura University, Egypt

For more information about this series, please visit: www.routledge.com/Mathematical-Engineering-Manufacturing-and-Management-Sciences/book-series/CRCIMIE

Artificial Intelligence Using Federated Learning

Fundamentals, Challenges, and Applications

Edited by Ahmed A. Elngar, Diego Oliva
and Valentina E. Balas

CRC Press
Taylor & Francis Group
Boca Raton London New York

CRC Press is an imprint of the
Taylor & Francis Group, an **informa** business

Designed cover image: Shutterstock—luchschenF

First edition published 2025
by CRC Press
2385 NW Executive Center Drive, Suite 320, Boca Raton FL 33431

and by CRC Press
4 Park Square, Milton Park, Abingdon, Oxon, OX14 4RN

CRC Press is an imprint of Taylor & Francis Group, LLC

ISBN: 978-1-032-77164-9 (hbk)
ISBN: 978-1-032-77246-2 (pbk)
ISBN: 978-1-003-48200-0 (ebk)

DOI: 10.1201/9781003482000

Typeset in Times
by Apex CoVantage, LLC

Contents

Preface

ARTIFICIAL INTELLIGENCE USING FEDERATED LEARNING: FUNDAMENTALS, CHALLENGES, AND APPLICATIONS

The rapid evolution of artificial intelligence (AI) has reshaped various aspects of our world, from healthcare to finance, and transportation to communication. Among the latest advancements, federated learning (FL) stands out as a revolutionary paradigm, enabling decentralized AI model training across multiple devices or organizations while preserving data privacy and security. This book, *Artificial Intelligence Using Federated Learning: Fundamentals, Challenges, and Applications*, aims to provide a comprehensive understanding of this cutting-edge technology, bridging the gap between theoretical foundations and practical implementations.

The inspiration for this book arises from the increasing need for privacy-preserving AI solutions in an era marked by growing concerns over data security and confidentiality. Traditional centralized AI models often require extensive data aggregation, posing significant risks of data breaches and privacy violations. Federated learning addresses these challenges by allowing models to be trained locally on edge devices, sharing only the necessary updates to a central server, thus ensuring that raw data remains secure and private.

This book is structured to cater to a diverse audience, including researchers, practitioners, and students. It is divided into three main sections:

1. **Fundamentals:** This section introduces the core concepts of AI and federated learning. We delve into the theoretical underpinnings of FL, explaining its architecture, key algorithms, and protocols. Readers will gain a solid foundation in the principles that make federated learning a viable and attractive approach to modern AI development.
2. **Challenges:** Despite its promising potential, federated learning faces several challenges that need to be addressed to fully harness its capabilities. This section explores issues such as data heterogeneity, communication efficiency, model accuracy, and security concerns. We discuss ongoing research and emerging solutions aimed at overcoming these obstacles, providing insights into the future directions of federated learning.
3. **Applications:** The final section showcases real-world applications of federated learning across various domains. Case studies and practical examples illustrate how FL is being utilized to solve complex problems in healthcare, finance, IoT, and beyond. Readers will learn about the benefits and limitations of FL in practice, along with strategies for successful implementation.

As the field of federated learning continues to evolve, this book serves as both a foundational text and a reference for advanced study. We hope to inspire and equip our readers with the knowledge and tools necessary to contribute to this exciting and impactful area of AI research and application.

We would like to express our gratitude to the many researchers and practitioners whose pioneering work in federated learning has paved the way for this book. Their contributions have been invaluable in shaping the content and direction of this work. We also extend our thanks to our colleagues, reviewers, and readers for their support and feedback.

We invite you to embark on this journey into the world of artificial intelligence and federated learning, exploring the fundamentals, confronting the challenges, and discovering the vast potential applications. May this book serve as a guide and inspiration in your endeavors within this dynamic and transformative field.

About the Editors

Dr. Ahmed A. Elngar is an associate professor and Head of the Computer Science Department at the Faculty of Computers and Artificial Intelligence, Beni-Suef University, Egypt. Dr. Elngar is also an associate professor of computer science at the College of Computer Information Technology, American University in the Emirates, United Arab Emirates. Also, Dr. AE is Adjunct Professor at School of Technology, Woxsen University, India. Dr. AE is the Founder and Head of the Scientific Innovation Research Group (SIRG). Dr. AE is a director of the Technological and Informatics Studies Center (TISC), Faculty of Computers and Artificial Intelligence, Beni-Suef University. Dr. AE has more than 150 scientific research papers published in prestigious international journals and over 35 books covering such diverse topics as data mining, intelligent systems, social networks, and smart environment. Dr. AE is a collaborative researcher. He is a member of the Egyptian Mathematical Society (EMS) and International Rough Set Society (IRSS). His other research areas include the Internet of Things (IoT), network security, intrusion detection, machine learning, data mining, and artificial intelligence. big data, authentication, cryptology, healthcare systems, and automation systems. He is an editor and reviewer of many international journals around the world. Dr. AE has won several awards, including the Young Researcher in Computer Science Engineering from the Global Outreach Education Summit and Awards 2019, as well as the Best Young Researcher Award (Male) (Below 40 years), Global Education and Corporate Leadership Award.

Dr. Diego Oliva In 2007, he obtained an electronics and computer engineering degree from the Centro de Enseñanza Técnica Industrial (CETI), the Industrial Technical Education Center (CETI) of Guadalajara, Mexico, and an MSc in electronic engineering and computer sciences from the Universidad de Guadalajara, Mexico in 2010. In 2015, he obtained a PhD in informatics from the Universidad Complutense de Madrid (UCM) in Spain. Since 2008, he has focused his research on developing, implementing, and designing metaheuristic algorithms. He has published more than 100 papers in international journals on topics related to optimization and its implementations. Since 2015, he has been a member of the Academia Mexicana de Computacion (AMEXCOMP), and since 2022, he has been a senior member of the Institute of Electrical and Electronics Engineers (IEEE). Since 2017, he has been a member of the Sistema Nacional de Investigadoras e Investigadores (SNII) in Mexico. In 2022, he obtained the distinction of Highly Cited Researcher by Clarivate-Web of Science. He is among the 2% most influential researchers worldwide, according to a report published by Stanford University and Elsevier in 2023. He has been the editor and author of several books at international publishing houses, and he is the associate editor and guest editor for several specialized journals with high impact factors. He is currently a professor and researcher at the Universidad de Guadalajara (CUCEI). He also collaborates with Mexican and foreign universities in several research projects. His main research interests are artificial intelligence, metaheuristic optimization algorithms, multiobjective optimization, parameter estimation in engineering, and image and digital signal processing.

Prof. Valentina E. Balas is currently Full Professor in the Department of Automatics and Applied Software at the Faculty of Engineering, "Aurel Vlaicu" University of Arad, Romania. She holds a PhD in applied electronics and telecommunications from Polytechnic University of Timisoara. Dr. Balas is the author of more than 300 research papers in refereed journals and international conferences. Her research interests are in intelligent systems, fuzzy control, soft computing, smart sensors, information fusion, modeling, and simulation. She is the Editor-in Chief of the *International Journal of Advanced Intelligence Paradigms* (IJAIP) and *International Journal of Computational Systems Engineering* (IJCSysE), member of the editorial board and member of several national and international journals, and an evaluator expert for national and international projects and PhD theses. Dr. Balas is the director of the Intelligent Systems Research Centre in Aurel Vlaicu University of Arad and Director of the Department of International Relations, Programs and Projects at the same university. She served as General Chair of the International Workshop Soft Computing and Applications (SOFA) in eight editions, 2005–2018, held in Romania and Hungary. Dr. Balas participated in many international conferences as organizer, honorary chair, session chair and member in steering, advisory or international program committees. She is a member of EUSFLAT and SIAM; a senior member of IEEE, and a member in TC–Fuzzy Systems (IEEE CIS), TC–Emergent Technologies (IEEE CIS), and TC–Soft Computing (IEEE SMCS). Dr. Balas was past Vice-President (Awards) of the IFSA International Fuzzy Systems Association Council (2013–2015) and is a joint secretary of the Governing Council of the Forum for Interdisciplinary Mathematics (FIM), A Multidisciplinary Academic Body, India. She is also director of the Department of International Relations, Programs and Projects and head of the Intelligent Systems Research Centre at Aurel Vlaicu University in Arad, Romania.

Contributors

Valli Bhasha Achukatla
Electronics and Communication
  Engineering Department
KSRM College of Engineering
Kadapa, Andhra Pradesh, India

Muhammad Mujtaba Asad
Sukkur IBA University
Sukkur, Pakistan

Luís Cardoso
Polytechnic Institute of Portalegre and
  Centre for Comparative Studies of
  the University of Lisbon
Lisbon, Portugal

Susanta Das
Ajeenkya D Y Patil University
Pune, Maharashtra, India

Jagjit Singh Dhatterwal
Department of Artificial Intelligence
  and Data Science
Koneru Lakshmaiah Education
  Foundation
Vaddeswaram, Andhra Pradesh, India

Abdelaziz El-hoshoudy
Egyptian Petroleum Research Institute
Nasr city, Cairo, Egypt

Ahmed A. Elngar
Faculty of Computers and Artificial
  Intelligence
Beni-Suef University
Beni-Suef City, Egypt

Utpal Ghosh
Department of Computer Science
Sarojini Naidu College for Women
Kolkata, West Bengal, India

Monali Gulhane
Symbiosis Institute of Technology
Symbiosis International (Deemed
  University)
Pune, Maharashtra, India

Nidhi Gupta
Department of Computer Science and
  Engineering
School of Engineering and Technology
Sharda University
Greater Noida, Uttar Pradesh, India

Walaa Hassan
Faculty of Computers & Informatics
Suez Canal University
Ismailia, Egypt
Faculty of Computer Science
Misr International University
Cairo, Egypt

Anupriya Jain
Professor School of Computer
  Applications MRIIRS Faridabad
New Delhi, New Delhi, India

Vamsi Krishna Kadiri
Electronics and Communication
  Engineering Department
KSRM College of Engineering
Kadapa, Andhra Pradesh, India

Balaram Yadav Kasula
University of The Cumberlands
Williamsburg, Kentucky, USA

Kuldeep Singh Kaswan
School of Computer Science and
  Engineering
Galgotias University
Greater Noida, India

Pavan Kumar Kattela
Electronics and Communication
  Engineering Department
KSRM College of Engineering
Kadapa, Andhra Pradesh, India

Daksh Khurana
Department of Computer Science &
  Engineering
Symbiosis Institute of Technology
  (SIT)
Pune, Maharashtra, India

Binesh Kumar
Research Scholar
Department of Chemistry
Guru Jambheshwar University of
  Science and Technology
Haryana, India

Shrabanti Kundu
Department of Computer Science and
  Engineering
University of Kalyani
Kalyani, West Bengal, India

Kiran Malik
Department of Computer Science and
  Engineering
Matu Ram Institute of Engineering and
  Management
Rohtak, Haryana, India

Pratap Singh Malik
Assistant Professor
Computer Science and Engineering
Guru Jambheshwar University of
  Science and Technology
Haryana, India

Ali Said Al Matari
A'Sharqiyah University
Oman
IAIN Curup
Sumatera, Indonesia

Appalaraju Muralidhar
Assistant Professor
Computer Science Engineering
VIT
Chennai, India

Thangavel Murugan
Assistant Professor
Department of Information Systems and
  Security
College of Information Technology
University of United Arab Emirates
Al Ain, United Arab Emirates

Muthmainnah
Universitas Al Asyariah Mandar
Polewali Mandar, Sulawesi Barat,
  Indonesia

Manjushree Nayak
Associate Professor
Department of Computer Science and
  Engineering
NIST University
Berhampur, Odisha, India

Akkarapon Nuemaihom
Buriram Rajabhat University
Buriram, Buriram, Thailand

Ahmad J. Obaid
University of Kufa
Najaf, Iraq

Debasish Padhi
Department of Computer Science and
  Engineering
NIST University
Berhampur, Odisha, India

Nitin Rakesh
Symbiosis Institute of Technology
Symbiosis International (Deemed
  University)
Pune, Maharashtra, India

Rathipriya Ramalingam
Assistant Professor
Department of Computer Science
Periyar University
Salem, India

Raveendra Reddy
Assistant Professor
Computer Science Engineering
Vignan University
Guntur, India

Habiba Mohamed
Faculty of Computer Science
Misr International University
Cairo, Egypt

Atharva Saraf
Ajeenkya D Y Patil University
Pune, Maharashtra, India

Prodhan Mahbub Ibna Seraj
American International University
 Bangladesh (AIUB)
Dhaka, Dnaka, Bangladesh

Sivabalan Settu
Assistant Professor
Computer Science Engineering
Vignan University
Guntur, India

Seema Sharma
Professor
School of Computer Applications
MRIIRS
New Delhi, New Delhi, India

Khushwant Singh
Research Scholar
Department of Computer Science and
 Engineering
University Institute of Engineering and
 Technology
Maharshi Dayanand University
Haryana, India

Yudhvir Singh
Professor
Department of Computer Science &
 Engineering
University Institute of Engineering &
 Technology
Maharshi Dayanand University
Haryana, India

Vikas Siwach
Assistant Professor
Computer Science and Engineering
UIET, MDU
Rohtak, India

Shaurya Sameer Talewar
Ajeenkya D Y Patil University
Pune, Maharashtra, India

Khushbu Trivedi
Ajeenkya D Y Patil University
Pune, Maharashtra, India

M. Venkatanarayana
Department of Electronics and
 Communication Engineering
KSRM College of Engineering
Kadapa, Andhra Pradesh, India

Pawan Verma
Department of Computer Science and
 Engineering
School of Engineering and
 Technology
Sharda University
Greater Noida, Uttar Pradesh,
 India

Arkas Viddy
Universitas Al Asyariah Mandar
Politeknik Negeri Samarinda,
 Kalimantan, Indonesia

Idi Warsah
Institut Agama Islam Negeri Curup
Curup, Sumatera, Indonesia

Pawan Whig
Vivekananda Institute of Professional
  Studies-TC
New Delhi, New Delhi, India

Mohit Yadav
Assistant Professor
Department of Mathematics
University Institute of Sciences
Chandigarh University
Mohali, India

Ramesh Kumar Yadav
Assistant Professor
Department of Data Science
Christ University
Bengaluru, India

Ahmad Al Yakin
Universitas Al Asyariah Mandar
Polewali Mandar, Sulawesi Barat,
  Indonesia

Nikhitha Yathiraju
University of The Cumberlands
Williamsburg, Kentucky, USA

Syed Zahiruddin
Department of Electronics and
  Communication Engineering
KSRM College of Engineering
Kadapa, Andhra Pradesh, India

1 Federated Learning
Overview, Challenges, and Ethical Considerations

Jagjit Singh Dhatterwal, Kiran Malik, Kuldeep Singh Kaswan, and Ahmed A. Elngar

INTRODUCTION TO MACHINE LEARNING

Machine learning (ML) is a subfield of artificial intelligence. The objective of ML is to exhibit to PCs new errands by breaking down existing information; along these lines, they might make forecasts or assessments all alone, with next to no assistance from people. The expression "AI" alludes to an umbrella term for different ways to deal with information investigation, design identification, and knowledge creation (Mitchell, 1997). Konečný et al. (2016) state that concentrated models are generally utilized by conventional AI draws near. To prepare and fabricate these models, conglomerating information from many sources and storing it in a focal area is fundamental.

The most common way of preparing models utilizing information gathered from a few hubs or sources and shipped off a solitary server is known as incorporated learning (Yang et al., 2019). Many individuals are stressed over this methodology, particularly about information security. The concentrated stockpiling of touchy information improves the probability of safety breaks. Also, while working with colossal data sets or information spread across numerous areas, brought together advancing frequently faces adaptability issues (Kairouz et al., 2019). Incorporated learning faces this impediment.

An option in contrast to the limits of unified learning has arisen with the coming of united learning (FL). This objective is achieved by remembering cooperative model preparation for scattered gadgets, as expressed by Bonawitz et al. (2017). The exploration of Yang et al. (2019) recommends that to make Combined Learning (FL) work better, information should be put away on nearby gadgets like cell phones, IoT gadgets, or edge servers. This strategy permits these contraptions to learn together while yet safeguarding buyer information. In contrast with unified techniques, decentralization eases tension in information transport networks while safeguarding the protection of clients' singular data (Kairouz et al., 2019).

In combined learning, client gadgets do calculations on neighborhood information and afterward send any refreshed models to a focal server or aggregator, as expressed by Bonawitz et al. (2017). Yang et al. incorporate these progressions into a

DOI: 10.1201/9781003482000-1

total model that shields client protection while gathering information from different sources in their 2019 exploration.

LITERATURE SURVEY

Unified Learning (FL) permits medical services associations to prepare models cooperatively using delicate clinical information while safeguarding patients' protection to an outrageous degree (Sheller et al., 2020). Florida (FL) has a few purposes in various fields, one of which is medication. Monetary associations have likewise researched information-secure forecast models utilizing united learning approaches. Sending crude information to unified servers in an IoT ecosystem isn't needed; edge gadgets might gain from another and make adjustments depending on the situation. Unified Learning (FL), first recommended in 2019 by Liu et al., makes this practical.

It is standard practice to utilize strategies like differential classification all through the learning stage to keep the model precise while safeguarding the protection of individual data sources (Shokri et al., 2015). The field of Unified Learning (FL) keeps on focusing on the assurance of clients' very own data. Moreover, unified learning (FL) frameworks frequently utilize safe conglomeration systems and encryption strategies to ensure the mystery of model changes during transmission (Bonawitz et al., 2017).

Similarly, as with each innovation, FL has its extraordinary arrangement of issues. As per concentrates finished by Li et al. in 2020, gadget heterogeneity is a significant impediment to arriving at an overall model union. This term suggests aberrations in information conveyance and handling power between gadgets. As per Hard et al., unified learning frameworks may not work as proficiently when confronted with correspondence issues like drowsy associations or confined transfer speed.

Konečný et al. (2016) state that there is constant work to further develop calculation correspondence proficiency, reinforce models to endure antagonistic attacks, and take care of the issues of security precision balance. There is a consistent expansion in the field's leap forward, and progressing research in unified learning is currently centred around specific points. Research on united strategies for advancement that permit contribution in decentralized learning while at the same time holding model proficiency is presently in progress by Smith et al. (2021).

UNDERSTANDING FEDERATED LEARNING

An inventive way to deal with AI is unified learning, which takes into account the solid and scattered creation of models on different edge devices (Yang et al., 2019). Combined learning is another method that is just entering the AI crowd. Devices such as cell phones, Web of Things (IoT) devices, and edge servers are working together to create a joint model without moving natural data (Kairouz et al., 2019). Coordinated efforts and shared information are the foundations of the decentralized learning worldview. The creation of AI models using decentralized information sources that guarantee information obfuscation was proposed by Google in a significant report in 2017 (Bonawitz et al., 2017).

This effort is responsible for the advancement of unified learning. The inherent advantages of this field have led to a huge expansion of innovative work aimed at exploring its potential uses in many areas (Kaswan et al., 2022a).

According to Li et al. (2020), the core idea of unified learning is to decentralize information processing so that models can be created on any teaming-up gadget or hub. According to Mohassel and Zhang (2017), privacy and security concerns are sufficiently addressed by storing sensitive customer information on the devices that create it instead of sending it to a central server. Konečný et al. (2016) state that new AI computations, strategies for mediation, and security-saving frameworks have contributed to the further development of combined learning. By integrating unified learning into a grounded system, this development effectively nullifies the disadvantages of concentrated learning. Unified learning is a significant advantage in situations where the protection of sensitive data is essential. Information breaches and unlawful access to individual data are less likely with this approach, as Yang et al. (2019) show.

United Learning smoothes the cycle by allowing models to be created at unique points. According to Hard et al. (2018), this strategy improves the performance and adaptability of AI models while simplifying the handling of different data sets distributed across multiple devices. The problems caused by information storage and policies could be better solved with the help of combined learning.

Table 1.1 Shows the summary of the federated Learning.

TABLE 1.1

History of Federated Learning

Year	Milestones and Key Developments in Federated Learning
2016	Google Brain introduces federated learning, aiming to enable on-device machine learning.
2017	Initial research publications on federated learning emphasize privacy-preserving aspects.
2018	Advancements in differential privacy techniques for federated learning emerge in research.
2019	Google releases federated learning Frameworks to facilitate decentralized model training.
2020	Exploration of secure and communication-efficient protocols for federated learning begins.
2021	Federated Learning applications expand into healthcare, IoT, finance, and telecommunications.
2022	Ongoing research focuses on federated transfer learning and lifelong federated learning.
2023	Innovations in federated meta-learning and adaptive aggregation techniques gain attention.
2024	Integration of federated learning with edge computing and federated reinforcement learning.
Present	Continued advancements in privacy-preserving methods and broader adoption in various domains.

KEY COMPONENTS OF FEDERATED LEARNING

Recently, a creative and free strategy for learning and teaching has developed. Achieving the ideal outcome depends on the targeted activities of some important elements (Kaswan et al., 2022b). At the center of consideration are client devices that often use edge registration. These devices can perform computations while in proximity. A wide range of electronic devices can utilize their information to create bounded models; this includes everything from cell phone information to IoT reviews. This technique limits the ability to share information with other devices. As described by Yang et al. (2019), each device is part of the organization and maintains its model nearby by sending adjustments to the central server, which in some cases is called an aggregator. The process of model accumulation is controlled by the aggregator, which can be set up on-premise or in the cloud. For this purpose, the framework collects and combines updates from numerous client devices. Bonawitz et al. (2020) refer to a combination of authors.

To guarantee a protected and smooth connection between the client's hardware and the specialist, Unified Learning utilizes the best correspondence advancements and security strategies. These innovations not only facilitate the exchange of model limits and updates, but likewise ensure the security of client information and data (Dhatterwal et al., 2022a). It is fundamental to use encryption techniques such as differential security and the interplay of learning and explicit secure accumulation to ensure the protection of information during model extraction. McMahan, H. B. and others authored the 2017 paper. To guarantee that nearby models match the focal PC and to diminish the probability of information mistakes, unified averaging is a great approach for correspondence security. According to Konečný et al. (2016), it is critical for the creation of neighborhood models that edge client gadgets have reasonable computational and capacity limits.

The research by Nishio and Yonetani designs an edge-driven method that effectively reduces idle state by reducing the repetition of information transmission to a focal server. By storing the basic data on the device, this system also increases security (Kaswan et al., 2021). All model updates from the devices can be merged and collected on the focal server. It also acts as an operational hub for collecting information and creating thoughts. As Yang et al. (2019) point out, it is possible to create a global model by merging knowledge data from different provincial data sets into a common data set.

To function appropriately, Unified Learning relies on several correspondence components. The business offers arrangements that enable client gadgets and the focal specialist to exchange information and synchronize models. The protection of information originating from individual devices can be ensured through the use of a shared secure total. This ensures that changes to the model can be incorporated securely. The 2017 paper was authored by McMahan, H. B., et al. Protecting clients' information gathering by concealing their data is one of the many correspondence-related uses for differential security measures. According to Bonawitz et al. (2020), the research refers to various authors. The security and well-being of the Unified Educational Experience is ensured through the execution of these aggregate safeguards.

HOW FEDERATED LEARNING WORKS

Helpful preparation of models across multiple edge gadgets or clients is enabled by Combined Averaging Calculation, a decentralized rationalization approach. For Combined Figuring out how to work, this system should be followed. In this technique, the model is refreshed locally before it is sent to a focal server for shuffling, which utilizes the neighborhood information to prepare the gadgets. Specialists found that the gadgets involved in the review made changes after getting the accumulated model and sending it back. The secrecy of the information is ensured by the self-reproducing learning system of this system (Dhatterwal et al., 2022).

Numerous means are utilized in decentralized model preparation. Deciding which gadgets are useful is the most important component. Each gadget puts away a specific data set. Rest assured that all data stored on these gadgets is kept secret, even if they could benefit from their information. The preferences of the learned models are sent after preparing interaction with the fundamental PC. It is the duty of the focal server to ensure that all gadgets update and gather these elements at the same time (Yang et al., 2019).

To create a global model, Unified Gaining uses the pooling technique to merge data from different devices. The use of a weighted typical method is common. The system integrates the device feedback, the objective result of the model, and other indicated measurements into the boundaries of the model.

The synchronization of subsequently accumulated information across all gadgets is important for the model refresh cycle in United Learning. From the main PC, each gadget in the process receives the latest adaptation of the worldwide model. The devices can then utilize the acquired information to work on their neighborhood models. As described by Kairouz et al. (2019), the iterative system plans to improve the execution and understanding of the model through repeated patterns of model selection, adaptation, and rearrangement.

The Unified Averaging Calculation tracks down the world's best arrangement by merging appropriate model updates with neighborhood calculations. As outlined in the 2020 study by Li et al., the understanding shown is achieved after a comprehensive conversation and the formation of a point-by-point model. Through this process, people can participate in cooperative learning while ensuring the classification of information collected from each device.

It is mandatory to create models openly and straightforwardly to reduce security issues related to the consistent maintenance and handling of information. Consequently, sensitive customer data is not yet stored on the devices in the customer area.

The acquisition systems utilized in unified learning influence the viability and exactness of the worldwide model. Exploring different acquisition methods should reduce biases and ensure a fair evaluation of the contributions of different devices. Next, some models are introduced: fundamental averaging, weighted averaging, and discrimination-based collection of protection data.

United Learning uses model updating methods to ensure that model boundaries remain reliable across multiple devices, regardless of whether those devices have different information boundaries and processing capacities. Unified Averaging with Force (FedAvgM) and Government Stochastic Angle Plunge (FedSGD) are two strategies used in appropriate detection to increase combination and robustness.

The central PC in Unified Advancing not only collects information but also manages various other functions. It controls the organizational connections of devices, monitors the progress of model updates, and ensures that the global model is accurate and predictable. As Yang et al. (2022) point out, the server simplifies the execution of United Advancing by organizing the compilation and appropriation of models for each of the affected devices.

Since combined learning is iterative, it includes some patterns of nearby preparation and worldwide gathering. As a result, the representation of the model can get better and better. According to Konečný et al. (2016), the worldwide model works on its representation with each cycle by utilizing different sources of information and consolidating the collected knowledge while ensuring the protection of the information.

Two protection techniques, secure collection, and differential security are stated by United Figuring to guarantee that delicate information stays secret throughout the model conglomeration. According to Mohassel and Zhang (2017), these strategies ensure that the merged model adheres to all guidelines for protecting individuals' data.

Due to its characteristic freedom, Combined Learning can continue with the tasks regardless of whether there are difficulties with the network or defective hardware. The collaborative cycle will continue in any case, even if devices fail or companies experience delays, as a 2020 study by Smith et al. shows. To achieve this goal, it could be very useful to include more real devices in the model accumulation process. Flexibility and resilience to disappointment are certainly of great importance in this context.

United Learning can effectively solve issues brought on by the absence of freedom and personality confirmation in information, changes in gadget types, and heterogeneous information circulation by utilizing enhancement options. Weighted inspection, customizable learning rates, and model transformation are among the techniques recommended by Li et al. (2020) to mitigate the effects of these problems. At the end of the day, these strategies lead to an easier demonstration of blending and a more generally talked about execution.

The features and pace of unification of unified learning procedures are the current focus of assembly research. The examination by Dinh et al. (2020) gives insight into how far the gadgets are occupied with cooperative improvement. The outcomes demonstrate that the technique is exact and dependable.

Unified learning has long struggled with the problem of productively dealing with the proliferation of non-ID information. Some research focuses on effectively addressing cooperative learning calculations that can deal with non-delegated information designs across different devices to address this issue. Yang et al. (2022) found that these procedures act on the trustworthiness and generalizability of the prepared models.

It is undoubtedly conceivable to circumvent the limits of performance and speed that can be encountered when utilizing combined learning in edge-figure environments. For the advancement of edge insights, combined learning is fundamental to streamline model preparation and updating on low-capital devices (Gao et al., 2021). We focus on expanding the utilization of existing PC and transmission innovations to achieve this goal.

To address the security issues with United Learning, we are looking for areas of strength. The utilization of harmful strategies in models and the portrayal of byzantine practices are two conspicuous examples of these limitations in practice. To make United Learning frameworks more secure, specialists are examining various methods. The aforementioned parts incorporate conventions for safe amalgamations, calculations that can survive Byzantine shortcomings, and differential protection-based shields.

The Combined Learning area is currently researching novel methods such as versatile collection and unified meta-learning. The goal is to improve model combination and execution while dealing with different gadgets and non-IID information. The goal of these complex pooling approaches is to work on cooperative learning in cluttered, information-rich environments.

Unified supportive learning, constant combined learning, and unified moving learning are dynamic areas of research. The commitment to collaborative learning is rapidly advancing the progress of these projects. According to researchers, the goal of these updates is to make cooperative methods relevant to a broader range of learning circumstances and certifiable challenges (e.g., Chen et al., 2023).

The stages involved in a typical FL process as shown in Figure 1.1:

- Initialization: The allocation of a centrally run first global model with random, initial weights is done by a central server.
- Client Selection: The server chooses a group of available equipment (participants) to be in the learning round among the various equipment that present themselves.
- Model Distribution: The server delivers the current global model's weight to clients which was the one chosen.
- Local Training: The client will proceed to perform the training using their local data and subsequently get the updated model weights.
- Model Updates: Through sending once-updated model weights to our server, clients perform this function.
- Aggregation: The server's job is to combine these updates normally, by averaging, to correct the global model.
- Model Evaluation: The server uses the global model to appraise it to figure out its efficiency.
- Iteration: After steps 2 to 7 are performed, the process is repeated until the model converges or the desired performance state is attained.

Algorithm 1: Federated Learning Algorithm

Data: Client devices $\{C_1, C_2, \ldots, C_n\}$, Central server S, Global model
        parameters θ
Result: Updated global model parameters θ
1 Initialization: Randomly initialize global model parameters θ ;
2 **for** *each communication round* $t = 1, 2, \ldots, T$ **do**
3   S broadcasts θ to all client devices;
4   **for** *each client* C_i **do**
5   C_i receives θ from S and trains a local model w_i using its own
        data;
6        $w_i \leftarrow$ LocalTraining(C_i, θ) ;
7   **end**
8     $W \leftarrow \{w_1, w_2, \ldots, w_n\}$;
9     $\theta \leftarrow$ Aggregate(W) ;
10   S receives θ from clients and updates the global model ;
11 **end**

Functions:
Function LocalTraining(C_i, θ):
    Data: Client C_i, Global model parameters θ
    Result: Trained local model parameters w_i
12   Perform local training on client C_i using its own data and θ $w_i \leftarrow$
    LocalUpdate(C_i, θ) **return** w_i

Function Aggregate(W):
    Data: Local models $\{w_1, w_2, \ldots, w_n\}$
    Result: Updated global model parameters θ
13   Aggregate local models using averaging or other aggregation methods
    $\theta \leftarrow \frac{1}{n} \sum_{i=1}^{n} w_i$ **return** θ

FIGURE 1.1   federated learning Algorithm

CHALLENGES AND LIMITATIONS

The security of information is crucial in the field of combined learning, as the information is only processed in a few places. It is essential to develop and apply sound protection strategies while acquiring and maintaining models to reduce the likelihood of information breaches and security violations.

The extra effort required to design connections is a major issue that arises when multiple devices participate in correspondence. Expanded correspondence costs and potential breaks in fixing issues could result from the standard exchange of model changes between the main PC and other devices. This could affect the student's data security, as indicated by an examination.

In situations where there are not many devices or the networks are not suitable for handling many information movements, there may be capacity issues in data transfer when transferring model updates between devices. According to Konečný et al. (2016), managing models with enormous scale or a large number of devices causes ongoing processes to stall. As a result, model synchronization becomes less productive and reliable, which increases the time required for interaction.

Since combined learning uses a wide range of devices and information sources, the execution is more complicated. Differences in processing capacity, information dispersion, and the steadfastness of gadgets could make it a difficult endeavor to coordinate information from many sources while keeping up the precision and impartiality of the prepared model (Karimireddy et al., 2021).

Regardless of whether encryption and other measures are utilized to improve protection in general, concerns about security and information insurance continue to emerge. Xie et al. (2023) emphasize the necessity of great strength areas for executing systems and secure association methods to shield the united profits from malicious endeavors like model harm or supposition assaults.

It is fundamental to execute model printing strategies and advance correspondence conventions to diminish information transfer rates between gadgets and the main PC. It is fundamental that you think about the transmissions by performing the expected steps. Quantization and differential pressure are two of the techniques used to enhance the cost-effectiveness of correspondence without compromising model execution).

It is fundamental to utilize imaginative common learning strategies that can effectively oversee fluctuating information designs and unmistakable qualities since gadgets and information indices are noteworthy by nature. Modeling the unification cycle as well as execution in different environments could be improved by effectively answering device factors through the use of versatile overall methods and customized combined learning calculations).

APPLICATIONS OF FEDERATED LEARNING

With the introduction of Combined Learning, an advanced strategy that enables the collaborative preparation of models while effectively regulating protection issues, medical care and testing in medicine has fundamentally changed. In addition, by utilizing diverse patient data, medical institutions can gain significant experience in protecting individual information about patients. By decentralizing the gathering of information and simplifying the development of expectation models, this goal is achievable.

HEALTHCARE AND MEDICAL RESEARCH

For healthcare and clinical research organizations, this is a top priority since it enables collaborative AI model training without compromising patient privacy. This technology makes it possible for medical organizations, research centers, and pharmaceutical companies to develop AI models jointly so that the privacy of the data is not compromised. With federated learning, models can access and learn from distributed data sets which are dispersed across multiple healthcare institutions while data security is preserved (Sheller et al., 2020). The use of this method can be seen in the improvement of disease detection and clinical imaging. Furthermore, organizations can follow strict privacy rules such as GDPR or HIPAA through federated learning where it is possible to just utilize useful insights from diverse data sets without directly sharing those sets of data (Yu et al., 2021). This enables the pharmaceutical people to run trials effectively.

INTERNET OF THINGS (IoT) AND SMART DEVICES

Federated Learning will be able to achieve the goals of the IoT ecosystem by performing complex algorithms on devices with limited resources. Several smart devices like sensors, wearables, and smart appliances may not have sufficient computing power and internet connection. By using the method of federated learning, models can be trained across multiple devices without transferring raw data to centralized servers, as reduced communication is a benefit of such an approach (Konečný et al., 2016). The applications span various areas such as improvement of anomaly detection in IoT systems, customization in smart homes, and enabling predictive maintenance in industrial IoT setups (Lim et al., 2020).

FINANCIAL SERVICES AND BANKING

The finance sector has been leading in predictive modelling, risk identification, and fraud detection while taking into account the adoption of data protection regulations, thereby ensuring compliance through the use of federated learning. Banks and other financial institutions may protect their customers' private data with the help of federated learning methods. This is achieved through a distributed network of branches and ATMs to retrain the A1 models. The feature of federated learning guarantees that customers preserve their privacy while information required for loan approval and credit scoring processes is gathered from different sources (Chen et al., 2019). The capability to facilitate collaboration among financial institutions in sharing insights on fraudulent behavior patterns is one of the advantages of adopting federated learning in fraud detection systems. As reported by Zhang et al., the cooperative way guarantees the privacy of highly confidential transactional data.

FEDERATED LEARNING IN INDUSTRY AND RESEARCH

Among the areas where federated learning has been applied widely are healthcare, telecommunications, and finance. It can give healthcare institutions the independence to create collaborative disease prediction models without depending on centralized data collection.

Telecommunication firms can improve network speed and customer satisfaction by deploying federated learning. This is done by jointly training models in this network that distribute the base stations. Gupta et al. raise the point that this method is useful in several areas such as resource allocation, spectrum management, and predictive maintenance which reduces data transmission overhead.

Research on federated learning is a relatively new but quickly evolving area. Scientists are involved in a quest for newly emerging technological applications, optimization methods, and mechanisms for privacy protection.

Further research in federated learning should focus on its potential use for other topics, including federated reinforcement learning, federated transfer learning, and lifelong federated learning. The main purpose of Chen et al. (2023) is to develop a federated learning technique that optimizes performance according to the different learning strategies and difficult real-world situations. At present, research is focused on advanced data patterns and techniques.

COMPARISON WITH OTHER LEARNING PARADIGMS

Centralized learning centralizes all data in one location for effective model training. The main advantage is the access to a huge set of data, which in turn leads to a more efficient model. On the other hand, the prevalence of data breaches and the violation of privacy become more feasible as centralized data storage increases the chances of privacy breaches.

Decentralized System Analysis: Distributed Learning provides model training amongst several nodes or servers to eliminate the need for a central aggregator. Single nodes work with smaller data components, while others coordinate to update a global model. Despite this approach being a distributional one, the problem of high communication costs has been caused by an excessive number of intra-node connections.

Federated Learning: To assist the learning of new information, federated learning takes into account the characteristics of both distributed and centralized learning models. It keeps the data secure, while the models are trained on scattered devices. This function lets one train their mode locally instead of transmitting the data to a remote server eliminating any security issues. Through the process of updating the model parameters but not the raw data, federated learning is suitable for privacy-sensitive applications.

Through decentralization, federated learning limits the susceptibility to data breaches. User privacy is kept as sensitive data resides on local devices. As a result of decentralized learning, there arises no need for centralized data distribution. The importance of this approach is twofold: firstly, it helps collaborative model learning by leveraging multiple devices with heterogeneous data sets. Through the use of distributed computing, federated learning takes advantage of multiple computing devices which in turn, leads to efficiency in processing while minimizing the costs of data transmission.

ETHICAL AND LEGAL CONSIDERATIONS

Training AI models based on distributed data across multiple devices is what federated learning does. Since data usually remains on consumer devices for a long time, it is difficult to tell the ownership of the aggregated model, which results in data ownership complications. The rights and responsibilities must be set as clear as possible among data contributors, model developers, and aggregators to achieve peace and the shared utilization of the data provided (Bagdasaryan et al., 2020). The consent of the users is critical in the case of Federated learning as well as when the data is retrieved from personal devices.

Users will be expected to have an extensive knowledge of data usage, model training objectives, and the problems of the federated learning projects before its full implementation. It is informed consent that allows users to make wise selections regarding their data submission and promotes transparency. Unlike some of the privacy standards, such as the General Data Protection Regulation (GDPR) of the European Union, the Health Insurance Portability and Accountability Act (HIPAA) of the healthcare sector, and a variety of local laws, federated learning is not met by their norms inherently. Data anonymization, encryption, and secure aggregation are some of the techniques that will be required to protect user privacy. Yang et al. mention in their 2022 paper that it is vital to implement such measures to meet the requirements.

The field of federated learning is based on ethical behaviour. The concepts of non-discrimination, transparency, accountability, should be at the core of everything that follows before the start of federated learning, according to Yang et al. (2022). Federated learning systems have to comply with existing regulatory settings. Naseri et al. (2022) indicate that this approach should be based on privacy-by-design principles, robust security measures, and the obtaining of the respective licenses or certificates. Bonawitz et al. (2020) suggested that federated learning systems need to be constantly observed and modified to keep up with shifting privacy laws, ethical standards, and legal requirements.

FUTURE DIRECTIONS AND EMERGING TRENDS

Strategies for Enhancing Privacy: As summarized by McMahan et al. (2018), robust security measures have been brought about by combining powerful privacy-preserving technique like homomorphic encryption, secure multi-party computation, and differential privacy to the federated learning.

More refined aggregation algorithms would give rise to better uniformity, fairness and adaptability to the different kinds of information and device capabilities. Hence, aggregation methods such as hierarchical aggregation, adaptive aggregation, and federated meta-learning have been suggested in the literature (Li et al., 2024).

Streamlined Communication Methods: Researchers currently work on the techniques of federated learning, which should minimize communication load while keeping model precision and convergence. Asynchronous communication protocols and decentralized development strategies are included in these tactics (Yang et al., 2023). The models are able to acquire new knowledge and the way they work can be changed dynamically to adjust to different situations. Among others, lifelong federated learning and federated transfer learning are the two areas where this phenomenon is predominant. These approaches in fact support a better information flow between devices, domains or tasks.

After the deployment of federated learning the process of training AI models will be changed in a radical way. Part of this transition involves decentralization from centralized data repositories to distributed and privacy-preserving technologies. The initiatives involved in raising the privacy and security standards might have a positive impact on subsequent advancements in federated learning. These regulations protect the privacy of individuals and ensure ethical data management, which will serve as the basis for both federated approaches and generalized artificial intelligence methods.

Federated Learning's Emphasis on Collaboration: The tendency towards collaboration and open-source principles in federated learning can kindle a more cooperative attitude and data sharing among the AI developers, thereby bringing the increased adoption of open-source tools and methodologies.

CONCLUSION

Federated Learning redesigns AI paradigms by combining data privacy with distributed model training. Thanks to decentralized data processing, it becomes possible to train the models here collaboratively across different devices and still ensure private information. The Federated Averaging Algorithm provides for the uncomplicated integration of the device-specific features into a centrally updated global model which also eliminates operational complexity. The various advantages of federated learning indicate that it has a promising future. The deployment of edge devices, on the other hand, can decrease the privacy issues of centralized systems. This paradigm shift allows us to utilize the wide variety of data sources without the need to violate any privacy laws. In effect, federated learning emerges as a very useful solution in the healthcare and the Internet of Things fields, where resources are limited and data is extremely sensitive. Having your data secured, you can go on to build your model without any concern. With research, many new achievements for federated learning are expected to be achieved in the future AI is about to witness a major change as a result of improved privacy protection, well-established communication protocols, and integration of today and tomorrow technologies such as Internet of things and Edge computing. Now, a new era has begun where these emerging phenomena are emphasized. This new era incorporates distributed model training, accountable data management, and collective intelligence. Federated learning is an example of novel AI solutions that are founded on privacy protection, collaborative sphere and always in compliance with ethical principles in dealing with private data. It could have a deep influence on how models are trained and used, and therefore, is critical in the development of a reliable and ethically sound AI. Its ability to skillfully handle privacy concerns makes this philosophy differ from other AI ways.

REFERENCES

Bagdasaryan, E., Veit, A., Hua, Y., Estrin, D., & Shmatikov, V. (2020). How to backdoor federated learning. *arXiv preprint arXiv:2007.14191*.

Bonawitz, K., Eichner, H., Grieskamp, W., Huba, D., Ingerman, A., Ivanov, V.,. . .& Moore, D. (2017). Practical secure aggregation for privacy-preserving machine learning. In CCS '17: *Proceedings of the 2017 ACM SIGSAC Conference on Computer and Communications Security (CCS)* (pp. 1175–1191). https://doi.org/10.1145/3133956.3133982

Bonawitz, K., Grieskamp, W., Huba, D., Ingerman, A., Ivanov, V., Kiddon, C.,. . .& Vasilaky, K. (2020). Towards federated learning at scale: A systematic comparison of privacy-preserving aggregation algorithms. *arXiv preprint arXiv:2007.10987.*

Chen, K., Lin, J., Yang, Q., Zhou, P., & Zheng, Y. (2023). Federated reinforcement learning: A comprehensive review. *arXiv preprint arXiv:2302.04567.*

Chen, M., Xu, X., Zhang, Z., & Cheng, S. (2019). Federated learning in mobile edge networks: A comprehensive survey. *IEEE Communications Surveys & Tutorials*, 21(3), 3262–3293.

Dhatterwal, J. S., Kaswan, K. S., Jaglan, V., & Vij, A. (2022a). Machine learning and deep learning algorithms for IoD. In *The Internet of Drones: AI Applications for Smart Solutions* (p. 237). CRC Press.

Dhatterwal, J. S., Kaswan, K. S., Preety, D., & Balusamy, B. (2022b). Emerging technologies in the insurance market. In *Big Data Analytics in the Insurance Market* (pp. 275–286). Emerald Insight.

Dinh, T., Tran, T., Pham, T., Lee, S., & Kim, Y. (2020). Convergence analysis of federated learning: A survey. *arXiv preprint arXiv:2005.07089.*

Gao, L., Wu, Q., Wei, Z., Zhang, L., & Zhang, H. (2021). Federated learning for edge intelligence: A comprehensive survey. *IEEE Communications Surveys & Tutorials*, 23(1).

Hard, A., Rao, K., Mathews, R., Ramaswamy, S., Beaufays, F., Augenstein, S.,. . .& Eichner, H. (2018). Federated learning for mobile keyboard prediction. *arXiv preprint arXiv:1811.03604.*

Kairouz, P., McMahan, H. B., Avent, B., Bellet, A., Bennis, M., Bhagoji, A. N.,. . .& Ramage, D. (2019). Advances and open problems in federated learning. *arXiv preprint arXiv:1912.04977.*

Karimireddy, S. P., Stich, S. U., Jaggi, M., & Hoffmann, M. (2021). Adaptive federated optimization. *arXiv preprint arXiv:2102.05203.*

Kaswan, K. S., Dhatterwal, J. S., & Kumar, K. (2021). Blockchain of internet of things-based earthquake alarming system in smart cities. In *Integration and Implementation of the Internet of Things Through Cloud Computing* (pp. 272–287). IGI Global.

Kaswan, K. S., Dhatterwal, J. S., Kumar, S., & Lal, S. (2022b). Cybersecurity law-based insurance market. In *Big Data: A Game Changer for Insurance Industry* (pp. 303–321). Emerald Publishing Limited.

Kaswan, K. S., Dhatterwal, J. S., Sharma, H., & Sood, K. (2022a). Big data in insurance innovation. In *Big Data: A Game Changer for Insurance Industry* (pp. 117–136). Emerald Publishing Limited.

Konečný, J., McMahan, H. B., Ramage, D., & Richtárik, P. (2016). Federated optimization: Distributed optimization beyond the datacenter. *arXiv preprint arXiv:1511.03575.*

Li, T., Sahu, A. K., Talwalkar, A., & Smith, V. (2020). Federated learning: Challenges, methods, and future directions. *IEEE Signal Processing Magazine*, 37(3), 50–60.

Lim, B., Kang, J., Jung, J. J., & Oh, S. (2020). Federated learning for Internet of Things: A comprehensive survey. *IEEE Access*, 8, 23537–23565.

McMahan, H. B., Abadi, M., Chu, A., Goodfellow, I., McMahan, B., Mironov, I.,. . .& Ramage, D. (2018). Advances and open problems in federated learning. *arXiv preprint arXiv:1812.09975.*

Mitchell, T. (1997). *Machine Learning.* McGraw Hill.

Mohassel, P., & Zhang, Y. (2017). Secureml: A system for scalable privacy-preserving machine learning. In *2017 IEEE Symposium on Security and Privacy (SP)*, San Jose, CA, USA (pp. 19–38). IEEE Xplore. doi: 10.1109/SP.2017.12

Sheller, M. J., Edwards, B., Reina, G. A., Martin, J., Pati, S., Kotrotsou, A., & Milchenko, M. (2020). Federated learning in medicine: Facilitating multi-institutional collaborations without sharing patient data. *Scientific Reports*, 10(1), 1–11.

Shokri, R., Shmatikov, V., & Smith, M. (2015). Privacy-preserving deep learning. In *Proceedings of the 22nd ACM SIGSAC Conference on Computer and Communications Security* (pp. 1310–1321). Association for Computing Machinery.

Smith, V., Chaudhuri, K., Sanjabi, M., & Talwalkar, A. (2021). Federated learning: Strategies for improving communication efficiency. *arXiv preprint arXiv:2106.03638.*

Xie, H., Li, Y., Xu, T., & Zhang, Z. (2023). Secure and robust federated learning: A comprehensive survey. *arXiv preprint arXiv:2305.01972.*

Yang, Q., Liu, Y., Chen, T., & Tong, Y. (2019). Federated machine learning: Concept and applications. *ACM Transactions on Intelligent Systems and Technology (TIST)*, 10(2), 1–19.

Yang, Q., Liu, Y., Chen, T., & Tong, Y. (2022). Federated machine learning: Concept and applications. *ACM Transactions on Intelligent Systems and Technology (TIST)*, 11(2), 1–19.

Yang, Q., Liu, Y., Chen, T., & Tong, Y. (2023). Communication-efficient federated learning: Recent advances and future directions. *ACM Transactions on Intelligent Systems and Technology (TIST)*, 10(2), 1–19.

Yu, K. H., Beam, A. L., & Kohane, I. S. (2021). Artificial intelligence in healthcare. *Nature Reviews Drug Discovery*, 20(5), 373–374.

2 In-Depth Analysis of Artificial Intelligence Practices

Robot Tutors and Federated Learning Approach in English Education

*Muthmainnah, Akkarapon Nuemaihom,
Ahmad Al Yakin, Prodhan Mahbub Ibna Seraj,
Muhammad Mujtaba Asad, and Ahmed A. Elngar*

INTRODUCTION

According to Khang et al. (2023a), integrating traditional education into complex systems is gaining increasing attention because of the integration of artificial intelligence (AI) with education. Advances in new education curricula and infrastructure systems have led to the widespread adoption of artificial intelligence in various fields, including administration, resource development, and more. Research on intelligent education has so far concentrated on methods to improve students' learning efficiency and intelligent education monitoring, but limited research has examined how to assess the efficacy and level of teaching provided by teachers utilizing AI robotics. Indeed, a variety of factors, such as students' efforts, the availability of contemporary learning tools, and the style of their instructors, affect their ability to learn (Muthmainnah et al., 2023). We present a smartphone-based robotics learning system based on blended learning for educators. By understanding the feelings of undergraduate students when they learn a foreign language, the realization of federated learning (FL) improves recognition performance and enables data sharing and collaborative modelling.

The new idea of Industry 6.0 highlights the importance of human workers in production with increasingly harmonious technological collaboration. The goal of Industry 6.0 is to combine the knowledge and experience of human adaptation with advanced technologies to make production processes more adaptive and collaborative (Chourasia et al., 2022). This 6.0 industrial revolution enables the sector to combine human and robotic workers by adopting the former key enablers such as the

DOI: 10.1201/9781003482000-2

Internet of Things (IoT), big data analytics, and artificial intelligence. The goal is to have a highly collaborative, adaptable, and agile team (Khalid et al., 2023).

By these statements, the robot tutor that AI-driven technology is one of the edge devices that can be used together with FL, where this collaboration emphasizes cooperation between human–robot interaction (HRI) to train and improve new machine-based skills. Through this practice, the virtual robot learning model adapts to the learning environment without having to send data and information to a remote server, which poses a security risk for certain bot programs (Papadopoulos et al., 2021). This FL approach includes safety measures and more efficient and effective human–robot collaboration, which is supported by robot facilitation that is consciously personalized by the surrounding environment (Lindblom et al., 2020). Subsequently, this concept is adopted for teaching modern English and individual production, where FL has an impact in the field of education at large. This statement is supported by the results of research conducted by Sadiku et al. (2022) and Driss et al. (2023), who suggest that the FL approach can help schools create AI-driven learning models or machine learning by protecting user data. Protecting user data is highly recommended so that the use of AI-driven learning applications can be adjusted, and manufacturing becomes more personalized and user-centered by utilizing insights gained by customizing specific products, processes, or services such as AR, VR, MR, or other educational technologies (Shakeer & Babu, 2024; Wu et al., 2023). Taking this route has the potential to increase the practicality and efficacy of AI, augmented reality, and virtual reality experiences in educational or teaching contexts by making them more immersive and realistic (de Moraes Rossetto et al., 2023; Khang et al., 2023a; Klimova et al., 2023). Moreover, the FL approach in English as a Foreign Language (EFL) can help students learn from each other and share what they know.

Adopting the federated learning paradigm as outlined by Angurala and Khullar (2023), our objective is to optimize English language teaching through decentralized machine learning models. Previously, in traditional language classrooms, all data and learning insights usually resided on a centralized server, like some previous learning platforms. In federated learning, this paradigm changes, where each student's device becomes a decentralized node, storing its data locally via their smartphone.

In our EFL class, we teach our students to use AI applications with a virtual agent approach, robot tutor, or language learning platform on their respective devices, smartphones, or laptops, thus enabling the creation of personalized learning pathways for each undergraduate student. As students engage with EFL language practice, the model adapts to their strengths and weaknesses. For example, if a student excels in vocabulary but struggles with pronunciation, the federated learning model will of course adapt the exercises to meet their specific needs in EFL, and vice versa. If an undergraduate student has difficulty with grammar, then the adjustments to the exercises are done practically. In our EFL scenario, students' interactions with language applications generate insights locally on their devices, where federated learning promotes collaborative learning experiences. Insights gained from each device contribute to collective model refinement. We present federated learning principles, such as grouping our students who excel in a particular language skill, so the federated learning model (FLM) will adapt to combine these successful strategies, thereby providing benefits for the entire class and according to the needs of each group.

AI virtual robotic tutor Lily and Elsa platforms are used in language learning that utilizes the federated learning model to provide adaptive content recommendations. As students progress, the model refines its understanding of their preferences and learning styles. We identify, for example, that when a student consistently engages in conversational practice, the model will suggest more interactive speaking activities to improve fluency, and the student can activate the audio feature to practice and continually refine its language proficiency predictions, ensuring adaptability to evolving classroom needs.

Therefore, we were glad to see the target and changes in language teaching by decentralizing the learning process, personalizing teaching, preserving privacy, encouraging collaboration, and providing real-time feedback. In our analysis, this fresh method heralds a sea change in language instruction and points the way toward the development of more effective and flexible classroom settings for English as a Foreign Language students. By working together to train machine learning models, academic institutions can protect students' personal information while also collecting and sharing expertise in a certain area. Industry 6.0 participants may work together more effectively and share more information. This promotes collective intelligence and innovation in various industries (Zeng et al., 2022). The framework can also make use of FLM to promote the ethical and responsible usage of AI. To promote AI models that are just transparent and accountable, collaboration is paramount. This will help to actively prevent any type of unethical behavior, such as prejudice or bias. This strategy can be used to create Industry 6.0 AI systems that are more trustworthy and responsible. These transitions in education also will help to make sure that this technology is used in an ethical way and following social norms.

Integrated human–robot FLM based on collaborative human–robot integration. One important part of working together is the ability to communicate and analyze data in real time within the language domain. More collaborative integration between humans and robots is a logical outcome of users being able to enhance their language abilities in a complex learning environment that is customized to their needs. It makes it easier to train machine learning models that can be adjusted to meet the unique requirements of each student, which in turn allows for the creation of highly customized and personalized learning. As the focus here is on actual 3D capabilities and other cutting-edge production technologies, humans and machines may work together to create one-of-a-kind products (Jia & Liu, 2019; Yakin et al., 2022).

FEDERATED LEARNING MODEL IN ENGLISH CLASS

Indonesia is one of the countries where people who do not speak English must now have a good understanding of English with the correct communication skills (micro and macro learning) because English is the language of global communication. One needs a reliable tool that can automate the training and identification process to improve their English skills. In the field of English education, the challenge is to be able to get tailored instruction to improve their English skills with tools that can help them determine where they are having difficulties. The tool must be accessible, easy to use, and embedded in students' smartphones and accommodate students with various needs whenever and wherever they are. It would be great if this tool could also provide instant feedback to users, allowing them to quickly correct their language errors independently. Recent technological advancements have brought forth a range of products designed to streamline English language learning through automation. According to Chen et al. (2019), federated learning is a method whereby small devices collaborate by sharing computing resources and local data to construct reliable models on distributed systems that operate on a large scale. The objective of FL is to discover answers to several issues that computers aren't capable of handling, including those involving users' private data, computing in real time, and AI robots embedded in devices (Shakeer & Babu, 2024). A global machine learning model can be trained using local weights from each user's data using federated learning, which enables numerous users to collaborate without revealing any sensitive information (Weller et al., 2022). This method is employed to achieve a high level of precision in the global model, as stated by Liu et al. (2020) and Narayan et al. (2016). Yin et al. (2021) demonstrated that Google Translate's federated learning (FL) approach in Florida effectively mitigates privacy risks and reduces data exchange costs while adhering to GDPR regulations, outperforming traditional centralized machine learning methods (Lyu et al., 2020). Several studies have compared deep neural networks (DNNs) with FL (Taik et al., 2024) due to the widespread use of DNN in the field of machine learning, but there are still not many articles comparing FL with several other learning techniques, including in the educational realm.

The encouraging findings on DNNs have increased their appeal in various applications. The integration of DNN into FL is not without problems, especially because of one obstacle. In their research, Mazzocca et al. (2024) explore potential developments and future perspectives for federated learning that protect privacy and data. Li et al. (2021) also compared several FL systems. In addition to providing an overview of FL, this paper also focuses on its application to image data security, collaborative AI, and tools in learning EFL. Only a few studies have examined possible barriers to the use of FL in EFL, even though previous studies have examined the advantages of FL over classical machine learning in terms of protection, privacy, and convergence efficiency (Sánchez et al., 2024). To address this gap, this chapter examines existing research on the subject, categorizes the challenges of FL in foreign language education and suggests potential solutions. Despite limited research on the application of FLM in EFL, this chapter aims to contribute to its exploration.

The AI robotics application can understand users' speech patterns and provide them with tailored recommendations by combining machine learning algorithms with voice recognition technology, among other features. In fields such as education, business, and customer service, where proper English communication is essential, the need for such tools is always increasing. Recent advances in deep learning have significantly increased the effectiveness and precision of systems that handle written and spoken language. The idea of end-to-end learning is one approach; this requires teaching a single neural network to translate between unprocessed audio data and literal text transcription (Hagiwara, 2021; Eslit, 2023). This method has shown encouraging results in tasks that involve understanding written language, namely vocabulary, grammar, and spoken recognition, with automatic speech recognition and speech synthesis features. One such idea is the attention mechanism, which allows the model to pay close attention to important parts of the input sequence. As a result, results in areas such as speech recognition and translation have improved. In addition, the capacity to represent long-term dependencies in a series of improving language skills has been enhanced using robotics-based teaching models, so that English language processing is increasingly modern and interactive (Wang, 2023).

To improve English language proficiency, robotic learning approaches are now being used in this study. Therefore, there is an urgent need for intelligent and real-time help systems to improve English language proficiency. The suggested work revolves around natural language processing. We present artificial intelligence and mobile applications with support for multiple users in this article. When training this model with a federated learning approach on the aggregated database, we also consider data privacy issues. Therefore, this work aims to develop privacy-preserving mobile and web multi-app training based on blended learning to assist students in improving their English language proficiency. One of the main reasons for suggesting blended learning-based systems is so that they can incorporate future technologies that will expand the possibilities of related applications.

FEDERATED LEARNING: A TAXONOMY

Federated learning is a framework for language training that emerged out of the intersection of machine learning and rising concerns about data privacy and security. The necessity to remedy the drawbacks of traditional machine learning models, particularly when handling complex and large data sets, led to this categorization. According to Zhang et al. (2021), there are three distinct types of FL solutions. Designed with privacy and computational efficiency in mind, these types are tailor-made for dealing with massive data sets. Let's have a look at the many forms of federated learning and describe them:

1. Horizontal federated learning. To define HFL, it is a process whereby numerous entities work together, with each entity having its local data set that contains identical features. It enables the simultaneous training of a model on many data sets independently, eliminating the necessity for data exchange. Although the model incorporates the collective wisdom of all involved parties, each one keeps full authority over its data (Beltrán et al., 2023). A common scenario in EFL classrooms involves students from diverse academic backgrounds employing a variety of mobile applications to enhance their grammar proficiency. The data, consisting of responses to similar questions, is scattered among various platforms such as Lily, Elsa-AI, Google translation, Google sites, and MindMeister© apps. By collecting knowledge from many different students, horizontal FL can improve the precision of grammatical models.
2. Vertical federated learning (VFL) is defined as a collection of practices that apply when two or more parties have complementary knowledge about a shared data set characteristic. This class makes it easier to train models on data sets that do not share any characteristics (Duan et al., 2022). By

limiting data sharing to what is essential, parties can protect the privacy of their data set components. In language learning programs, for example, users can choose their vocabulary and save it on their device by using the MindMeister© application. Without revealing users' specific word choices, vertical FL can be used to improve language models collaboratively, for example, by collecting information from various applications, and the platform can propose customized vocabulary exercises.

3. Last, TFL stands for federated learning transfer. Efficiency gains during model training through the incorporation of previously trained global model information into locally targeted models is one possible interpretation of transfer learning. With the use of a large data set, TFL trains a global model. Subsequently, local models refine their performance using their data sets. Maximizing learning efficiency without compromising data security is possible with this strategy (Chen et al., 2023). Sentiment analysis engines are trained to correctly detect the tone of English text by initially exposing them to data in that language. This is because these engines are designed to improve learning performance on multiple distributions by transferring knowledge from one distribution to another. To train a model on English data, one can use FL transfer. To optimize learning across different language distributions, knowledge learned from English data is transferred and adapted to improve sentiment analysis performance on English knowledge.

FL is related to NLP theory and the theory of human language acquisition, namely that humans convey a large amount of information through every word and syllable they pronounce when speaking. Natural language communication is enhanced and influenced by various factors, such as the subject matter and the sophistication of the vocabulary used. To enhance language acquisition efficiency and reduce the emotional strain on teachers and students, it is crucial to explore the potential of AI. Students can learn independently, even without a teacher, because AI is set to be available 24/7. The traditional concept of remembering and evaluating the structure, language, meaning, and unique tone of each sentence is an impractical idea (Liu & Lu, 2023).

As can be seen, one branch of artificial intelligence offers machines that can read, understand, predict, and extract meaning from human spoken language, and this intelligence is of great benefit to EFL students (Celik, 2023). One common application of this field is sentiment analysis, which involves identifying and classifying the emotional tone of text, such as positive, negative, or neutral. Information is provided regarding factors that influence student preferences. With the growing popularity of sentiment analysis on the web, individuals can learn from multiple perspectives and stories. Through social media, more and more people are sharing their opinions with strangers by tweeting about a variety of subjects (such as politics, consumer preferences, vacations, and other topics where the app excels).

Expanding to support the collaboration of FLM for EFL, data translation is known as machine translation (Benbada & Benaouda, 2023), which uses machines to translate different languages as students become more adept at translating using technology. Google Translate is the most well-known example of an application that uses statistical machine learning (SML) language processing for EFL, and this engine was very popular for students before the advent of AI applications for language practice such as Lily Tutor and Elsa AI. The basic premise is to collect a set of possible texts, looking for similar texts in opposing languages, with the possibility of them being translated into another language. Computer speech recognition, or autonomous speech recognition (ASR), is another name for speech recognition such as Elsa AI. It involves carrying out verbatim transcription of a speech input stream through a program programmed into a computer system. With proper pre-processing, some of these uses of natural language processing can become plain text that can be parsed in a way that any computer can parse. This includes describing a sequence of numeric words as input. There needs to be a numerical representation that conveys multiple verbal meanings while maintaining semantic integrity for each syllable. There is a contemporary paradigm in NLP where the main method is to achieve this.

METHOD AND PARTICIPANTS

The research methodology adopted in this research is an experimental method, specifically using a pre-experimental design (Marsden & Torgerson, 2012). This categorization is due to the initial nature of the design, which does not have the characteristics of a true experiment due to the presence of external variables that can influence the dependent variable. Consequently, the experimental results, which represent the dependent variable, are not exclusively governed by the independent variables. This limitation arises due to the absence of control variables and non-random sample selection.

The specific research design used is the one-group pretest-posttest design. In this design, a pretest is given before implementing the treatment, which aims to provide a baseline measure. This approach increases the precision of the assessment of treatment effects by facilitating comparison with pretreatment conditions. The design is characterized by giving a pretest (initial test) and posttest (final test) sequentially to a single group. The research procedure steps use a one-group pretest and posttest design, which includes measuring the dependent variable (pretest), providing treatment (X) for six weeks involving students by implementing a federated learning approach using the Tutor Lily and Elsa AI applications, as well as subsequent measurement of the dependent variable after treatment, as shown in Table 2.1.

To operationalize the one-group pretest-posttest design, a pretest (O_1) was given to students in semester 1 of the 2023 academic year to ensure their initial knowledge before implementing the treatment. Next, treatment was carried out in the form of implementing federated learning in AI-integrated classes to improve learning outcomes on the concept of English language skills. The difference in initial and

TABLE 2.1

Pre-Experimental Design

O_1	X	O_2
Pretest	Treatment	Post Test

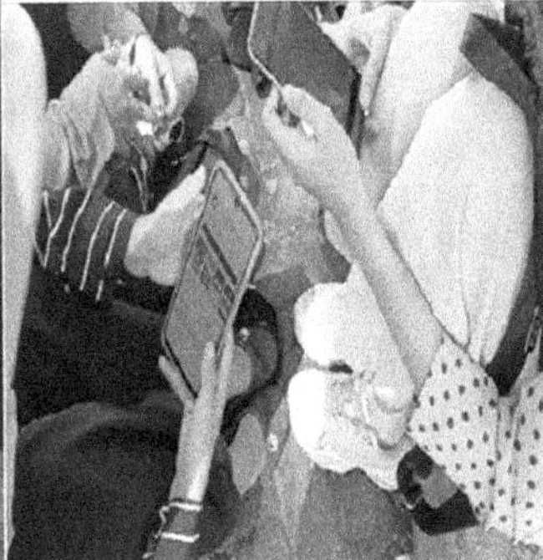

FIGURE 2.1   FLM approach.

final test scores (O_1 and O_2) is assumed to represent the effect of treatment (X). This design is in line with the research objective, assessing improvements in student learning outcomes after implementing the AI-based FL model.

To describe the research context, the population is undergraduate students at Al Asyariah Mandar University, totaling 219 students. The sample selection involved 13 study programs: Indonesian language education, mathematics education, Pancasila and citizenship education, public health, sharia economics, Islamic management, agrotechnology, agribusiness, animal husbandry, communication science, government science, informatics engineering, and informatics systems. The samples exposed to AI were the communication study program, animal husbandry study program, public health study program, and agrotechnology study program for each learning model, so a total of four experimental classes were obtained with a sample of 74 undergraduate students. The research took place in the first (odd) semester of the 2023 academic year, and data collection was carried out six times in the experimental class, as shown in Figure 2.1.

ADOPTING FL AI-DRIVEN PEDAGOGY

Among a sample of first-semester undergraduate students combining two AI tutors, our main aim was to identify the impact of robotics learning activities with a federated learning approach on English language skills. Participants were given treatment for eight meetings. The population of this study was 219 undergraduate students with beginner-level English skills. The sample consisted of 74 students with an age range of 18–21 years, 11 males and 63 females. Participants attended eight English lectures in class with AI interaction by the learning syntax, as shown in Table 2.2. The design of this research was quantitative with a pre-experiment or did not use a control class.

This English language teaching model provides an innovative and technology-infused language teaching approach, utilizing blended learning and AI robot tutors with a focus on English language skills. Undergraduate students engage in interactive and personalized activities, thereby fostering a comprehensive and effective language learning experience that can enhance learning outcomes.

TABLE 2.2

Teaching Model Incorporating FL

Step 1. Warmup	In the first stage, undergraduate students are introduced to the concepts of robotics-based learning and federated learning, which involve conversations with AI robot tutors to develop conversation skills. Undergraduate students are asked to engage in interactive dialogue with an AI tutor, with a focus on pronunciation, fluency, and contextual communication. Start by downloading the Robot Tutor application on the student's smartphone.
Step 2. Federated	Undergraduate students are asked to form groups and discuss with their group friends the conversation theme, the weather. They discussed determining weather keywords by combining robot tutor Lily and Mind Mapping. At this stage, undergraduate students are asked to access Tutor Lily, which helps them discover vocabulary related to weather. After that, they organized their group's vocabulary according to the weather conditions that Tutor Lily shared. Undergraduate students will practice pronunciation of vocabulary for those who are not yet able to pronounce the vocabulary, practice accents, and practice correct writing.
Step 3. Composition	Under Tutor Lily's guidance, students collaborate with their peers to create practice scenarios for writing dialogues. These exercises are designed to improve grammar, coherence, and vocabulary, with AI providing feedback. In this stage, undergraduate students access L2 using Tutor Lily's translation features and Google Translation. To perfect the dialogue, lecturers guide undergraduate students and facilitate with Grammarly. This activity trains collaborative grammar that utilizes blended learning, ensuring collective improvement in grammatical accuracy. In addition, at this stage, undergraduate students structurally enrich their vocabulary through AI-guided exercises and contextual usage exercises.
Step 4. Practice	Practice stage: after the dialogue is complete, it continues with personalized reading with AI with the aim of improving reading comprehension through a personalized and AI-guided reading experience. A dialogue script about the presented material (weather) makes up the reading material that the AI has curated. Together with their group mates, they practice with each other to sharpen their listening skills with AI-generated audio content. In addition to Tutor Lily, undergraduate students access Elsa-AI. This stage of AI integration aims to train pronunciation, practice listening, and practice pronunciation with the help of AI and assess understanding.
Step 5. Performance	In the performance stage, undergraduate students and their group friends display their English language skills by recording a video presentation and uploading it to YouTube.

RESEARCH INSTRUMENTS

This research uses a combination of test and non-test instruments to collect data for comprehensive analysis. The testing component consists of a multiple-choice questionnaire consisting of 25 questions given at the initial stage. This test had a dual purpose, assessing the initial proficiency of both experimental groups and establishing the homogeneity of their abilities. This careful assessment aims to ensure that, at the start of the study, the two groups do not show statistically significant differences in their capacities, as shown in Figure 2.2.

To assess whether the experimental groups had made any improvements in their learning outcomes, they were both given an identical multiple-choice questionnaire. To fully grasp the efficacy of the interventions used in the study, this cyclical testing method is employed.

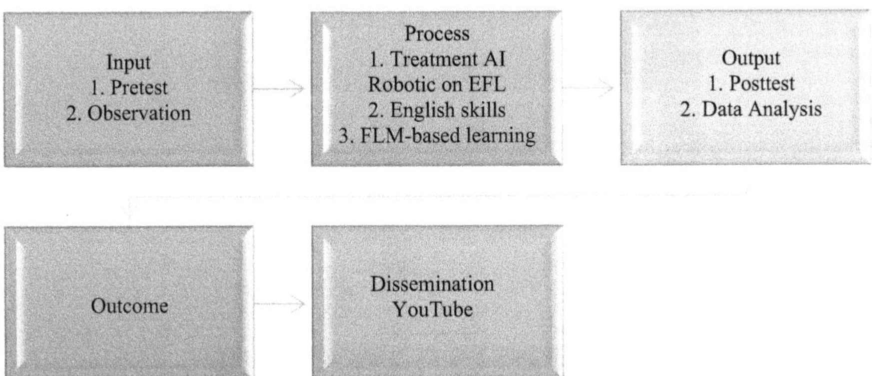

FIGURE 2.2   Research procedure.

A psychomotor evaluation, which is not a test but is administered at the same time as the test instrument, is also conducted. The goal of this all-encompassing strategy is to gain a thorough knowledge of the many facets of student learning outcomes by integrating test and non-test instruments. After carefully administering the instruments and collecting the data, as well as making sure that the instruments were valid and appropriate for the research purpose, a series of advanced data analyses were carried out. To conduct a thorough and rigorous review of the treatments used and their effect on the experimental group's learning outcomes, this study seeks to acquire new insights into the complexities of the research findings shown in Figure 2.3.

DATA ANALYSIS

After a comprehensive verification process and extensive tests for validity and compliance with set standards, the next crucial step is to quantify the research results. Two sophisticated techniques used in the analysis process are quantitative descriptive analysis and inferential statistics, which are both made possible by SPSS version 26. Thoroughly analyzing and making sense of the numerical data produced by the instrument is essential for quantitative descriptive analysis. Using statistics like the mean, median, and standard deviation, one can get a good idea of the dataset's central tendency and dispersion and then conclude the set's key properties. At the same time, relevant inferences are drawn using inferential statistics and dispersion in the collected data. Simultaneously, inferential statistics are used to draw meaningful conclusions and infer patterns in datasets. The use of SPSS version 26, powerful statistical software, allows the application of various statistical tests.

RESULTS AND DISCUSSION

Using descriptive statistics, we were able to better understand the distribution and central tendencies of performance indicators in the data set that included pre-and posttest scores from 74 participants. Participants' baseline performance on the

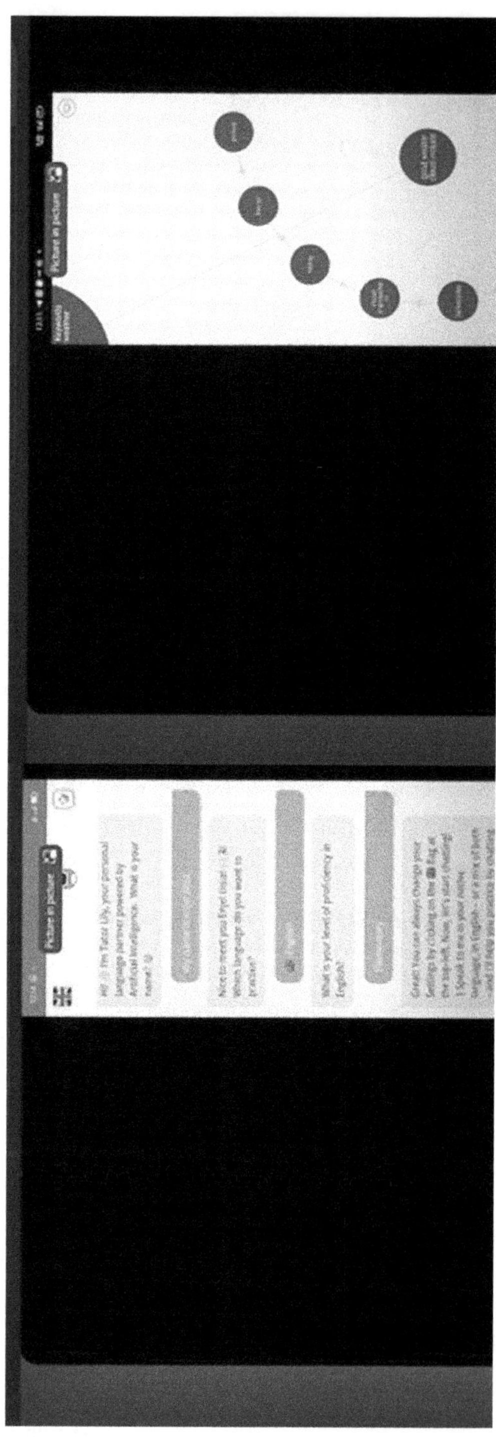

FIGURE 2.3   Extracting performance in YouTube.

TABLE 2.3
Pre- and Post-Test Results

	N	Minimum	Maximum	Mean	Std. Deviation
Pretest	74	30.00	70.00	54.3243	8.96433
Posttest	74	60.00	95.00	79.6622	9.52038
Valid N (listwise)	74				

TABLE 2.4
Chi-Square Test

	Pretest	Posttest
Chi-square	58.811[a]	41.459[b]
df	8	7
Asymp. sig.	.000	.000

[a] 0 cells (0.0%) have expected frequencies less than 5. The minimum expected cell frequency is 8.2.

[b] 0 cells (0.0%) have expected frequencies less than 5. The minimum expected cell frequency is 9.3.

pretest showed a substantial amount of variability, with scores ranging from 30 to 70 on the scale in Table 2.3.

Table 2.3 shows the mean score was 54.3243, and the standard deviation was 8.96433. Afterward, on the posttest, the scores varied from 60.00 to 95.00, showing an increasing tendency. There was a 9.52038 standard deviation and a significant increase in the mean posttest score to 79.6622. Both evaluations use the same valid N (listwise) value of 74, which shows that the data set is comprehensive.

The relationship between pretest and posttest scores was assessed using statistical tests, specifically the chi-square test. The results in Table 2.4 show that there are statistically significant consequences. With 8 degrees of freedom, the pretest chi-square value is 58.811, resulting in a p-value of 0.000, which is considered highly significant. Apart from that, the posttest shows a p-value of 0.000, 7 degrees of freedom, and a chi-square value of 41.459. It is worth mentioning that in both cases, zero cells (0.0%) meet the minimum predicted cell frequency requirement, with an expected frequency below five. For the pretest, the lowest predicted cell frequency was recorded at 8.2, and for the posttest, it was 9.3.

Table 2.5 shows the mean difference of 54.32432 from the test results is statistically significant, according to the pretest, which has a t-statistic of 52.131 with 73 degrees of freedom and a p-value of 0.000. The 95% confidence interval of the difference (52.2475, 56.4012) further supports these findings. The posttest also shows a t-statistic of 71.980 with 73 degrees of freedom and a significant p-value of 0.000, which means the mean difference of 79.66216 from the test value of 0 is statistically significant. The 95% confidence interval accompanying the difference (77.4565, 81.8679) further indicates the dependence of this effect.

TABLE 2.5
One Sample Test

| | | | | Test Value = 0 | | |
| | | | | Mean | 95% Confidence Interval of the Difference | |
	t	df	Sig. (2-Tailed)	Difference	Lower	Upper
Pretest	52.131	73	.000	54.32432	52.2475	56.4012
Posttest	71.980	73	.000	79.66216	77.4565	81.8679

TABLE 2.6
ANOVA

	Sum of Squares	df	Mean Square	F	Sig.
Between Groups	1707.894	7	243.985	3.872	.001
Within Groups	4158.322	66	63.005		
Total	5866.216	73			

TABLE 2.7
Paired Sample Correlations

		N	Correlation	Sig.
Pair 1	Pretest and posttest	74	.403	.000

To determine whether the pretest scores varied significantly between categories, we used analysis of variance (ANOVA), as shown in Table 2.6. We found that the between-group variability, which represents differences among the groups, accounted for 1707.894 sums of squares with 7 degrees of freedom, giving us a mean square of 243.985 when we examined the sources of variation. The estimated F-statistic of 3.872 and its related p-value of.001 among the groups demonstrated a statistically significant difference. Alternatively, there was a mean square of 63.005 and a sum of squares of 4158.322 for the within-group variability, which reflects individual differences within each group. There were 66 degrees of freedom in this analysis. With a total of 73 degrees of freedom, the sum of squares for the pretest scores was 5866.216. The statistical significance of the between-group F-ratio indicates that there are obvious group-wise variations in the pre-test results.

Table 2.7 shows the results of a paired sample correlation analysis conducted on 74 individuals to examine the relationship between pretest and posttest scores. Calculation of the correlation coefficient (r) between the pretest and posttest scores shows that there is a quite positive relationship, with a value of 0.403. The statistical significance of the correlation, with a p-value of 0.000, indicates that it is very unlikely that the observed relationship occurred by chance, as shown in Figure 2.4.

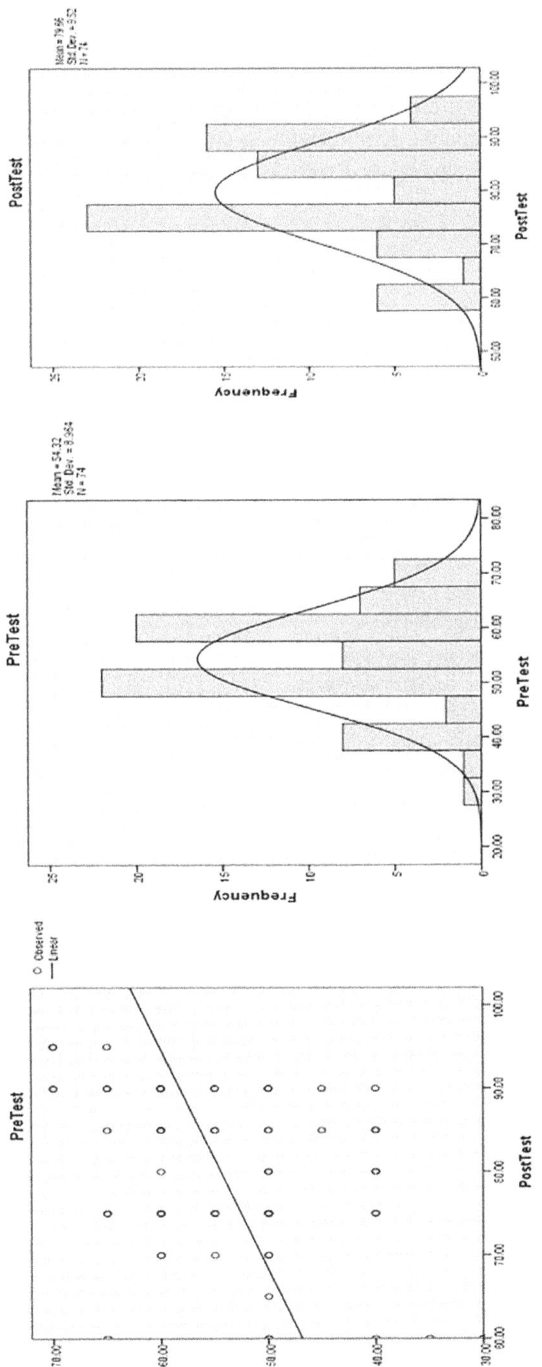

FIGURE 2.4 The histogram pretest and posttest.

Table 2.8 shows the survey data set includes descriptive statistics for variables related to EFL classroom pedagogical considerations, AI practices, robotic tutor involvement, blended learning implementation, AI practices, and blended learning integration, as well as general pedagogical considerations. Primary statistical measures were applied to participants' responses after each variable was evaluated using various statements. For AI practices, participants demonstrated a strong understanding of the subject (mean = 4.5270) and actively engaged with AI-enabled technologies (mean = 4.4189).

TABLE 2.8
Questionnaire Results

Descriptive Statistics

	N	Minimum	Maximum	Mean	Std. Deviation
AI Practices					
1. I understand the practice of artificial intelligence (AI) in education, and I am happy	74	3.00	5.00	4.5270	.68692
2. I use AI-powered tools or applications in English as a Foreign Language (EFL) learning very well.	74	2.00	5.00	4.4189	.77648
3. From my perspective, AI-driven tools are a promising avenue for optimizing language acquisition.	74	2.00	5.00	4.1892	.87077
4. AI Tutor Lily and Elsa's technology or practice integrated into EFL lessons motivates interest in learning English compared to before without using AI.	74	1.00	5.00	4.1486	.85500
5. I believe that AI-based practices can contribute to individualized learning experiences in EFL classrooms.	74	2.00	5.00	4.3378	.83218
Robot Tutor					
Using a robot tutor in EFL class is very interesting.	74	2.00	5.00	4.3243	.74223
I have found that incorporating a robot tutor into my English learning has been very beneficial.	74	2.00	5.00	4.3649	.82053
I am more comfortable learning when lecturers use Robot Tutor Lily and ELAS compared to traditional teaching methods.	74	2.00	5.00	4.2027	.84367
My language skills can be improved effectively with the help of AI.	74	1.00	5.00	4.1216	.92088
Implementation of federated learning					
I enjoy learning English with the concept of federated learning (FL) in the context of language education.	74	1.00	5.00	4.1351	.95551
Learning federation in EFL class can help my deep learning of English.	74	2.00	5.00	4.2297	.86875
I feel the potential benefits when lecturers use federated learning for language teaching.	74	1.00	5.00	4.3919	.82458
I am excited about the federated learning model in an EFL environment.	74	1.00	5.00	4.2162	.86437
I believe that federated learning can address privacy issues in language education.	74	2.00	5.00	4.2973	.83960

TABLE 2.8 *(Continued)*
Questionnaire Results

Descriptive Statistics

	N	Minimum	Maximum	Mean	Std. Deviation
Integration of AI practices and federated learning					
Integrating AI practices and federated learning in EFL teaching helps me to analyze language, from vocabulary to designing practical sentences and dialogues.	74	1.00	5.00	4.2162	.89550
I felt the synergy between AI practices and federated learning as being able to optimize language learning outcomes.	74	2.00	5.00	4.1622	.92198
I like to learn English by using the AI and FL approaches.	74	1.00	5.00	4.0811	1.03045
Pedagogical considerations:					
I am sure that the integration of robot tutoring, artificial intelligence (AI) practices, and federated learning can achieve maximum learning outcomes.	74	2.00	5.00	4.1486	.90179
I believe in the impact of AI practices, especially robot tutoring, on encouraging student engagement and participation in EFL lessons.	74	2.00	5.00	4.1757	.88144
In my opinion, educators should adapt their teaching strategies to accommodate the incorporation of AI and federated learning practices in EFL teaching.	74	1.00	5.00	4.1486	.96063
I recommend that educators incorporate AI practices and federated learning into teaching strategies for EFL and other courses.	74	2.00	5.00	4.1757	.86576
Valid N (listwise)	74				

DISCUSSION

As stated in Table 2.3, the data of descriptive statistics provide a detailed picture of the distribution of participants' performance, which can be used as a starting point for inferential studies and to guide possible educational initiatives. Based on Table 2.4, the results of the chi-square test showed that the association test was strong, indicating a correlation between pre- and posttest scores. This can help us understand educational interventions or experimental conditions better. Using a test value of 0, a one-sample test was performed on the pretest and posttest data sets to see whether the mean differences in each group were statistically significant. The results showed that there was a statistically significant change between the mean scores before and after the test, which explains how well the intervention of the FL teaching method worked. A substantial drop in performance between the two sets of assessments suggests the need for additional research into the intervention's efficacy or the effectiveness of the teaching method. Consequently, the coefficient value, which shows the strength of the link, reveals that participants with higher pretest scores also tend to receive higher posttest scores. This study sheds light on people's performance on both exams and can provide light on the kinds of educational interventions or experimental conditions that may have produced the observed correlations.

Based on Table 8, the mean scores of students' responses from the survey ranged from 4.1216 to 4.3919. This indicated that undergraduate students were generally positive in terms of their experiences with Tutor Robot and the application of federated learning in the EFL classroom. Positive feedback was also received by AI practice integration and federated learning, highlighting the real capacity of AI practice integration and federated learning model to improve language learning outcomes such as Bowser's added-on (Farahrani et al., 2023), which trains algorithms using distributed data sets stored on edge devices or servers. This proposed method eliminates the need to communicate data, thereby bridging the privacy utility gap introducing a new personalized clustered FL (PCFL) method that trains models by leveraging shared features in the personal data and by capturing unique features in each client's data. Although the proposed methodology successfully overcomes privacy concerns, experimental findings show that it hardly loses performance when compared with centralized model training methods.

Figure 2.2 shows the teaching model taxonomy by integrating the FL concept, which is based on the FL classification, namely horizontal federated learning, vertical federated learning, and federated learning transfer, and the AI application domain. Segmentation of the language data obtained is distributed to test EFL English skills, and the personalized approach is part of the teaching model that is tested in the classroom. Some of the main obstacles include quota costs, network constraints, the diversity of AI systems in use, and concerns about privacy and security. Analysis of survey paper topics and their correlation with language learning outcomes suggest potential improvements in FL implementation through categorization. Moreover, a majority of participants advocated for the integration of AI into English as a Foreign Language classrooms, echoing the positive findings on student engagement reported by Braines et al. (2017).

In addition to the challenges faced, the study revealed that undergraduate students perceived FLM as a valuable tool for enhancing their EFL language skills, consistent with previous research (Chhikara et al., 2020; Purgina, 2019). Their studies showed that the use of various artificial intelligence applications can support language training speed, efficiency, generalization ability, and suitability for various domains, such as the Internet of Things (IoT), Google Translation collaboration, storing data on vocabulary, grammar, sentences obtained on smartphone devices, notes, and security data for each undergraduate student. Integrating FL in class has its own challenges. In this study, this is the first time this has been done and required repeated explanations so that undergraduate students could understand each learning syntax structure and adapt well. Priorities, including optimal internet networks and different types of smartphones, certainly do not hinder FL acceleration in class.

When designing learning scenarios with the integration of Tutor Lily and Elsa AI (customized models depending on the learning environment), personalization is an additional component to consider. Barriers to implementing an FL-based learning system are undergraduate students' high internet costs and the need for data storage space on their respective smartphones. Because of the large number of devices that may form a combined network, data transfer over the network is much slower than computing performed locally. As a result, effective networking has been the subject of much research in Indonesia and the Asia Pacific in general. The diversity of underlying systems also means that devices in a combined network may have different communication capabilities. Different devices may also have varying amounts of storage and processing power, with some downloading quickly and others being slow. In many circumstances, network limitations do not mean that only some devices can take part in an iteration. Therefore, the application of learning techniques with the FL approach can overcome and find solutions to these obstacles.

The adoption of combined learning methods, similar to those used in education, is still uncommon in fields like engineering, architecture, IoT, and health, where the term "FL" is predominantly used. Several elements contribute to improving learning outcomes with FL techniques; apart from technology and teaching models, conditions and consideration of students' learning needs are also considered. The time required for blended learning to come together considers not only the processing power of language acquisition but also the time required to adapt to the technique. Therefore, to implement and optimize the FL approach, it is important to consider the quality of the internet, student motivation, and learning experience so that this learning model works well and can be applied to other courses. To optimize the combined deep learning model for AI applications, it can be carried out adaptively, considering the availability of devices and their readiness to participate in the learning process.

CONCLUSION

This research incorporates a federated learning approach in language learning by using three FL characters in learning activities to see how federated mobile learning can help overcome the problem of low English language skills, interest, and learning motivation, especially those related to the anywhere and anytime learning paradigm. As an innovative decentralized learning solution, the FL approach was tested in language classes to overcome the problems of decentralized EFL language learning. We introduce the concept of federated learning and the idea of blended learning and its main characteristics in a way that is easy for everyone to understand. The results of our study outline the steps for implementing blended learning in language classes and the possibility that it can be adapted for other courses. Moreover, this research shows that blended learning can be used with simulations that have been run for eight meetings, and our results show that blended learning techniques can improve learning outcomes, motivation, and interest in learning English. Additional studies are needed to address the many unresolved concerns and barriers.

The application of FL in the field of language can be done with the concept of a collaborative human–robot integration approach. Collaboration occurs by emphasizing real-time features in data sharing and analysis. A more collaborative integration of humans and robots is of course supported by a more flexible learning environment, allowing users to improve their language skills according to their needs in a sophisticated learning environment. Strategies based on collaborative human–robot integration can be used to implement FL in the language domain. Real-time data exchange and analysis is a key component of collaboration. Users can improve their language skills in a complex learning environment tailored to their needs, which of course supports more collaborative integration between humans and robots.

REFERENCES

Al Yakin, A., Obaid, A. J., Muthmainnah, R. S. A. M., Khalaf, H. A., & Al-Barzinji, S. M. (2022). Bringing technology into the classroom amid Covid 19, challenge and opportunity. *Journal of Positive School Psychology, 6*(2), 1043–1052.

Angurala, M., & Khullar, V. (2023). Federated learning-based privacy preserved English accent training ecosystem for people with Indian language accent. *Entertainment Computing, 46*, 100572.

Beltrán, E. T. M., Pérez, M. Q., Sánchez, P. M. S., Bernal, S. L., Bovet, G., Pérez, M. G.,. . .& Celdrán, A. H. (2023). Decentralized federated learning: Fundamentals, state of the art, frameworks, trends, and challenges. *IEEE Communications Surveys & Tutorials, 25*(4), 2983–3013. http://dx.doi.org/10.1109/COMST.2023.3315746

Benbada, M. L., & Benaouda, N. (2023). *Investigation of the Role of Artificial Intelligence in Developing Machine Translation Quality. Case Study: Reverso Context and Google Translate Translations of Expressive and Descriptive Texts. Language Combination: Arabic-English/English-Arabic* (Doctoral dissertation, Faculty of Letters and Languages-Department of English).

Braines, D., O'Leary, N., Thomas, A., Harborne, D., Preece, A. D., & Webberley, W. M. (2017). Conversational homes: A uniform natural language approach for collaboration among humans and devices. *International Journal on Advances in Intelligent Systems, 10*(3/4), 223–237.

Celik, I. (2023). Towards intelligent-TPACK: An empirical study on teachers' professional knowledge to ethically integrate artificial intelligence (AI)-based tools into education. *Computers in Human Behavior, 138*, 107468.

Chen, D., Jiang, X., Zhong, H., & Cui, J. (2023). Building trusted federated learning: Key technologies and challenges. *Journal of Sensor and Actuator Networks, 12*(1), 13.

Chen, M., Suresh, A. T., Mathews, R., Wong, A., Allauzen, C., Beaufays, F., & Riley, M. (2019). Federated learning of n-gram language models. *arXiv preprint arXiv:1910.03432.*

Chhikara, P., Singh, P., Tekchandani, R., Kumar, N., & Guizani, M. (2020). Federated learning meets human emotions: A decentralized framework for human–computer interaction for IoT applications. *IEEE Internet of Things Journal, 8*(8), 6949–6962.

Chourasia, S., Tyagi, A., Pandey, S. M., Walia, R. S., & Murtaza, Q. (2022). Sustainability of industry 6.0 in global perspective: Benefits and challenges. *Mapan, 37*(2), 443–452.

de Moraes Rossetto, A. G., Martins, T. C., Silva, L. A., Leithardt, D. R., Bermejo-Gil, B. M., & Leithardt, V. R. (2023, November). An analysis of the use of augmented reality and virtual reality as educational resources. *Computer Applications in Engineering Education, 31*(6), 1761–1775. https://doi.org/10.1002/cae.22671

Driss, M. B., Sabir, E., Elbiaze, H., & Saad, W. (2023). Federated learning for 6G: Paradigms, taxonomy, recent advances and insights. *arXiv preprint arXiv:2312.04688.*

Duan, Q., Hu, S., Deng, R., & Lu, Z. (2022). Combined federated and split learning in edge computing for ubiquitous intelligence in internet of things: State-of-the-art and future directions. *Sensors, 22*(16), 5983.

Eslit, E. R. (2023). Elevating language acquisition through deep learning and meaningful pedagogy in an AI-evolving educational landscape. *Preprints.* https://doi.org/10.20944/preprints202309.0658.v1

Farahani, B., Tabibian, S., & Ebrahimi, H. (2023). Towards a personalized clustered federated learning: A speech recognition case study. *IEEE Internet of Things Journal, 10*(21), 18553–18562. doi: 10.1109/JIOT.2023.3292797

Hagiwara, M. (2021). *Real-World Natural Language Processing: Practical Applications with Deep Learning.* Simon and Schuster.

Jia, T., & Liu, Y. (2019, October). Words in kitchen: An instance of leveraging virtual reality technology to learn vocabulary. In *2019 IEEE International Symposium on Mixed and Augmented Reality Adjunct (ISMAR-Adjunct)* (pp. 150–155). IEEE.

Khalid, U., Iqbal, H., Vahidian, S., Hua, J., & Chen, C. (2023). CEFHRI: A communication efficient federated learning framework for recognizing industrial human–robot interaction. *arXiv preprint arXiv:2308.14965.*

Khang, A., Muthmainnah, M., Seraj, P. M. I., Al Yakin, A., & Obaid, A. J. (2023a). AI-aided teaching model in education 5.0. In *Handbook of Research on AI-Based Technologies and Applications in the Era of the Metaverse* (pp. 83–104). IGI Global.

Khang, A., Shah, V., & Rani, S. (Eds.). (2023b). *Handbook of Research on AI-Based Technologies and Applications in the Era of the Metaverse.* IGI Global.

Klimova, B., Pikhart, M., Polakova, P., Cerna, M., Yayilgan, S. Y., & Shaikh, S. (2023). A systematic review on the use of emerging technologies in teaching English as an applied language at the university level. *Systems, 11*(1), 42.

Li, Q., Wen, Z., Wu, Z., Hu, S., Wang, N., Li, Y.,. . .& He, B. (2021). A survey on federated learning systems: Vision, hype and reality for data privacy and protection. *IEEE Transactions on Knowledge and Data Engineering, 35*(4), 3347–3366. doi: 10.1109/TKDE.2021.3124599

Lindblom, J., Alenljung, B., & Billing, E. (2020). Evaluating the user experience of human–robot interaction. In *Human–robot interaction: Evaluation Methods and Their Standardization* (pp. 231–256). Springer.

Liu, B., & Lu, Z. (2023). Design of spoken English teaching based on artificial intelligence educational robots and wireless network technology. *EAI Endorsed Transactions on Scalable Information Systems, 10*(4), e12–e12.

Liu, Y., Yuan, X., Xiong, Z., Kang, J., Wang, X., & Niyato, D. (2020). Federated learning for 6G communications: Challenges, methods, and future directions. *China Communications, 17*(9), 105–118.

Lyu, L., Xu, X., Wang, Q., & Yu, H. (2020). Collaborative fairness in federated learning. Federated learning*: Privacy and Incentive*, 189–204.

Marsden, E., & Torgerson, C. J. (2012). Single group, pre-and post-test research designs: Some methodological concerns. *Oxford Review of Education, 38*(5), 583–616.

Mazzocca, C., Romandini, N., Montanari, R., & Bellavista, P. (2024). Enabling federated learning at the edge through the IOTA tangle. *Future Generation Computer Systems, 152*, 17–29.

Muthmainnah, Obaid, A. J., Al Yakin, A., & Brayyich, M. (2023, June). Enhancing computational thinking based on virtual robot of artificial intelligence modeling in the English language classroom. In *International Conference on Data Analytics & Management* (pp. 1–11). Springer Nature.

Narayan, R., Chakraverty, S., & Singh, V. P. (2016). Quantum neural network based machine translator for English to Hindi. *Applied Soft Computing, 38*, 1060–1075.

Papadopoulos, G. T., Antona, M., & Stephanidis, C. (2021). Towards open and expandable cognitive AI architectures for large-scale multi-agent human-robot collaborative learning. *IEEE Access, 9*, 73890–73909.

Purgina, M. (2019). *Mobile Technology for Gamification of Natural Language Grammar Acquisition* (Unpublished doctoral dissertation, The University of Aizu). https://uaizu. repo.nii.ac.jp.

Sadiku, M. N., Musa, S. M., & Chukwu, U. C. (2022). *Artificial Intelligence in Education.* iUniverse.

Sánchez, P. M. S., Celdrán, A. H., Xie, N., Bovet, G., Pérez, G. M., & Stiller, B. (2024). Federated trust: A solution for trustworthy federated learning. *Future Generation Computer Systems, 152,* 83–98.

Shakeer, S. M., & Babu, M. R. (2024). A study of federated learning with internet of things for data privacy and security using privacy preserving techniques. *Recent Patents on Engineering, 18*(1), 1–17.

Taik, A., Abouaomar, A., & Cherkaoui, S. (2024). Green federated learning-based models and protocols. In *Green Machine Learning Protocols for Future Communication Networks* (pp. 63–102). CRC Press.

Wang, T. (2023, January). Innovative strategies for the development of international Chinese language education based on deep learning models. *Applied Mathematics and Nonlinear Sciences, 9*(1), 1–5. doi: 10.2478/amns.2023.2.00138

Weller, O., Marone, M., Braverman, V., Lawrie, D., & Van Durme, B. (2022). Pretrained models for multilingual federated learning. *arXiv preprint arXiv:2206.02291.* https://doi.org/10.48550/arXiv.2206.02291

Wu, D., Yang, Z., Zhang, P., Wang, R., Yang, B., & Ma, X. (2023). Virtual-reality inter-promotion technology for metaverse: A survey. *IEEE Internet of Things Journal, 10*(18), 15788–15809. doi: 10.1109/JIOT.2023.3265848

Yin, X., Zhu, Y., & Hu, J. (2021). A comprehensive survey of privacy-preserving federated learning: A taxonomy, review, and future directions. *ACM Computing Surveys (CSUR), 54*(6), 1–36.

Zeng, R., Zeng, C., Wang, X., Li, B., & Chu, X. (2022). Incentive mechanisms in federated learning and a game-theoretical approach. *IEEE Network, 36*(6), 229–235.

Zhang, C., Xie, Y., Bai, H., Yu, B., Li, W., & Gao, Y. (2021). A survey on federated learning. *Knowledge-Based Systems, 216,* 106775.

3 Enabling Federated Learning in the Classroom

Sociotechnical Ecosystem on Artificial Intelligence Integration in Educational Practices

Ahmad Al Yakin, Arkas Viddy, Idi Warsah,
Ali Said Al Matari, Luís Cardoso,
Ahmed A. Elngar, Ahmad J. Obaid,
and Muthmainnah

INTRODUCTION

The introduction of AI technology can affect teacher–student relationships. Understanding the sociological aspects of this relationship is critical to ensuring that implementing blended learning enhances, rather than hinders, the educational experience (Muthmainnah et al., 2023a). Teachers play an important role in mediating the impact of AI on the classroom environment and fostering a positive sociotechnical ecosystem. This chapter explores the integration of blended learning into educational practice, focusing on sociological aspects to see the impact of its implementation in the classroom. Federated learning, a decentralized machine learning approach, has great potential to improve educational outcomes. However, its successful implementation requires a different understanding of the social dynamics in the educational environment. To investigate the sociological implications of the introduction of artificial intelligence (AI) through blended learning in the classroom, examining the potential benefits and challenges from a sociotechnical perspective with the basic assumption that artificial intelligence has become an integral part of various sectors and their sectors. Its application in education is receiving increasing attention.

Blended learning, a decentralized machine learning paradigm, offers a unique approach to incorporating AI in the educational landscape that focuses on the sociological dimensions that enable blended learning in the classroom, highlighting the complex interactions between technology and social structures Ng et al. (2023). To understand the implications of AI integration, it is important to take a sociological perspective. Sociological theories, such as symbolic interactionism, social

DOI: 10.1201/9781003482000-3

constructivism, and sociology of education, provide valuable frameworks for analyzing the dynamics between students, teachers, and AI technologies in learning environments (Pedro et al., 2019). The benefits of federated learning in education address concerns regarding data privacy and security by allowing model training on decentralized devices. This approach empowers educational institutions to leverage the benefits of AI without compromising sensitive student information (Rajbongshi et al., 2022). Additionally, blended learning facilitates a personalized learning experience by adapting to individual student needs and improving overall academic performance. Despite its potential benefits, the integration of blended learning in classrooms poses challenges with sociological implications. Issues such as the digital divide and ethical considerations surrounding the application of AI require careful consideration. A sociological perspective helps address these challenges by examining power dynamics, social stratification, and their impact on diverse student populations Donnelly (2017). Rapid progress on the Internet of Things (IoT) has been observed in recent years. This network enables extensive sensing and computing capabilities to connect various objects to the Internet (Čolaković and Hadžialić, 2018). An approach to developing intelligent and privacy-preserving IoT systems called federated learning has recently been proposed.

The recently proposed federated learning (FL) technique has attracted the interest of many scientists who want to learn more about its possibilities and uses (Zhang et al., 2022). The main goal of FL is to answer the following question: Is it possible to train a model locally, without transferring data to a central repository? Traditional machine learning algorithms do not always provide the desired results when working in an FL framework that emphasizes labor. Another benefit of FL over more conventional machine learning approaches is that it allows the underlying algorithm to learn from its errors (Ouyang et al., 2021). FL has been used in various fields and industries, including healthcare, the Internet of Things, transportation, and defense, and this chapter will investigate symbolic interaction on mobile applications used in education.

We were motivated to write this chapter due to the increased interest in this sector since FL emerged in 2016, with a wide variety of applications, challenges, and concerns related to this new paradigm. As a result, there is a paucity of recent survey papers and preprints that discuss the FL field from various viewpoints. AbdulRahman et al.'s (2020) research on federated learning for mobile-edge networks highlights challenges in privacy, security, resource management, and communication overhead. Additionally, it shows some uses of FL for edge networks. Various FL settings, including horizontal FL, vertical FL, and compound transfer learning, were categorized and described by AbdulRahman et al. (2020) and Aledhari et al. (2020) according to the characteristics of the data distribution. Focusing on these four key areas—communication, system heterogeneity, statistical heterogeneity, and privacy—can lead to more effective federated learning (FL) implementations. Wei et al. (2020) concentrate on implementation issues and its current techniques. Wireless communication was highlighted by Konečný et al. (2016) as a potential application area for FL.

Given the rapid growth, wide range of applications, and significant impact of FL in various fields of study, it is surprising that there has not been a literature review that covers all the bases regarding the modeling, implementation, and technical aspects of FL in terms of sociotechnical aspects, details, and deployment and directs researchers to relevant contributions. This fact prompted us to conduct comprehensive survey research, identify key issues, and develop a survey that incorporates the new taxonomy. Therefore, to overcome the gaps that occur regarding the form of implementing integrated learning in the classroom, the integration of artificial intelligence in education holds great promise for personalized learning and enhanced educational experiences by leveraging student data centers and concerns regarding data privacy and security by facilitating learning spaces with a federated learning approach, offering a secure and collaborative approach to leveraging AI in the classroom with sociotechnical ecosystems in educational practice.

SMART SOCIOTECHNICAL ECOSYSTEM THROUGH FEDERATED LEARNING MODEL

In general, the integration of artificial intelligence in educational practices has great potential for personalized learning and enhancing more meaningful educational experiences Al (Yakin et al., 2023). It cannot be denied that surfing using technology raises concerns about privacy and data security, which results in significant challenges (Bhutoria, 2022). Federated learning is emerging as a promising solution, enabling the development of collaborative AI in the classroom while maintaining the privacy of student data while interacting with this AI. There are several benefits of FL in learning, such as that with FL, you can overcome data privacy problems by storing student data on their devices, such as smartphones, tablets, or laptops. This decentralization eliminates the need to share raw data with a central server, thereby reducing the risk of breaches, misuse, and dissemination of personal data (Yuan et al., 2023).

According to Southworth et al. (2023), collaborative FL by integrating AI can foster collaboration between institutions or classrooms by allowing them to participate in model training without compromising the privacy of existing student data. Each device automatically trains a local model based on its data, and only model updates, not raw data, are shared or disseminated. This collaboration facilitates the development of more robust and generalizable AI models that can learn from diverse educational contexts. The most popular FL capabilities are training models on local data and facilitating personalized learning experiences in the learning space, offline or online. Therefore, by utilizing individual student data, customized models can be created to meet specific needs and learning styles (Ji et al., 2021). Based on these conditions, adjusting educational content, suggesting relevant learning resources, and providing differentiated feedback can be maximized. Traditional AI models often require large computing resources, creating a bottleneck for smaller institutions or those with limited resources. FL distributes the training workload across participating students' devices, reducing resource burden, making AI-supported education more accessible, and creating learning convenience (Dhananjaya et al., 2024).

To better understand the role of FL (Han and Zhang, 2020), it is necessary to study its interaction with the broader sociotechnical ecosystem of AI integration in education. This ecosystem includes a variety of stakeholders, including students, teachers, developers, and policymakers, who interact with and influence technology in the context of existing educational practices and social norms. Studies highlight the importance of considering the pedagogical implications of AI integration (Hess et al., 2021). Integrating FL successfully requires careful alignment with existing curricula and teaching philosophies, ensuring that AI tools complement and enhance, not replace, traditional pedagogy. In addition, the human element remains important, where teacher training and support are essential to efficiently maximize the potential of FL in the classroom (Lazarus et al., 2023).

Accelerated AI and federated learning provide promising avenues in education while addressing data privacy concerns. However, it is important to acknowledge the existing challenges and consider FL within the broader sociotechnical ecosystem. By fostering collaboration, addressing ethical considerations, and thoughtfully integrating FL into existing pedagogical practices, the educational community can harness FL's unique potential to create personalized and engaging learning experiences for all students in higher education (Bienkowski et al., 2012).

The digitalization and development of Mixverse in the educational landscape are undergoing transformative change as educators and institutions leverage innovative technologies to revolutionize learning and teaching practices. The rapid advancement of artificial intelligence has opened the door to new applications in various sectors, including education. Although AI's potential for personalized learning and improving educational experiences has been widely recognized, effectively integrating AI into the classroom environment still presents ongoing challenges. Artificial intelligence has great potential to personalize learning experiences, meet individual student needs, and provide educators with valuable insights into student progress. However, integrating AI into a complex and dynamic classroom environment requires careful consideration of factors beyond the purely technical aspects. Existing research primarily focuses on the technical aspects of applying AI in educational settings, with limited exploration of the broader sociotechnical ecosystem. These discovered gaps require a deeper understanding of the social and technical factors that influence the successful integration of AI in the classroom, considering the diverse stakeholders and their interactions (Lazarus Cowin et al., 2023).

This chapter aims to address this critical gap by exploring the sociotechnical ecosystem surrounding the integration of AI in educational practices, particularly in higher education. We will examine the social and technical factors that influence the successful implementation of AI-powered learning in the classroom environment, with a focus on the interactions between technology, pedagogy, and human actors.

More and more research is being conducted to investigate the potential applications of AI in education. Research has explored AI-powered tutoring systems, personalized learning platforms, and intelligent feedback mechanisms, showing promising results in various educational contexts. Experts emphasize the importance of considering the pedagogical implications of AI integration. Inderawati et al. (2024) argue that successful implementation requires careful alignment with the existing curriculum and teaching philosophy. Additionally, the human element remains important, and Aeni et al. (2024) highlight the need for educators to have the necessary skills and knowledge to effectively utilize and integrate AI tools into their teaching practices.

Additionally, the research underscores the social complexity involved in AI integration (Sadek et al., 2024) and raises concerns about potential biases embedded in

AI algorithms, necessitating responsible development and implementation practices. Additionally, Le-Nguyen and Tran (2024) emphasize the importance of addressing ethical considerations around data privacy, student agency, and the potential for AI to exacerbate existing educational disparities. By exploring the sociotechnical ecosystem, this research aims to bridge the gap between the technical development of AI and its successful integration in the complex social settings of the classroom. By examining the interactions between technology, pedagogy, and human actors, we seek to contribute valuable insights to guide the responsible and effective implementation of AI-powered learning for the benefit of all stakeholders in the educational process.

This chapter provides insights into the integration of blended learning in educational practice from a sociological perspective. Although blended learning promises to revolutionize education, its successful implementation requires a different understanding of social dynamics in the classroom. By combining theory and sociological frameworks, educators and policymakers can address challenges and harness the benefits of AI to drive academic excellence in diverse learning environments. Although they cover important and fundamental issues in many FL subjects and research domains, such as resource management, privacy and security, and application areas, as well as core system models and design, George and Wooden's paper (2023) was published at the same time as this survey, but the approach is different. We anticipate that this chapter will provide a comprehensive picture of existing issues, categorize them, and help researchers develop effective strategies for promoting a variety of new technologies and current topics. The following are the main points of this work compared to previous surveys.

1. We detail the development of students' knowledge construction with artificial intelligence-based FL approaches in the classroom, examine FL subjects and research fields in detail, classify the contributions and efforts of FL paradigms currently trending in research and industry, and provide a detailed overview and core analysis of the model FL and system design, covering important technical aspects. Additionally, we discuss the difficulties and interesting unanswered questions that lay the foundation for future FL solutions in the educational landscape.

2. To cover all domains that offer FL techniques so far, we create a taxonomy of FL application areas in the field of education and analyze sociological aspects, which are of course very different from researchers in the fields of Internet of Things, edge computing, networking, robotics, the world of networks, models, recommendation systems, cybersecurity, online classes (learning management system/LMS), wireless communications, and electric vehicles.

3. We go further by reviewing the literature to determine whether important contributions to the FL paradigm have highlighted the complex interactions between technology and social structure.

4. We classify FL as a term learning approach with optimization methods based on the objective function and calculated parameters, and we provide an in-depth evaluation of the recommended resource management strategies for FL scenarios.

DECENTRALIZATION OF EDUCATION THROUGH AI FOR THE SOCIOTECHNICAL ECOSYSTEM

An innovative method for protecting student information while incorporating artificial intelligence into classroom activities is presented by federated learning (Tan et al., 2022). By tackling current issues and encouraging teamwork, FL may revolutionize classroom instruction, opening the door to individualized learning and providing teachers with innovative resources to meet the needs of their students from all backgrounds. Future research and development efforts could position FL to play a pivotal role in the ethical and successful integration of AI into classrooms, paving the way for more engaging and tailored teaching for all students.

Integrating new technologies like artificial intelligence is currently driving a fundamental transition in the field of education. But getting students to use technology in class is only half the battle. To guarantee successful technological integration, a sociotechnical approach is necessary, which highlights the intricate interplay between social and technical elements and algorithms. The sociotechnical theory posits that the success of any system including educational systems depends on the degree to which its social and technical components are interdependent on one another. Acknowledging the interplay between social elements like pedagogy, student demands, and school culture and technological elements like the purpose, applications, and constraints of educational technology (e.g., platforms driven by artificial intelligence) is essential in educational taxonomy. To integrate technology smoothly and successfully, understanding these interconnections is crucial (Linderoth et al., 2024).

Educators must be provided with the information and training to make successful use of technology, according to multiple studies. This calls for an appreciation of technology's possibilities, its incorporation into the educational process, and the resolution of any obstacles that may arise for the benefit of students (Denny et al., 2024). Gill et al. (2024) conducted a study to explore optimal methods for integrating technological tools into traditional educational structures. Adaptations to teaching methods may be necessary, for instance, if certain technologies necessitate a change to a student-centered learning style.

Other studies look at serious problems like algorithmic bias, student agency, and data privacy. For technology to be used fairly and ethically in education, it must be developed and used responsibly. In addition, research by Kaliraj et al. (2024) investigated ways technology might help close achievement disparities and make education accessible to everyone. This necessitates attending to details like the availability of devices and dependable internet connections for students. Technology offers numerous advantages in education when a sociotechnical approach is used. To ensure that technology is seamlessly integrated into the classroom, taking into account both social and technical factors is crucial. This will lead to an engaging and productive learning environment for students, as technology can be utilized to create personalized learning paths that cater to each student's unique needs and learning style. By doing so, we can effectively support a diverse range of students. According to research (Goel et al., 2024), teachers can spend more time fostering meaningful student engagement and cooperation if they use technology to streamline administrative work.

There are no longer clear boundaries between the two phases of learning: the classroom and the workplace (Southworth et al., 2023). The average person today has more knowledge than they can process, and the workforce of the future will need to know more than that. Promoting high-quality educational opportunities across life's transitional contexts—from the family home, classroom, workplace, and beyond—is critical to fostering lifelong learning. Rather than starting with a certain level of education, professionals need to find ways to incorporate learning into their daily work. Learning also occurs in many environments, such as families, clubs, and online communities, and occurs at all ages and in almost every occupation. The findings from these specific cases should inform more generalizable learning theories and more creative and perceptive systems, practices, and evaluations in a variety of fields. With a focus on continuous learning, the benefits of learning in the classroom, in the community, at home, and work can all be combined.

Increasing professional specialization, market changes (local and global are no longer distinguishable), and new technologies for information sharing, communication, and collaboration, made possible by Internet technologies and specialized applications, are characteristics of today's society (Abulibdeh et al., 2024). The skills taught in school, such as reading, writing, and arithmetic, are not sufficient for situations like this. Some examples of new capabilities include the ability to work remotely, communicate effectively in diverse teams, build consensus, adapt knowledge artifacts, address problems with unclear or missing specifications, and utilize online and specialized collaboration.

According to Ali et al. (2024), the following are some characteristics of the sociotechnical ecosystems that have emerged throughout the past 10 years. The deep and lasting digitalization and transformation of our time are rooted in social and cultural change as well as technology. Rather than technology being the primary driver of change in complex situations, Sriram (2024) argues that human behavior and social organization undergo gradual change. He said that this includes not only hardware, networks, and software but also people, processes, regulations, laws, and the movement of raw materials and finished products, as well as many other components. In addition, these conditions require the co-design of social and technical systems using models and concepts that consider the artifacts and the social context in which the system will be used. Meta-design between AI, social interaction, and federated learning is an important part of this system because it empowers users to shape and adapt the technical system to their own needs. As we know, this system is designed for humans, not just one person. They must be able to support both individuals and groups, but groups achieve their goals through the combined efforts of all members and involve collaboration. Therefore, the question of how to create an environment that encourages individual and social efforts without disrupting either is of paramount importance.

According to Bednar and Welch (2020), we need new perspectives and methods to address the fundamental problems associated with distributed intelligence and the development of sociotechnical systems to help people live better and more fulfilling lives, so we researched to find ways of exploiting the potential of new technology based on a widespread and reliable computing environment and increasing people's proficiency in using technology in the mixverse era. Rather than conveying pre-digested information, new media and technologies can play an important role in facilitating meaningful activities, social debate and discussion, stakeholder understanding, and the formulation and resolution of real problems. Based on this worldview, the following sociotechnical arrangements are necessary to facilitate self-directed learning and learning communities in higher education. These systems should facilitate the enhancement of individual and collective knowledge; this means that they not only help individuals when they work alone but also when they collaborate with others and are members of several groups simultaneously, whether through digital interactions or interacting with objects or people in the real world. This seems to be an excellent strategy to improve the learning experience.

Improvisation, evolution, and creativity are nothing extra in a world where nothing is guaranteed. Lecturers must use it as an opportunity to find better and original answers to a problem. Sociotechnical arrangements that enable new collaborative designs are needed to empower self-directed learners and reinvigorate social innovation in learning communities. To empower users to take on the role of designers and unleash their creativity, meta-design outlines goals, methods, and procedures (Tabo, 2020). AI-enabled technologies can support self-directed learning by shifting the focus from mere consumption of information to active participation in its production. AI-based federated learning goes beyond user-centered and participatory learning design in many ways and is encouraged to share ideas and goals.

METHODOLOGY

This research used quantitative methods. We used Google Forms to create our online surveys. This survey consisted of five parts. In the first part, we tried to collect some basic information from participants, including basic demographics such as student age, gender, and self-reported level of technology competency.

A Likert scale consisting of 5 points is used to assess technology skills such as: strongly agree, agree, somewhat agree, disagree, and disagree. In the second part of the chapter, we look at how people understand the basics of AI (such as learning using AI bots); how they utilize AI in learning (such as accessing information); and how they process, share, and discuss information. Asking whether participants were aware that AI and deep learning are widely discussed in the educational community, the third subsection sought to explore the understanding of AI as a topic in learning. In the fourth part, we asked students to rate their feelings and attitudes towards artificial intelligence, considering their answers to the following questions: (1) how they view AI about humans and society; (2) how they feel towards artificial intelligence as a tool for socializing information and knowledge; (3) how they feel about the development of AI in the federated learning model; (4) how they feel about AI regarding its potential to improve learning outcomes and its integration into the teacher professional education curriculum. Using five different 5-point Likert scales, these sections were scored: (5) strongly agree, (4) agree, (3) somewhat agree, (2) disagree, and (1) strongly disagree. The five components included questions about respondents' opinions about the use of artificial intelligence in social interactions in the classroom and their hopes for future education in this area Figure 3.1.

FIGURE 3.1 Playing AI on federated learning model.

STATISTICAL ANALYSIS

The results of data analysis, namely the average value and standard deviation of continuous variables, are displayed. Data presented as percentages and frequencies are known as categorical variables. We described "strongly agree" and "agree" as agreement for the descriptive statistics, and "agree" and "strongly agree" as disagreement. The results of this research data were processed using the Statistical Package for Social Sciences (SPSS version 20) used for data entry and analysis. Data was analyzed using descriptive statistics to obtain percentages, frequencies, standard deviations, and means. The reliability and validity tests were determined by calculating Cronbach's alpha, which produced 0.82.

RESULTS: SURVEY RESPONDENTS

A total of 43 respondents from professional teacher education took the time to respond and were considered for analysis. Upon completion of the study, precise data will be accessible concerning the enrollment figures and demographic profile of professional teacher education (PPG) students for the 2023 and 2024 academic years. Preliminary findings indicate that there were 20 first-year teachers and 13 second-year students, with a gender distribution of approximately 35% male and 65% female. The majority of participants were aged 22 to 25, with 24 respondents falling within this age range and 19 specifically being 22 to 25 years old. In terms of device ownership, all respondents owned smartphones and laptops. Most respondents had prepaid internet subscriptions using smartphone data to access the internet.

OVERALL PERCEPTION AND CATEGORY OF RESPONDENTS' RESPONSES TOWARD SOCIOTECHNICAL AI

Studying new technology courses in the first year and studying design thinking in the second year, this research intends to investigate how professional teacher education students view AI in education with a federated learning model. AI-based education is strongly supported by the results of this poll. Learning with AI is often seen as preferable by students in higher education, and they state that taking this class is their first experience interacting with AI and getting to know AI.

This leads the researchers to the confident conclusion that AI learning with the FL model is very attractive for professional teacher education students at Al Asyariah Mandar University. Even though inadequate internet conditions due to having to prepare quotas do not provide a big obstacle, this has a positive impact on the atomic habits of those who adapt to AI. This activity certainly uses creative ways of teaching and learning so that the educational process continues and is meaningful. In many cases, students perceive learning with AI integration as more comfortable and safer. In conclusion, the findings of recent research support students' positive attitudes toward AI learning with the federated learning model described in Table 3.1.

The results of survey data analysis regarding respondents' opinions regarding the integration of AI with federated learning in the sociotechnical ecosystem model described in Table 3.1 show that students generally have a positive attitude towards artificial intelligence in education, especially with federated learning. In both statements, students demonstrated a reasonable level of understanding regarding the benefits of AI in education (Statement 1) and their belief that AI-driven learning tools are becoming more common (Statement 2). As stated in Statement 3, they enjoy leveraging AI with FL, and as stated in Statement 4, they recognize its potential in accessing learning materials. Many people feel comfortable disclosing personal information to AI systems (Statement 5). Students demonstrated a greater level of comfort in terms of AI information processing, while they generally felt comfortable using AI tools (Statement 6). Statement 8 stated that students had a positive impression of sharing and debating material in class using AI tools. Personalized learning (Statement 9) and collaboration (Statement 10) are two areas that this research suggests could be beneficial.

In the "mixverse era" (Statement 11), students recognize the importance of AI and deep learning, and they feel comfortable discussing AI in class (Statement 12). According to Statement 13, socio-technical skills can be improved by incorporating AI-based learning into the curriculum. The broader impact of AI on our ability to understand the universe may require further investigation (Statement 14). Finally, the survey shows that students support the development of AI in federated learning (Statement 18), believe that AI can improve learning outcomes (Statement 19), feel comfortable relying on AI for learning (Statement 17), have positive perceptions of the role of AI in society (Statement 16), and strongly support integrating AI into teacher education (Statement 20). The results of this research show a bright future for educational technology that utilizes artificial intelligence in the sociotechnical ecosystem, as shown in Figure 3.2.

Using the descriptive statistics in Table 3.2, this study investigated student sentiment regarding educational approaches involving federated learning and artificial intelligence. By providing a brief overview of the data, descriptive statistics help us understand the data's central tendency, variability, and the shape of the distribution of students' answers. There is only one variable in Table 3.2, and the descriptive statistics aimed at this variable might be a survey question asking students' opinions about FL and AI models. This question uses a scale with a range of 39 points, which

TABLE 3.1

Student Survey

Statements	N	Minimum	Maximum	Mean	Std. Deviation
Part 1: Understanding AI basics through FL model (5 items)	43	3.00	5.00	4.4651	.59156
1. I understand the functions and benefits of artificial intelligence (AI) in education through the federated learning model.					
2. I interact with AI-powered tools in class and outside of class to access information about education or teaching.	43	3.00	5.00	4.2093	.67465
3. I enjoy learning to use AI with federated learning in class and outside of class.	43	2.00	5.00	4.3721	.65550
4. I believe that AI can be used to access information for learning purposes (e.g., searching websites, and finding relevant resources).	43	3.00	5.00	4.2791	.66639
5. I feel comfortable sharing information and interacting socially with AI-powered learning tools.	43	2.00	5.00	4.2558	.84777
Part 2: Utilizing AI in learning (5 items)					
6. I feel comfortable using AI-powered tools in the learning process.	43	2.00	5.00	3.9535	.78539
7. I believe AI helps process the information to learn.	43	3.00	5.00	4.3023	.67383
8. I believe that it is easy to share information with others through AI-powered tools (e.g., online discussion forums) and discuss in the classroom.	43	3.00	5.00	4.3023	.63751
9. I find it useful to use AI to personalize the learning experience for different students.	43	3.00	5.00	4.3953	.62257
10. I believe AI can be a valuable tool for collaborating and learning with others.	43	3.00	5.00	4.4186	.62612
Part 3: AI as a learning topic (5 items)					
11. I believe AI and deep learning essential in the mixverse era.	43	3.00	5.00	4.1860	.69884
12. I am comfortable carrying out discussion activities using AI in education.	43	2.00	5.00	4.4419	.70042
13. I believe that AI-based learning as part of the regular curriculum is useful for improving socio-technical skills.	43	2.00	5.00	4.3256	.74709
14. Learning with AI can help students better understand the world around them.	43	2.00	5.00	3.5581	.82527
15. Students should be involved in discussions about the use of AI in education.	43	3.00	5.00	4.3721	.61811
Part 4: Attitudes toward AI (5 items)					
16. I see the role of AI as very effective in relation to humans and society in general.	43	3.00	5.00	4.2791	.62965
17. I feel comfortable relying on AI tools to learn and disseminate information and knowledge.	43	3.00	5.00	4.3488	.68604
18. I agree with the development of AI in a federated learning model (where the data remains on the user's device).	43	3.00	5.00	4.1163	.69725
19. I feel the potential of AI to efficiently improve student learning outcomes.	43	3.00	5.00	4.1628	.75373
20. I believe that the integration of AI into teacher professional education curricula is very important.	43	3.00	5.00	4.3488	.68604
Valid N (listwise)	43				

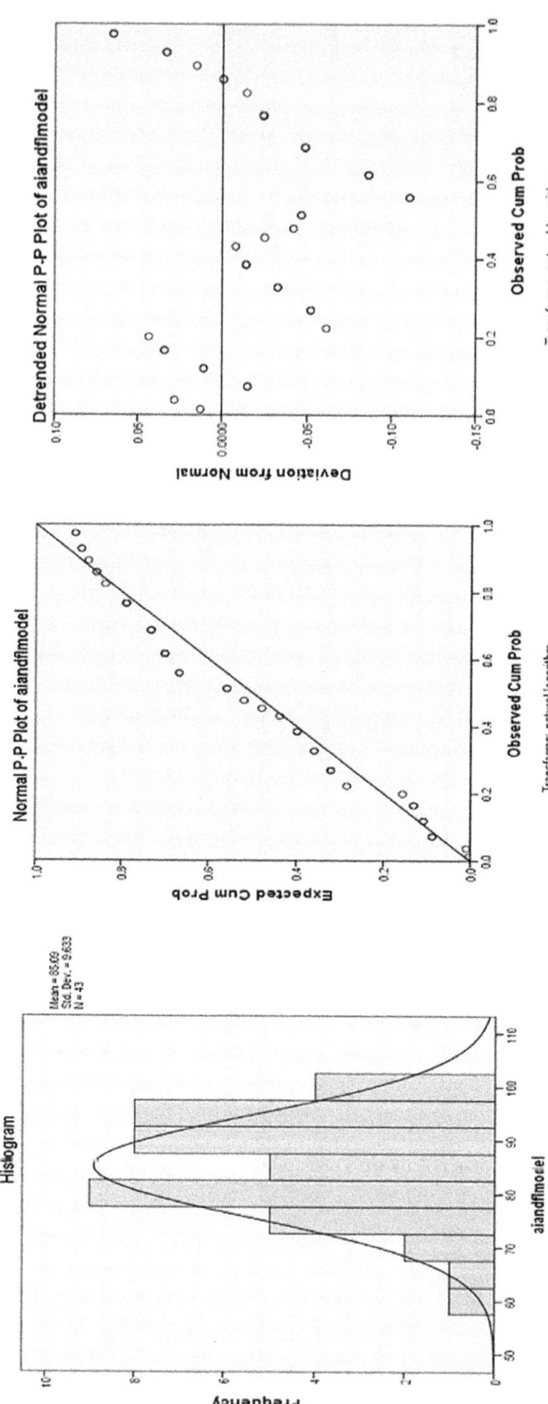

FIGURE 3.2   Histogram and normal P-P plot.

TABLE 3.2

Description of Statistics

	N	Minimum	Maximum	Mean	Std. Deviation	Skewness		Kurtosis	
	Statistic	Statistic	Statistic	Statistic	Statistic	Statistic	Std. Error	Statistic	Std. Error
AI-FL model	43	60	99	85.09	9.633	–.554	.361	–.194	.709
Valid *N* (listwise)	43								

TABLE 3.3

Test Statistics

Algorithm Test	AI-FL model
Chi-square	8.884[a]
DF	22
Asymp. sig.	.994

Twenty-three cells (100.0%) have expected frequencies less than 5. The minimum expected cell frequency is 1.9.

is indicated by a minimum score of 60 and a maximum score of 99. With this scale being a Likert scale, survey statements regarding FL and AI models can cover an alternative range from strongly disagree (60) to strongly agree (99). Table 3.2 shows a mean score of 85.09 and a standard deviation of 9.633. Students generally seem to have a positive view of FL and AI models. However, there is a lot of variation in these sentiments, as seen in the standard deviation. In other words, some children have a more optimistic viewpoint than others. Kurtosis (.361) and skewness (–.554) values: A slight negative skew indicates a slight bias towards higher scores, which could be an indication of a positive skew attitude. If the kurtosis value is close to 3, it means the distribution is almost normal.

Data results on the chi-square test (Table 3.3) summarize the chi-square test statistics applied to the AI variable with the federated learning model on the social-technical ecosystem (likely representing student responses to a survey about attitudes towards AI and the federated learning model). It is known that the data shows that the chi-square test statistic (8.884a) is relatively low. The degrees of freedom (DF) are 22, indicating the number of categories used to classify student responses. Asymptotic significance (asymp. sig.) is a very high value (0.994), exceeding the common threshold of 0.05 for statistical significance.

The findings of the reliability analysis for the AI variable with the federated learning model on the social-technical ecosystem, which uses Cronbach's alpha, are displayed in Table 3.4. This variable is a collection of 20 survey questions intended to measure how students feel about educational approaches that use federated learning

---

TABLE 3.4
Reliability Statistics

Cronbach's Alpha	N of Items
.943	20

---

and artificial intelligence. Survey data results with 20-item questions measuring opinions on AI and FL models and Cronbach's alpha show how well these questions measure the same basic concepts. The very high result of 0.943 exceeds the generally recognized criterion of 0.7 for satisfactory internal consistency. This suggests a strong correlation between survey items and their ability to measure cohesive ideas. With such a high alpha value, researchers can be confident that this research variable is a reliable indicator of student views.

The research results showed that there was greater confidence that the survey could reflect students' true opinions.We can now more accurately identify instances where students' beliefs differ across categories or contexts. Although a high alpha implies that the survey is internally consistent, it does not guarantee content validity. The results of this research survey have a very high level of internal consistency, as indicated by a Cronbach's alpha value of 0.943. Due to its high alpha value, this instrument is overall reliable for measuring students' opinions about the use of FL and AI in the classroom.

DISCUSSION

Table 3.1 examines a student survey asking about their knowledge, use, and feelings towards AI in education, specifically as it relates to federated learning. According to the survey results presented in Table 3.1, students who have received FL instruction demonstrated a moderate level of comprehension regarding the characteristics and advantages of artificial intelligence in the educational setting (Statement 1, mean = 4.47). They also believe that the use of AI-powered tools for educational purposes is increasing, both inside and outside the classroom (Statement 2, mean = 4.21). The next statement claimed that using FL and AI was a pleasant experience for students, according to Statement 3 (mean = 4.37). Furthermore, from the data, it is known that there is a high consensus regarding AI's ability to access learning material (Statement 4, mean = 4.28). Additionally, factors influencing the ease with which individuals share data and interact with AI-powered learning tools vary (Statement 5, mean = 4.26).

Statement 7, mean = 4.30, indicates a higher level of comfort for students regarding the involvement of AI in processing learning information, while Statement 6, mean = 3.95, indicates that students feel comfortable using AI tools overall. Statement 8, mean = 4.30, indicates that students like sharing information through AI-powered tools and discussing it in class. According to the data results in Statement 9, the average score is 4.39, which shows that the possibility of AI for individualized educational experiences is recognized. According to students in the teacher education profession, AI has the potential to improve learning and collaboration (Statement 10, mean = 4.42). In Statement 11, the mean is 4.19, recognizing the importance of AI and deep learning in the "mixverse era" (possibly referring to hybrid physical-virtual reality settings). Statement 12, mean = 4.44, indicates that students feel comfortable talking about AI in class. Then the poll shows that socio-technical skills can be improved by incorporating AI-based learning into the curriculum, which can be seen in Statement 13 (mean = 4.33).

A more thorough investigation of this subject may be needed, although there is an encouraging trend in the use of AI to learn about the world (Statement 14, mean = 3.56). Then, Statement 15, mean = 4.37, shows that students support participation in class debates regarding the use of AI in the classroom. Most of the students expressed positive views regarding the usefulness of AI for humans and society (Statement 16, mean = 4.28). Based on Statement 17, the average score of students' self-confidence in using AI technology for learning and disseminating knowledge is 4.35.

There is a consensus on the need to develop AI in FL, where user data is stored on the device, according to Statement 18 (mean = 4.12). Statement 19, mean = 4.16, shows that the public has a positive impression of AI's ability to improve learning outcomes. Then, according to Statement 20 (mean = 4.35), this research shows that there is great support for incorporating AI into teacher professional education. Based on the survey, it is known that students generally have a favorable opinion of AI in education. This is especially true when it comes to federated learning. The positive impact of AI on education, teamwork, and individualization has been recognized by students. This poll opens the door to further research and the potential use of AI-powered learning aids in the classroom, although this may require further investigation in certain areas, such as the global impact of AI.

Additional details regarding the appropriate choice of survey questions and answers are needed for a more thorough interpretation, as this analysis relies on a single variable. Furthermore, descriptive statistics by themselves cannot reveal group differences and causal relationships. Students' perspectives on the use of FL and AI models in the classroom can be better understood with the use of descriptive statistics (Tables 3.1 and 3.2). Although every respondent's answers are different, the data shows an optimistic future. We can look at the relationships between these attitudes and other student traits or learning outcomes in future studies. We can learn more about students' views on the pros and cons of using FL and AI models in the classroom by looking at the survey questions themselves and using inferential statistics.

This chapter provides insights into the integration of blended learning in educational practice from a sociological perspective. Although blended learning promises to revolutionize education, its successful implementation requires a different understanding of social dynamics in the classroom. By combining theory and sociological frameworks, educators and policymakers can address challenges and harness the benefits of AI to drive academic excellence in diverse learning environments. Although they cover important and fundamental issues in many FL subjects and research domains, such as resource management, privacy and security, and application areas, as well as core system models and design, Yu et al. (2023) and Sartori and Theodorou (2022) were published at the same time as this survey, but the approach is different. We anticipate that the suggested survey will provide a comprehensive picture of existing issues, categorize them, and help researchers develop effective strategies for promoting a variety of new technologies and current topics.

The following are the main points of this work compared to previous surveys. (1) We detail the development of students' knowledge construction with artificial intelligence–based FL approaches in the classroom, examine FL subjects and research fields in detail, classify the contributions and efforts of FL paradigms currently trending in research and industry, and provide a detailed overview and core analysis of the model FL and system design, covering important technical aspects. Additionally, we discuss the difficulties and interesting unanswered questions that lay the foundation for future FL solutions in the educational landscape. (2) To cover all domains that offer FL techniques so far, we create a taxonomy of FL application areas in the field of education and analyze sociological aspects which are of course very different from researchers in the fields of Internet of Things, edge computing, networking, robotics, the world of networks, models, recommendation systems, cyber security, online classes, wireless communications and electric vehicles. (3) We go further by reviewing the literature to determine whether important contributions to the FL paradigm have highlighted the complex interactions between technology and social structure. (4) We classify FL as a term learning approach with optimization methods based on the objective function and calculated parameters, and we provide an in-depth evaluation of the recommended resource management strategies for FL scenarios.

Based on the results of this research, we realize that the inseparability of social and technological components in work is important from a socio-technical systems perspective, as has been done by Farrow (2023) and Schoenherr (2024), which is similar to this research, in which humans, tasks, technology, AI, physical space, and organization are the five pillars underlying the work system model. Workflow refers to the interactions between various components of a working system that develop over time as individuals, aided by tools and technology, perform activities in a physical setting within an educational framework. The sociotechnical system with the federated learning model is only one part where the application of AI technology influences changes in the relationship between elements of the work system in the Merdeka class. The new system's integration will significantly alter how teacher professional education students interact with technology, tasks, and their peers within the learning environment. This involves ensuring that technology seamlessly fits into their workflow, considering factors like task sequencing and the timing of technology usage. Although the results of the data from respondents did not claim that negative consequences (such as distraction and increased workload) can result from failure to integrate technology into the workflow during the learning process, good outcomes (such as acceptance and utilization) can occur when technology is implemented. Therefore, AI must be built to complement, not hinder, workflows in education by integrating it with other system components. For AI to be successfully integrated into these workflows, all aspects of the work system and the various levels and dimensions of workflow integration need to be considered.

This research is different from other research in that implementing an AI system not only emphasizes the development of AI technology but also confirms it with students' real-world environments, such as information obtained from AI that remains confirmed with other scientific sources. Of course, the results of this study refute the arguments of Salwei and Carayon (2022), which state that artificial intelligence is often created in a vacuum, ignoring the real world in which it will be used. Artificial intelligence has the potential to improve and support the learning process; however, the success of this technology depends on the acceptance of students and lecturers who teach with AI (Muthmainnah et al., 2023). Especially in decision-making and criticizing information, this happens. A lack of confidence and reluctance to utilize technology can arise from unclear communication about both the technology and its recommendations.

CONCLUSION

Our definition and operationalization of the sociotechnical scope that focuses on AI technology with federated learning models shows significant promise. Based on these results, we can better understand masking in general and find more socially and technically oriented methods in the learning process in Merdeka classes or the independent curriculum (KURMA). We have briefly reviewed the capabilities and limitations of various sociotechnical approaches to addressing this problem as a starting point. This work is purely the result of the opinions of professional teacher education students. We base our considerations on the premise that AI systems are getting a positive response and developing as part of the digitalization of education. Therefore, the application of AI in the classroom and the social and sociotechnical interactions formed with the federated learning model requires human intervention. Additionally, even if AI systems develop further in the future, we still think that humans will play an important role. Because our complementary approach views humans and technology as fundamentally different, we operate on this premise. Therefore, they did not compete to see which of the two had quantitatively comparable superior abilities. The reality is that their different strengths work better together. Computers excel in computing, but humans excel in thinking, critical thinking, systems thinking, and collaboration. Therefore, no matter how good human intelligence or technological capabilities are, an intelligent human–technology combination will always outperform both, as has been proven through this research. When it comes to AI systems, we strongly believe that this is especially the case in fields where understanding and responsibility are critical human traits, where combined human AI systems will work best. Rather, it is a human need for the conditions necessary for education to have a high level of involvement and higher-order thinking skills. The reason is that by involving technology, showing initiative, showing dedication, or accepting responsibility cannot be done simply by direction or assignment, but proper work design is essential so that human contribution can be motivated authentically by the design of socio-technical systems, i.e., the level of allocation of human functions to machines in the mixverse class.

REFERENCES

AbdulRahman, S., Tout, H., Ould-Slimane, H., Mourad, A., Talhi, C., & Guizani, M. (2020). A survey on federated learning: The journey from centralized to distributed on-site learning and beyond. *IEEE Internet of Things Journal, 8*(7), 5476–5497.

Abulibdeh, A., Zaidan, E., & Abulibdeh, R. (2024). Navigating the confluence of artificial intelligence and education for sustainable development in the era of industry 4.0: Challenges, opportunities, and ethical dimensions. *Journal of Cleaner Production*, 140527.

Aeni, N., Khang, A., Al Yakin, A., Yunus, M., & Cardoso, L. (2024). Revolutionized teaching by incorporating artificial intelligence chatbot for higher education ecosystem. In *AI-Centric Modeling and Analytics* (pp. 43–76). CRC Press.

Aledhari, M., Razzak, R., Parizi, R. M., & Saeed, F. (2020). Federated learning: A survey on enabling technologies, protocols, and applications. *IEEE Access*, *8*, 140699–140725.

Ali, M., Wood-Harper, T., & Wood, B. (2024). Understanding the technical and social paradoxes of learning management systems usage in higher education: A sociotechnical perspective. *Systems Research and Behavioral Science*, *41*(1), 134–152.

Al Yakin, A., Khang, A., & Mukit, A. (2023). Personalized social-collaborative iot-symbiotic platforms in smart education ecosystem. In *Smart Cities* (pp. 204–230). CRC Press.

Bednar, P. M., & Welch, C. (2020). Socio-technical perspectives on smart working: Creating meaningful and sustainable systems. *Information Systems Frontiers*, *22*(2), 281–298.

Bhutoria, A. (2022). Personalized education and artificial intelligence in the United States, China, and India: A systematic review using a human-in-the-loop model. *Computers and Education: Artificial Intelligence*, *3*, 100068.

Bienkowski, M., Feng, M., & Means, B. (2012). Enhancing teaching and learning through educational data mining and learning analytics: An issue brief. *Office of Educational Technology*. US Department of Education.

Čolaković, A., & Hadžialić, M. (2018). Internet of Things (IoT): A review of enabling technologies, challenges, and open research issues. *Computer Networks*, *144*, 17–39.

Denny, P., Gulwani, S., Heffernan, N. T., Käser, T., Moore, S., Rafferty, A. N., & Singla, A. (2024). Generative AI for education (GAIED): Advances, opportunities, and challenges. *arXiv preprint arXiv:2402.01580*.

Dhananjaya, G. M., Goudar, R. H., Kulkarni, A., Rathod, V. N., & Hukkeri, G. S. (2024). A digital recommendation system for personalized learning to enhance online education: A review. *IEEE Access*, *12*, 34019–34041. doi: 10.1109/ACCESS.2024.3369901

Donnelly, R. (2017). Blended problem-based learning in higher education: The intersection of social learning and technology. *Psychosociological Issues in Human Resource Management*, *5*(2), 25–50.

Farrow, R. (2023). The possibilities and limits of XAI in education: A socio-technical perspective. *Learning, Media and Technology*, *48*(2), 266–279.

George, B., & Wooden, O. (2023). Managing the strategic transformation of higher education through artificial intelligence. *Administrative Sciences*, *13*(9), 196.

Gill, S. S., Xu, M., Patros, P., Wu, H., Kaur, R., Kaur, K.,.. .& Buyya, R. (2024). Transformative effects of ChatGPT on modern education: Emerging era of AI chatbots. *Internet of Things and Cyber-Physical Systems*, *4*, 19–23.

Goel, P. K., Singhal, A., Bhadoria, S. S., Saraswat, B. K., & Patel, A. (2024). AI and machine learning in smart education: Enhancing learning experiences through intelligent technologies. In *Infrastructure Possibilities and Human-Centered Approaches with Industry 5.0* (pp. 36–55). IGI Global.

Han, Y., & Zhang, X. (2020, April). Robust federated learning via collaborative machine teaching. In *Proceedings of the AAAI Conference on Artificial Intelligence* (Vol. 34, No. 4, pp. 4075–4082). AAAI Press.

Hess, D. J., Lee, D., Biebl, B., Fränzle, M., Lehnhoff, S., Neema, H.,.. .& Sztipanovits, J. (2021). A comparative, sociotechnical design perspective on responsible innovation: Multidisciplinary research and education on digitized energy and automated vehicles. *Journal of Responsible Innovation*, *8*(3), 421–444.

Inderawati, R., Apriani, E., Jaya, H. P., Saputri, K., Wijayanti, E., & Muthmainnah, M. (2024). Promoting students' writing by using essay writing GPT: A mix method. In *Advanced Applications of Generative AI and Natural Language Processing Models* (pp. 249–264). IGI Global.

Ji, S., Saravirta, T., Pan, S., Long, G., & Walid, A. (2021). Emerging trends in federated learning: From model fusion to federated x learning. *arXiv preprint arXiv:2102. 12920.*

Kaliraj, P., Singaravelu, G., & Devi, T. (Eds.). (2024). *Transformative Digital Technology for Disruptive Teaching and Learning.* CRC Press.

Konečný, J., McMahan, H. B., Yu, F. X., Richtárik, P., Suresh, A. T., & Bacon, D. (2016). Federated learning: Strategies for improving communication efficiency. *arXiv preprint arXiv:1610.05492.*

Lazarus, M. D., Truong, M., Douglas, P., & Selwyn, N. (2024). Artificial intelligence and clinical anatomical education: Promises and perils. *Anatomical Sciences Education, 17*(2), 249–262.

Lazarus Cowin, J., Oberer, B., Lipuma, J., Leon, C., & Erkollar, A. (2023, September). Accelerating higher education transformation: Simulation-based training and AI coaching for educators-in-training. In *International Conference on Interactive Collaborative Learning* (pp. 532–541). Springer Nature Switzerland.

Le-Nguyen, H. T., & Tran, T. T. (2024). Generative AI in terms of its ethical problems for both teachers and learners: Striking a balance. In *Generative AI in Teaching and Learning* (pp. 144–173). IGI Global.

Linderoth, C., Hultén, M., & Stenliden, L. (2024). Competing visions of artificial intelligence in education—A heuristic analysis on sociotechnical imaginaries and problematizations in policy guidelines. *Policy Futures in Education*, 14782103241228900.

Muthmainnah, M., Khang, A., Al Yakin, A., Oteir, I., & Alotaibi, A. N. (2023a). An innovative teaching model: The potential of metaverse for English learning. In *Handbook of Research on AI-Based Technologies and Applications in the Era of the Metaverse* (pp. 105–126). IGI Global.

Muthmainnah, M., Obaid, A. J., Al Yakin, A., & Brayyich, M. (2023b, June). Enhancing computational thinking based on virtual robot of artificial intelligence modeling in the English language classroom. In *International Conference on Data Analytics & Management* (pp. 1–11). Springer Nature.

Ng, P. H., Chen, P. Q., Sin, Z. P., Jia, Y., Li, R. C., Baciu, G.,. . .& Li, Q. (2023, November). From classroom to metaverse: A study on gamified constructivist teaching in higher education. In *International Conference on Web-Based Learning* (pp. 92–106). Springer Nature.

Ouyang, X., Xie, Z., Zhou, J., Huang, J., & Xing, G. (2021, June). Clusterfl: A similarity-aware federated learning system for human activity recognition. In *Proceedings of the 19th Annual International Conference on Mobile Systems, Applications, and Services* (pp. 54–66). Association for Computing Machinery. https://doi.org/10.1145/3458864. 3467681

Pedro, F., Subosa, M., Rivas, A., & Valverde, P. (2019). Artificial intelligence in education: Challenges and opportunities for sustainable development. In *Working Papers on Education Policy*, 7 [17], 46 pages. UNESCO Digital Library.

Rajbongshi, S. K., Sarmah, K., & Sarmah, S. (2022). Unveiling the tapestry: Federated learning challenges and opportunities in the Indian educational landscape. *Mathematical Statistician and Engineering Applications, 71*(2), 747–762.

Sadek, M., Kallina, E., Bohné, T., Mougenot, C., Calvo, R. A., & Cave, S. (2024). Challenges of responsible AI in practice: scoping review and recommended actions. *AI & Society*, 1–17.

Salwei, M. E., & Carayon, P. (2022). A sociotechnical systems framework for the application of artificial intelligence in health care delivery. *Journal of Cognitive Engineering and Decision Making, 16*(4), 194–206.

Sartori, L., & Theodorou, A. (2022). A sociotechnical perspective for the future of AI: Narratives, inequalities, and human control. *Ethics and Information Technology, 24*(1), 4.

Schoenherr, J. R., Chiou, E., & Goldshtein, M. (2024). Building trust with the ethical affordances of education technologies: A sociotechnical systems perspective. In *Putting AI in the Critical Loop* (pp. 127–165). Academic Press.

Southworth, J., Migliaccio, K., Glover, J., Reed, D., McCarty, C., Brendemuhl, J., & Thomas, A. (2023). Developing a model for AI Across the curriculum: Transforming the higher education landscape via innovation in AI literacy. *Computers and Education: Artificial Intelligence, 4*, 100127.

Sriram, G. K., Malini, A., & Santhosh, K. M. R. (2024). State of the art of artificial intelligence approaches toward driverless technology. *Artificial Intelligence for Autonomous Vehicles*, 55–74.

Tabo, G. (2020). Designing infrastructures for learning: Technology and human praxis: A sociotechnical and sociocultural perspective to designing IT infrastructures in a resource constrained settings. Aalborg University's Research Portal (aau.dk).

Tan, A. Z., Yu, H., Cui, L., & Yang, Q. (2022). Towards personalized federated learning. *IEEE Transactions on Neural Networks and Learning Systems, 34*(12), 9587–9603. doi: 10.1109/TNNLS.2022.3160699

Wei, K., Li, J., Ding, M., Ma, C., Yang, H. H., Farokhi, F.,. . .& Poor, H. V. (2020). Federated learning with differential privacy: Algorithms and performance analysis. *IEEE Transactions on Information Forensics and Security, 15*, 3454–3469.

Yu, X., Xu, S., & Ashton, M. (2023). Antecedents and outcomes of artificial intelligence adoption and application in the workplace: The socio-technical system theory perspective. *Information Technology & People, 36*(1), 454–474.

Yuan, L., Sun, L., Yu, P. S., & Wang, Z. (2023). Decentralized federated learning: A survey and perspective. *arXiv preprint arXiv:2306.01603*.

Zhang, T., Gao, L., He, C., Zhang, M., Krishnamachari, B., & Avestimehr, A. S. (2022). Federated learning for the internet of things: Applications, challenges, and opportunities. *IEEE Internet of Things Magazine, 5*(1), 24–29.

4 Real-Time Implementation of Improved Automatic Number Plate Recognition Using Federated Learning

*M. Venkatanarayana, Syed Zahiruddin,
and Ahmed A. Elngar*

INTRODUCTION

An invention that came out of the UK for ANPR systems was developed by the police scientific development branch in 1976. Computer vision systems have real-world applications, such as the automatic detection and recognition of number plates. The objective is to establish a framework that can digitally record cars going past a specific point and then electronically identify them by finding the license plate in the picture, separating the letters from the plate, and identifying them. Some basic applications for number plate recognition systems are toll collection, traffic congestion control, personal security, visitor management systems, and tracking of stolen vehicles. However, it has gained much attention during the past 10 years in tandem with advancements in digital cameras and reduced computational complexity. It can easily extract and recognize vehicle number plate characters from a captured image. Essentially, it requires a digital camera that can capture an image, then the location of the number plate in the image, and then extract the characters using an open-source Python library to translate the pixels into numerically readable characters. The system is low cost compared to other existing ANPR systems.

In addition to not being very resilient, the previous solutions either used computationally expensive feature-based approaches like Hough transform or edge detection, or they used artificial neural networks, which need enormous amounts of training data. Lightweight operation and real-time number plate recognition are the goals of the proposed ANPR system. There are three stages to how an ANPR system works. The first is finding and taking a picture of a vehicle. The second is finding and extracting the license plate from that picture. Step three involves breaking down the image into its component characters using image segmentation and then using optical character recognition (OCR) to identify every character using a database containing information about each alphanumeric character.

Here, a federated learning model is proposed to extract a number plate from the captured image. Characters and numbers are extracted from tiny number plates using

DOI: 10.1201/9781003482000-4

OCR with the easy OCR/tesseract Python library. The proposed system is tested in real time, and improved results are observed. The section on "Literature Survey" presents a literature survey; procedural steps to train YOLO models are illustrated in the section titled "Procedural Steps to Do ANPR"; and methodology for federated learning is introduced in the section on "Federated Learning", followed by a results and discussion section which is followed by a conclusion.

LITERATURE SURVEY

Automatic number plate detection is the subject of various author-proposed methods. In 2013, the team of Patel et al. put out a machine learning-based solution for automatic license plate recognition. An infrared camera records the input, which is then pre-processed with noise reduction and contrast enhancement. Next, image features are extracted using contour tracing. Canny's edge detection is applied to detect the character edges of the number plate. Isolating the numbers and characters is the goal of segmentation. To identify the text, artificial neural networks (ANNs) compare it to previously seen patterns. The MATLAB software was used to simulate the entire system [1].

Arsenovic et al., in 2017, exhibited a review based on ANPR that possesses a camera module, sensor, control unit, GSM, and cloud server. The captured images are converted to grayscale and enhanced by changing the histogram. Sobel's edge detection method is used to detect edges. The next step is to process the morphological images, and after that, the edge-detected images are segmented. Finally, the machine learning approach is utilized to recognize the characters [2].

Using a deep learning network, the approach for identifying the number plate for the Turkish data set was proposed in 2018 by Kilic et al. TensorFlow, the Keras library for deep learning, and MATLAB simulation were utilized. Training used 75% of the photos, testing 25%, and validation 5%. Adaptive Gaussian thresholding, morphological modifications, and median blur smoothening were among the image processing methods utilized in this approach. The next step was to train the convolutional neural network (CNN) model using the altered images. After the decryption method, the extracted picture characteristics are fed into an LSTM network. This method offered good accuracy for the detection of numbers, letters, and all the characters [3].

A survey done by Ghadage et al. in 2020 on ANPR used Canny's edge detection for plate detection. To achieve better results, character segmentation was applied utilizing image binarization, CCA, and vertical and horizontal projection. Optical character recognition, template matching, and artificial neural networks for character recognition came next [4].

Gnanaprakash et al., in 2020, recommended a method for ANPR. Before anything else, you have to turn the video into still images and find the car in each one. After the cars are spotted, the following stage is to identify their license plates, and then the last step is to recognize the characters. The suggested DL model streamlines training with the help of the Image AI library. For vehicle identification, the accuracy reached 97%, for license plate localization it was 98%, and for character recognition, it was 90% [5].

Many researchers contributed their work to propose ANPR using various methods. This chapter focuses on capturing a car image using a Raspberry Pi camera. The number plate image is detected using the open-cv Python library, and using OCR libraries, the text is extracted. In real time, segmentation using open-cv to crop a tiny number plate region is a challenging task [6–11]. To overcome this, advanced YOLO models are utilized to isolate the desired number plate more precisely. This chapter proposes the YOLOv7 model to improve accuracy in car identity verification, vehicle license plate tracking, and text recognition.

ANPR LIMITATION FACTORS

Several factors that influence the performance of the ANPR system [12–17] are as follows,

- Poor resolution due to hardware limitations.
- Distorted images due to motion blur.
- Because of overexposure and shadow reflection, the lighting is dull and there is little contrast.
- A dust particle or other item blocking the plate.
- Reading license plates from the front and the back are different.
- To scan license plates, the vehicle lane can shift the camera's vertical orientation.
- Countries or states do not adhere to the same standards. No two vehicles from the same nation or state have license plates that are identical in style and typeface.
- The vehicle's speed.

These limitations can be rectified using the proposed method.

PROCEDURAL STEPS TO DO ANPR

Images are pre-processed, number plates are located, characters are segmented, and characters are recognized in the system depicted in Figure 4.1 [18–25].

1. Pre-processing:
   a) Import a color image, convert it to grayscale, and show it to the user.
   b) Gaussian Smoothing: It reduces noise from the imported image and improves the image quality. It enhances the character recognition by doing this.
   c) Plate Localization: Edge detection and federated learning are prominent approaches for localizing number plates. Edge detection minimizes image data except for edges. Next, we resize a pre-made slide window to find the spot with the most white pixels to isolate the number plate area. Then, we display the divided sub-image on the localization panel.
   d) ML Segmentation: A data set of car front and back images, annotated with number plate information, used to train and evaluate a model.

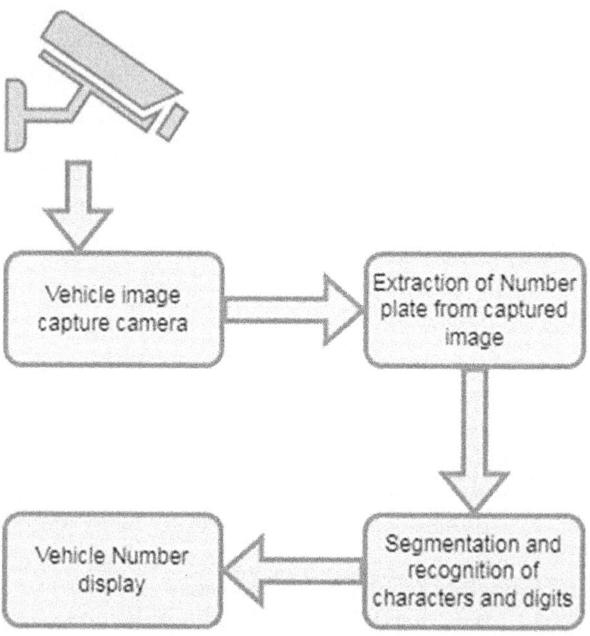

FIGURE 4.1   Block diagram of the proposed ANPR system.

2. Character Segmentation:
   a) Image binarization: Create a proportional image for the original license plate, encode it, and display it in the localization panel.
   b) Then, count the number of white pixels in each column of the binary sub-image using vertical projection.
   c) Character segmentation with vertical projection creates several components from a binary sub-image. The segmentation panel displays each segment, which should have one character.
3. Character Recognition: To identify the character therein, it applies character recognition to segmented regions of the sub-image.

FEDERATED LEARNING

i) Introduction to Federated Learning

Federated learning is decentralized machine learning. In the field of machine learning, it is common practice to train data by gathering information from various edge devices such as mobile phones and laptops. This data is then consolidated and sent to a centralized server. Machine learning algorithms acquire and process data, autonomously train themselves, and subsequently make predictions for newly generated data. Google, Amazon, and Microsoft are the leading players in the AI business, providing cloud-based AI products and APIs. Traditional AI methods include the transmission of sensitive user data to servers where models are trained.

ii) Types of Federated Learning

The three primary types of federated learning are:

1. **Horizontal Federated Learning** involves the distribution of data samples among devices or servers, where the model is trained jointly.
2. **Vertical Federated Learning** involves the partitioning of features among several devices, and the model is then trained using these complimentary features.
3. **Federated Transfer Learning** involves the fine-tuning of pre-trained models using decentralized data to perform specific tasks, hence minimizing the requirement for large amounts of local data.

iii) Working Principles of Federated Learning

- A centralized machine learning program will be distributed to all devices, allowing users to utilize them as required.
- The model will now incrementally acquire knowledge and improve its abilities by processing the data provided by the user, thereby enhancing its intelligence over time.
- Subsequently, the devices are permitted to transmit the training outcomes, derived from the localized version of the machine learning application, back to the central server.
- This identical procedure occurs on several devices that own a local replica of the application. The findings will be consolidated on the centralized server, this time excluding user data.
- The centralized cloud server now updates its central machine learning model using the aggregated training data. The current version of the model is far superior to the prior version that was deployed.
- The development team now upgrades the model to a more recent version. Users are then able to update the application with the improved model, which has been generated using their data.

To summarize, by implementing the six aforementioned stages, federated learning will establish a system that employs an encryption key that is not under the control of your centralized cloud server to encrypt user-sensitive data.

iv) How Federated Learning Works

Federated learning works by training a central model across decentralized devices or servers. Instead of moving all data to a central location, the model is trained locally on each device, and only the model updates are shared. This maintains privacy and allows collaborative learning without sharing raw data.

Advantages of Using Federated Learning

Federated learning in automatic number plate recognition systems offers several advantages, especially when dealing with sensitive data and resource-constrained environments. Here are some key benefits, as shown in Figure 4.2:

- **Localized Data Processing:** ANPR systems often involve processing sensitive information such as license plate numbers. Federated learning allows the model to be trained locally on each device without transmitting raw data, thus preserving the privacy of individual license plate information.
- **Reduced Latency:** Federated learning leverages the computational capabilities of edge devices, such as Raspberry Pi, allowing ANPR models to be trained and updated directly on the devices. This reduces the need for sending data to a central server, reducing latency in the recognition process.
- **Reduced Data Transfer:** Instead of sending large amounts of raw data to a central server, only model updates are transmitted during the FL process. This results in more efficient use of bandwidth and reduces the amount of data transferred over the network.
- **Scalability:** ANPR systems can benefit from the scalability of federated learning. As more devices join the network, the model can be updated collaboratively, making the system scalable without the need for centralizing data.
- **Device-Specific Learning:** ANPR systems often face variations in license plate styles, lighting conditions, and camera angles. Federated learning allows devices to adapt to local variations by training on their specific data sets, leading to more robust and context-aware models.
- **Reduced Data Exposure:** Since raw data remains on the local devices, the risk of exposing sensitive information is minimized. This can be crucial in ANPR applications where privacy and security are paramount.

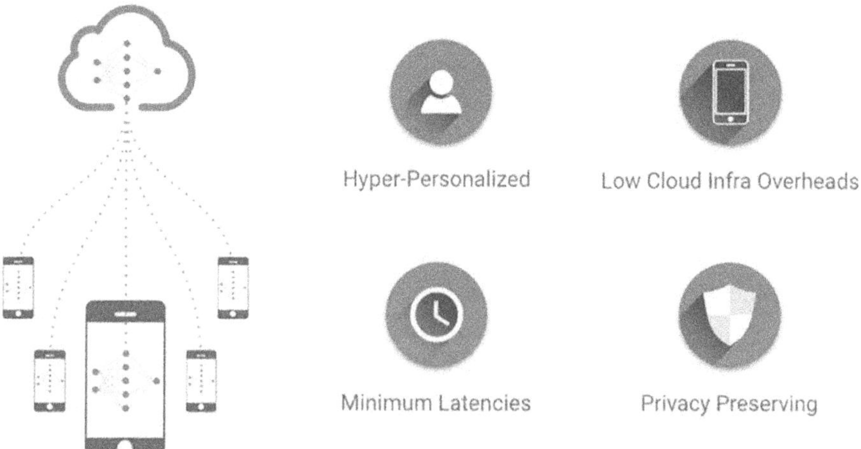

FIGURE 4.2   Federated learning and benefits.

Federated learning offers these advantages; its implementation should consider the specific requirements and constraints of ANPR systems, including the computational capabilities of edge devices and the need for real-time processing. Additionally, robust security measures should be in place to protect the federated learning process and the privacy of the data involved.

RESULTS AND DISCUSSION

The hardware setup is made following the above-defined steps, as shown in Figure 4.1. Sensors that detect a vehicle's presence, a Raspberry Pi camera that takes the picture, a personal computer that runs the algorithm, and a microcontroller that regulates all of the ANPR system's hardware make up the hardware model. Automatic number plate recognition is done by using two methods. Method 1 is the OpenCV segmentation, and Method 2 is based on federated learning.

Figures 4.3 (a&b) show the results specifying the number plate recognition. The accuracy of the reproduction may be affected by slight font variations on the original plates and the limitations mentioned earlier. The accuracy obtained for car detection, number plate recognition, and number extraction for both methods is shown in Table 4.1. Using federated learning, an accuracy of 98% for car detection, 97% for number plate recognition, and 94% for character extraction are achieved. Thus, by adopting federated learning, good accuracy is obtained, and the contour problems are reduced.

CONCLUSION

In this chapter, the real-time implementation of improved automated number plate identification using a federated learning approach is proposed. The implementation of this system comprises a Raspberry Pi camera, open-cv Python library, and OCR library to identify the license number plate and, in turn, extract text. The contour problem that occurred in the open-cv in real time number plate identification is

FIGURE 4.3 (a&b) Federated learning–based automated license plate identification.

minimized using deep learning techniques such as YOLOv7. The results obtained are based on OpenCV segmentation and federated learning methods. Using federated learning, accuracy is improved. The proposed system can be applied in a real-time environment with high overall and instant server response.

REFERENCES

[1] Atul Patel, Chirag Patel, and Dipti Shah, "Automatic Number Plate Recognition System (ANPR): A Survey," *International Journal of Computer Applications*, vol. 69, no. 9, 2013.

[2] Marko Arsenovic, Sran Sladojevic, Andras Anderla, and Darko Stefanovic, "Deep Learning Driven Plates Recognition System," *XVII International Scientific Conference on Industrial Systems*, Novi Sad, Serbia, 2017.

[3] Irfan Kılıc, Galip Aydin, and Turkish Vehicle, "License Plate Recognition Using Deep Learning," *International Conference on Artificial Intelligence and Data Processing*, pp. 1–5, IEEE, 2018. doi: 10.1109/IDAP.2018.8620744.

[4] Shraddha S. Ghadage, and Sagar R. Khedkar, "A Review Paper on Automatic Number Plate Recognition System using Federated learning Algorithms," *International Journal of Engineering Research & Technology (IJERT)*, vol. 8, no. 12, 2020, http://doi.org/10.17577/IJERTV8IS120398

[5] V. Gnanaprakash, N. Kanthimathi, and N. Saranya, "Automatic Number Plate Recognition Using Deep Learning," *IOP Conference Series: Materials Science and Engineering ICCSSS*, 2020, http://doi.org/10.1088/1757-899X/1084/1/012027

[6] R. Laroca, E. Severo, L. A. Zanlorenzi, et al., "A Robust Real-Time Automatic License Plate Recognition Based on the YOLO Detector," in *Proceedings of the 2018 International Joint Conference on Neural Networks (IJCNN)*, pp. 1–10, Rio de Janeiro, Brazil, July 2018.

[7] P. Bhogale, A. Save, V. Jain, and S. Parekh, "Vehicle License Plate Detection and Recognition System," *International Journal of Computer Applications*, vol. 137, no. 9, pp. 31–34, 2016.

[8] G. Sharma, "Performance Analysis of Vehicle Number Plate Recognition System Using Template Matching Techniques," *Journal of Information Technology & Software Engineering*, vol. 8, no. 2, 2018.

[9] K. Tejas, K. A. Reddy, D. P. Reddy, K. Bharath, R. Karthik, and M. R. Kumar, "Efficient License Plate Recognition System with Smarter Interpretation Through IoT," in *Soft Computing for Problem Solving*, pp. 207–220, Springer, 2019.

[10] N. Gupta, T. Sandep, P. Gupta, D. Goyal, and M. Goyal, "A Review: Recognition of Automatic License Plate in Image Processing," *Advances in Computational Sciences and Technology*, vol. 10, no. 5, pp. 771–779, 2017. ISSN 0973-6107.

[11] I. Ullah, and H.J. Lee, "License Plate Detection Based on Rectangular Features and Multilevel Thresholding," In *Proceedings of the International Conference on Image Processing, Computer Vision, and Pattern Recognition (IPCV)* (p. 153). The Steering Committee of the World Congress in Computer Science, Computer Engineering and Applied Computing, 2016.

[12] A. Sharma, A. Kumar, K.V.S. Raja, and S. Ladha, "Automatic License Plate Detection," *CS771 Course Project, Winter Semester 2015–16*, Indian Institute of Technology, Kanpur, 2015.

[13] P. Ravi Kiran Varma, Srikanth Ganta, B. Hari Krishna, and S.V.S.R.K. Praveen, "A Novel Method for Indian Vehicle Registration Number Plate Detection and Recognition Using Image Processing Techniques Detection and Recognition Using Image Processing Techniques," *International Conference on Computational Intelligence and Data Science (ICCIDS 2019)*, vol. 167, pp. 2623–2633, Elsevier, Procedia Computer Science, 2020.

[14] P. William, A. Shrivastava, N. Shunmuga Karpagam, T.A. Mohanaprakash, K. Tongkachok, and K. Kumar, "Crime Analysis Using Computer Vision Approach with Machine Learning," in N. Marriwala, C. Tripathi, S. Jain, and D. Kumar (Eds.), *Mobile Radio Communications and 5G Networks, Lecture Notes in Networks and Systems*, vol. 588, Springer, 2023, http://doi.org/10.1007/978-981-19-7982-8_25

[15] Ibtissam Slimani, Abdelmoghit Zaarane, Wahban Al Okaishi, and Issam Atouf, "Abdellatif Hamdoun An Automated License Plate Detection and Recognition System Based on Wavelet Decomposition and CNN," *Array*, vol. 8, pp. 1–7, Elsevier, 2020.

[16] K.B. Sathya, S. Vasuhi, and V. Vaidehi, "Perspective Vehicle License Plate Transformation Using Deep Neural Network on Genesis of CPNet," in *Third International Conference on Computing and Network Communications*, vol. 171, pp. 1858–1867, Elsevier, Procedia Computer Science, 2020.

[17] K. Tejas, K. Ashok Reddy, D. Pradeep Reddy, and M. Rajesh Kumar, "Efficient License Plate Detection by Unique Edge Detection Algorithm and Smarter Interpretation through IoT," in *7th International Conference on Soft Computing and Problem Solving*, pp. 1–11, Springer, 2020, https://doi.org/10.1007/978-981-13-1595-4_16

[18] Choong Young Jung, Keong Lee Kim, and Tan Chye Cheah, "License Plate Number Detection and Recognition Using Simplified Linear-Model," *Journal of Critical Reviews*, vol. 7, no. 3, pp. 55–60, 2020.

[19] Olamilekan Shobayo, Ayobami Olajube, Ohere Nathan, and Modupe Odusami, "Obinna Okoyeigbo Development of Smart Plate Number Recognition System for Fast Cars with Web Application," *Applied Computational Intelligence and Soft Computing Hindawi*, vol. 2020, pp. 1–7, 2020.

[20] Weihong Wang, "Jiaoyang Tu Research on License Plate Recognition Algorithms Based on Deep Learning in Complex Environment," *IEEE Access*, vol. 8, pp. 91661–91675, 2020.

[21] P. William, A. Shrivastava, P.S. Chauhan, M. Raja, S.B. Ojha, and K. Kumar, "Natural Language Processing Implementation for Sentiment Analysis on Tweets," in N. Marriwala, C. Tripathi, S. Jain, and D. Kumar (Eds.), *Mobile Radio Communications and 5G Networks, Lecture Notes in Networks and Systems*, vol. 588, Springer, 2023, http://doi.org/10.1007/978-981-19-7982-8_26

[22] S. Mishra, S. Choubey, A. Choubey, N. Yogeesh, J. Durga Prasad Rao, and P. William, "Data Extraction Approach Using Natural Language Processing for Sentiment Analysis," *International Conference on Automation*, vol. 2022, Computing and Renewable Systems (ICACRS), 2022, pp. 970–972, http://doi.org/10.1109/ICACRS55517.2022.10029216

[23] Deepak Narayan Paithankar, Abhijeet Rajendra Pabale, P. William, Rushikesh Vilas Kolhe, and Prashant Madhukar Yawalkar, "Framework for Implementing Air Quality Monitoring System Using LPWA-based IoT Technique," *Measurement: Sensors*, vol. 100709, 2023, http://doi.org/10.1016/j.measen.2023.100709

[24] K. Gupta, S. Choubey, Y. N, P. William, V.T. N, and C.P. Kale, "Implementation of Motorist Weariness Detection System Using a Conventional Object Recognition Technique," *International Conference on Intelligent Data Communication Technologies and Internet of Things*, vol. 2023, pp. 640–646, IDCIoT, 2023, http://doi.org/10.1109/IDCIoT56793.2023.10052783

[25] P. William, Y. N, V.M. Tidake, S. Sumit Gondkar, C. R, and K. Vengatesan, "Framework for Implementation of Personality Inventory Model on Natural Language Processing with Personality Traits Analysis," *International Conference on Intelligent Data Communication Technologies and Internet of Things*, vol. 2023, pp. 625–628, IDCIoT, 2023, http://doi.org/10.1109/IDCIoT56793.2023.10053501

5 Fake Currency Identification Using Artificial Intelligence and Federated Learning

Syed Zahiruddin, Vamsi Krishna Kadiri,
Valli Bhasha Achukatla, Pavan Kumar Kattela,
and Ahmed A. Elngar

INTRODUCTION

Identifying fake currency using artificial intelligence involves developing systems that can analyze various features of bank notes to distinguish genuine from counterfeit notes. A diverse data set of genuine and counterfeit currency notes is gathered. Ensuring the data set covers different denominations, countries of origin, and the physical condition of the notes. The data set obtained consists of several images of real and fake currencies. The different categories of Indian currencies differ in value estimation and color usage, separated by the quality of printing, the material used for printing, and other traits, which makes for simple visual distinguishing proof. To ensure accurate and reliable results, it is essential to clean and preprocess the images of currency notes before further analysis. This involves resizing, normalization, and noise reduction. Noise reduction reduces unwanted artifacts or random variations in the image, image resizing adjusts the image dimensions to meet specific requirements, and contrast enhancement adjusts the image to improve the visibility of objects or features. The relevant features are extracted from the currency notes through this approach. Features include texture, color, watermark characteristics, security thread patterns, and other distinctive elements. A suitable machine learning or deep learning model is chosen. For picture classification tasks, convolutional neural networks (CNNs) are employed; in the case where the input contains sequential patterns, recurrent neural networks (RNNs) or transformer models are helpful. In this work, CNN is preferred.

The model was trained with the preprocessed data set. During training, the model learns to recognize patterns and features that distinguish genuine from counterfeit notes. Assessing the model's ability to predict accurately on data, it hasn't been trained on. Fine-tuning the model is done based on validation results to improve its performance. To integrate the trained model into a system designed for real-time processing of currency note images. This could be integrated into ATMs, cash

DOI: 10.1201/9781003482000-5

processing machines, or other systems where currency authenticity needs to be verified. Regularly providing the model with new counterfeit examples to improve its ability to recognize and detect them. Regularly monitor its performance and retrain if necessary. To use AI models in real-time, they should be connected to devices like cameras or scanners that can take pictures of currency notes. Implementing advanced security protocols to safeguard the AI system from malicious intrusions that could undermine its accurate detection of counterfeit currency.

The quality and representativeness of the data set, in addition to the features chosen for examination, are important determinants of the model's performance. Furthermore, working closely with experts in the field of currency and security can provide valuable insights into the specific characteristics that differentiate genuine from counterfeit notes.

The second section is the literature survey, the third illustrates the methodology related to the proposed work, the fourth is about the stochastic gradient descent with momentum (SGDM) optimizer, the fifth describes the fundamentals of federated learning, the sixth illustrates the results and discussion of the proposed method, the seventh is a case study, and the eighth is the conclusion.

LITERATURE SURVEY

Until recently, it was surely true that "a picture is worth a thousand words", but the exploding research interests in the area of digital image processing during the period changed the researchers' opinion.

In the literature, several approaches exist for paper currency identification and recognition. Object identification and recognition based on shape, size, and different parameters are determined to recognize the pattern using Gabor wavelets. Gabor wavelet grids are localized within the area using region of interest (ROI) and are capable of identifying different image classes in the study [1]. An improved edge detection technique was used for detecting various edges that are important in identifying different objects. This research proposes a novel edge detection algorithm that utilizes a gray prediction model of first-order one variable, denoted as GM(1,1). The primary objective of this approach is to address the shortcomings associated with existing edge detection methods, which often result in undetected objects and the introduction of artificial information. It also applies various edge detection steps to improve the final output, which is then used for various applications [2].

The global search perspective of the genetic algorithm and the local search perspective of the back propagation (BP) algorithm were integrated into this study on the recognition of RMB (renminbi) numbers using a back propagation ANN [3]. It is based on a genetic algorithm that is trained. Renminbi is the legal tender used in China, and each is distinct. Another technique is called local binary pattern (LBP) [4] for extracting features from currency, and it is largely used in the banking system with good accuracy and fast recognition of features. Later a new approach using a microcontroller [5] for identifying the serial numbers of currency for automated banking systems was introduced. This method can automatically identify currency based on serial number data stored in government currency printing stations.

Each note of Indian currency has a unique serial number. To locate the serial number, various image processing techniques are used. A technique [6] is implemented for currency authentication systems using a Sobel operator for edge detection and considering three standard security features: Identification marks, security threads, and watermarks for feature extraction were chosen as the three qualities of Indian paper currency that help in the detection of counterfeits. In the work in [7], the SVM algorithm is used for identifying RMB numbers in currency. The work in [8] is for recognizing currency by detecting invisible marks. The study in [9] is about gesture recognition of images using CNN. In [10], biometric iris recognition based on neural networks is covered. In [11], a study about optical character recognition techniques is described for identifying and recognizing text and writing it in a machine-readable format and its applications. The study mentioned in [12] is for the recognition of currency, whether it is authentic or fake based on reserved security features that can easily identify the currency note.

The security elements of the currency are listed in the following, as shown in Figure 5.1:

1. Security thread
2. Intaglio printing
3. See-through register
4. Watermark and electrotype watermark
5. Color-shifting ink
6. New numbering pattern
7. An increase in the size of the identification markers and angular bleed lines

This chapter suggests a novel method for classifying and identifying cash using convolutional neural networks and digital image processing techniques.

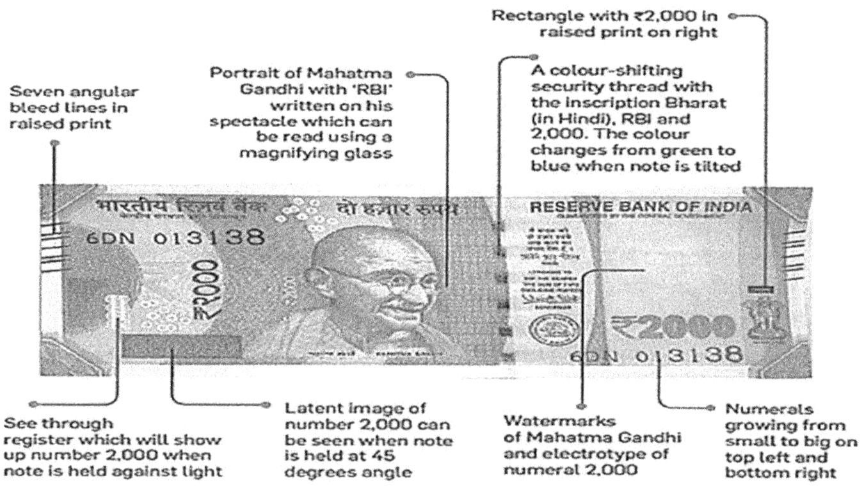

FIGURE 5.1 Security features of Indian currency.

METHODOLOGY

The proposed technique aims to identify and classify the currency, whether it is original or fake, and then give the denomination of the currency. For both identification and classification, the denomination of the currency is identified using the CNN classifier. Here the identification of the currency is done by pre-processing the image and extracting the features by training the CNN network with different features of original and fake notes. Then classifying the denomination of every note is done by training the CNN with the data set available.

DEEP LEARNING ALGORITHMS

Deep learning algorithms are derived from the ANNs, and they are illustrated in the classification of images and identification of objects. Compared to ANNs, CNNs create image characteristics using fewer parameter computations. Convolutional neural networks are used for data classification and feature detection in images [13–18].

CONVOLUTIONAL NEURAL NETWORKS

Convolutional neural networks learn from input directly, doing away with the requirement for manual feature extraction. Without the assistance of a human, it automatically recognizes the key elements of an image. It develops unique characteristics for each class on its own [19–23].

CNNs are popular because of the following factors

1. The network will learn the feature directly from the CNN; manual feature extraction is no longer necessary.
2. A CNN may be trained to do additional recognition tasks, enabling it to build on current networks.
3. CNNs offer highly accurate recognition results.

CNNs compare the features of input with the trained data and classify the currency. Classification using a CNN classifier is done to find whether it's an authentic or duplicate note. Therefore, the identification is done as follows:

If the note is original displays output as "Original" and also gives the accuracy of the classified model. If the note is a duplicate, "Fake" is displayed, as shown in Figure 5.2, and it classifies the note as a duplicate note and also gives the accuracy of the classified model. In this study, two CNN classifiers are used. The first CNN classifier will give the currency that is the original/duplicate with an accuracy of the classified model, and another CNN classifier will give the denomination of the currency with an accuracy of the classified model.

ARCHITECTURE

Any intermediary levels in a feed-forward neural network are referred to as hidden layers, as the activation function and final convolution obscure their inputs and outputs. A convolutional neural network is composed of multiple layers. Convolutional

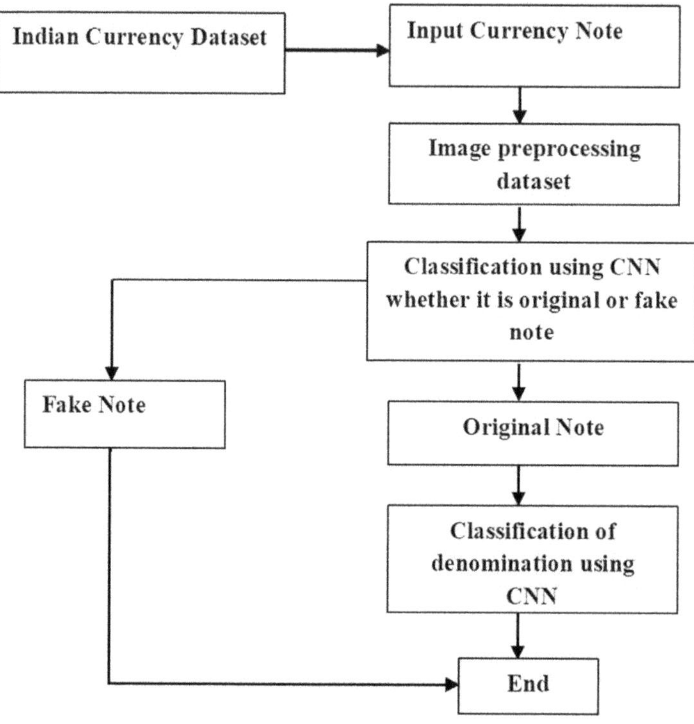

FIGURE 5.2 Block-level representation of the proposed method.

layers are located in the hidden layers of a convolutional neural network. This typically consists of a layer that performs a dot product between the convolution kernel and the layer's input matrix [24–27]. Two CNN networks are used in the suggested method, and Figure 5.2 displays the proposed method's block diagram.

1. For training a CNN, the data set is needed.
2. An image obtained from the data set undergoes preprocessing steps and subsequently used as input for the convolutional neural network classifier.
3. The CNN is divided into the following layers:
   a. Image input layer
   b. Convolutional layer
   c. Batch normalization layer
   d. ReLU layer
   e. Max pooling layer
   f. Fully connected layer
   g. Softmax layer
   h. Classification
4. For training a CNN, options are to be specified among the following:
   a. Optimizers
   b. Initial learning rate

    c. Maximum epochs
    d. Shuffle
    e. Validation frequency
    f. Verbosity
    g. Plots

LAYERS AND TRAINING

For any network to work properly, the first step is to outline the architecture of the network and then select data sets to train the network with distinct features. The functions of each layer of the CNN network are as follows:

Layers

a. *Image input layer:* The network receives images by creating an image input layer.
b. *Convolutional layer:* This layer conducts convolution operations. For convolving a picture first, the size of the filter, that is, for changing the neurons' connectivity to the next, is changed without changing the image properties. To calculate the size of a filter, use the formula ($h * w * c$), where h, w, and c represent the filter height, filter width, and number of channels in an image., respectively.

    A CNN network can be built with a single convolution layer or multiple convolutional layers. The number of convolutional layers in a CNN network is based on the volume and depth of the data. Three convolutional layers are utilized in this network. The filter is moved to the right by a specific stride value before filtering the entire width. Once the entire picture has been traversed, the same stride value returns to the image's start (left) and executes the operation [28–32].
c. *Batch normalization layer:* Observation for each channel is separately normalized by this layer. The batch normalization layer normalizes a small set of data among the total set alone for fastening the training process of a CNN network and minimizing the sensitivity of the network. This batch normalization is used in between the Convolution and ReLU layers. The ReLU layer is the rectified linear unit, which removes the nonlinearities present in the network.
d. *ReLU layer:* Its function is to perform the thresholding operation for every element and replace the element values that are below zero or negative with zero.
e. *Max pooling layer:* To accomplish downsampling, the max-pooling layer divides the input data as rectangular pooling regions and then calculates the max value of the particular region. The activation map's spatial resolution is mostly decreased by the pooling or downsampling layer.
f. *Fully connected layer:* This layer, as its name suggests, is the final destination for all neurons in the network. It joins all the features gathered by the preceding layers.
g. *Soft max layer:* This layer returns the result of the softmax function applied to the raw data.

h. *Classification:* This layer calculates the cross-entropy loss for classification tasks, accommodating both standard and weighted classification scenarios involving mutually exclusive categories.

Training

The next stage is to build the network's training options after defining the CNN's layers and defining the global training settings with the training options function. The first stage in training a network is to use CNN network optimizers to create a collection of options. The maximum number of training epochs is then defined, after which the learning rate for each sum of epochs is defined. The shuffle rate and validation frequency are configured. Also, verbose and plot settings are used to display the training and graphing progress.

STOCHASTIC GRADIENT DESCENT WITH MOMENTUM

The stochastic gradient descent with momentum optimizer is utilized in this training to establish the momentum value for getting suitable outcomes from predicted and actual outputs. It utilizes the segmentation maps that correspond with the input images. When paired with the cross-entropy loss equation, the energy function over the resulting feature map is represented by a pixel-wise soft-max.

SGD with momentum is an optimization algorithm commonly used to train deep neural networks. It is an improvement on the basic stochastic gradient descent (SGD) technique that aids in dampening oscillations and accelerating convergence. The SGD method updates the model's parameters by considering the loss's negative gradient about the parameters. However, this can lead to oscillations, especially in the presence of noisy or sparse gradients. Momentum is introduced to address this issue [32–38].

The key idea behind momentum is to maintain a moving average of the gradients and use it to update the parameters. This moving average is called the momentum term. The update rule for the parameters in SGD with momentum is expressed as:

$$v_t = \beta \cdot vt - 1 + (1 - \beta) \cdot \nabla J(\theta_t) \tag{1}$$

$$\theta_{t+1} = \theta_t - \alpha \cdot v_t \tag{2}$$

where,

v_t represents the momentum term at iteration t,

β represents the momentum coefficient (usually set between 0 and 1),

$\nabla J(\theta_t)$ represents the gradient of the loss with correspondence to the parameters at iteration t,

θ_t is the current set of parameters,

α is the learning rate.

The momentum term helps the optimization process by accumulating gradients over time and damping oscillations. It enables the optimizer to "roll" through flat regions and navigate more efficiently along narrow, curved valleys in the optimization landscape.

Typical values for the momentum coefficient (β) are in the range of 0.9 and 0.99. Usually, the learning rate (α) is set to a modestly positive number.

Due to its capacity to quicken convergence, SGD with momentum is a well-liked option for neural network training. In this work SGD is also employed, particularly in situations involving high-curvature, noisy, or sparse gradients.

FUNDAMENTALS OF FEDERATED LEARNING

Federated learning (FL) is a machine learning technique that allows models to be trained without data exchange among decentralized devices or servers that store local data samples. This collaborative learning paradigm allows for model training without centralizing data, addressing privacy concerns, reducing communication costs, and accommodating situations where data cannot be easily moved.

The fundamentals of federated learning are:

1. Decentralized Training: In traditional machine learning, data is typically centralized, and models are trained on that centralized data. In federated learning, the training process occurs on local devices or servers without aggregating raw data in one location.
2. Privacy Preservation: One of the main advantages of federated learning is privacy preservation. The only data that is transmitted are model updates (gradients), while raw data stays on local devices. As a result, there is less chance of sensitive data being revealed.
3. Communication Efficiency: Federated learning uses model updates (parameters or gradients) rather than raw data to be sent to a central server for model training. As a result, the network may communicate more effectively by requiring less data to be transferred over it.
4. Iterative Model Updates: The federated learning process typically involves multiple rounds of model updates. Local models are trained on raw data in each cycle, and only model updates or modifications to the model parameters are passed to the aggregator or central server.
5. Aggregation Techniques: The central server combines the model updates received from different devices to update the global model. Various aggregation techniques, such as simple averaging or more sophisticated methods, are used to combine these updates.
6. Heterogeneous Devices: Federated learning works well in situations when devices have different features and data distributions. It can adapt to different device capabilities and accommodate variations in local data sets.
7. Secure and Trusted Aggregation: Ensuring the security and trustworthiness of the aggregation process is crucial. Methods like secure multi-party computation (SMPC) or homomorphic encryption can be employed to protect the privacy of model updates during aggregation.
8. Applications: Federated learning is exceptionally useful in situations where data is distributed across many devices, resembling mobile phones, IoT devices, and edge devices. It has applications in healthcare, finance, smart cities, and other fields where privacy and data locality are important considerations.
9. Challenges: Federated learning introduces several issues such as dealing with non-identically distributed (IID) data, handling stragglers (devices with sluggish computation), and addressing potential biases in the aggregated model.

10. Open Source Frameworks: Several open-source federated learning frameworks exist, including Tensor Flow Federated (TFF), PySyft, and Flower, which facilitate the execution of federated learning protocols.

Federated learning is a rapidly evolving field with ongoing research to address its challenges and extend its applicability to various domains. Federated learning in fake currency identification is a scheme that permits models to be trained among decentralized devices or servers sharing local data samples without restoring them. This concept is applied to fake currency detection to enhance the overall model's performance and accuracy while preserving privacy.

The following is the procedure for fake currency identification using federated learning:

1. Data Distribution: A data set of fake currency images is distributed across various devices, including smartphones and local servers. The images are stored locally on each device. The central server holds the initial model parameters and coordinates the training process.
2. Initialization: The central server initiates a global method with some base parameters for fake currency detection.
3. Local Training: Each device processes the model locally using its data set without exchanging the data with the base server. Training is performed on the device using convolutional neural networks to identify patterns and features indicative of fake currency.
4. Model Update: After local training, only the updated model parameters (not the raw data) are returned to the base server.
5. Aggregation: The base server collects the model updates from all devices, adjusting the global model based on the received updates.
6. Iterative Process: The steps from 3–5 are frequently iterated. The model continues to improve based on the collective knowledge of all devices without raw data leaving the local devices.

Benefits of federated learning in fake currency detection:

1. Privacy Preservation: Since raw data remains on local devices, privacy concerns are minimized.
2. Decentralized Learning: Federated learning allows for the utilization of diverse data sets from different sources, improving the model's robustness.
3. Reduced Communication Overhead: Only model updates are communicated, reducing the amount of data shared over the network.
4. Adaptability: The model can adapt to new fake currency patterns that may be region specific, as it learns from various local data sets.

It's important to note that federated learning is an important approach, and the effectiveness lies in the quality and variety of the local data sets, as well as the coordination of the central server. Additionally, security measures must be in place to prevent adversarial attacks or malicious behavior during the federated learning process.

RESULTS AND DISCUSSION

The following figures are the experimental outputs.

The data set of various currency notes is used as the input figure, as shown in Figure 5.3. The image is supplied to the CNN layers after being resized to $512 \times 512 \times 3$.

The training progress is for determining whether the input image is authentic or fake, together with accuracy and training loss in CNN, which is illustrated in Figure 5.4, which also includes the denomination of the input image.

FIGURE 5.3   Sample of input image specifying Indian currency of 10 rupees 10.

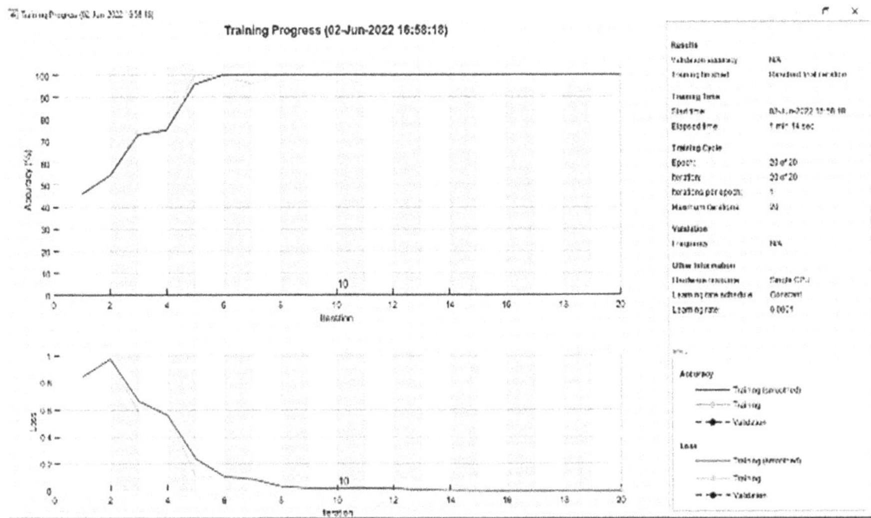

FIGURE 5.4   Training progress for verifying whether the currency is original/fake.

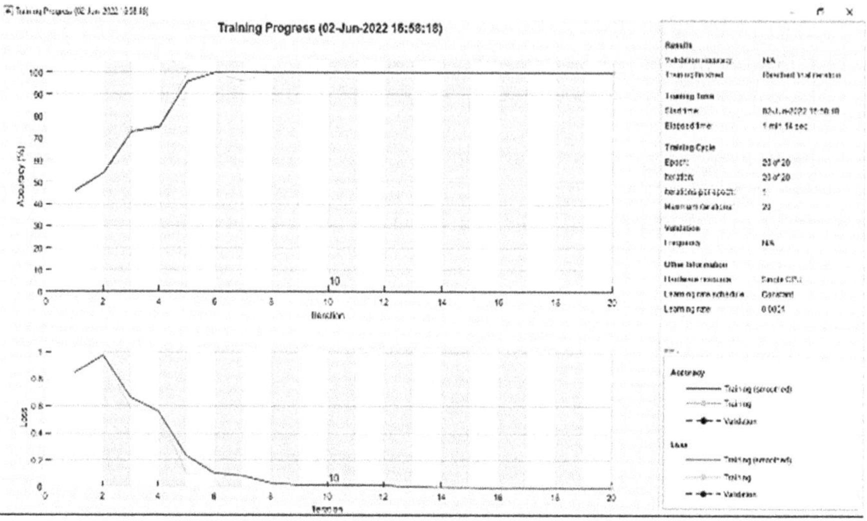

FIGURE 5.5 Training progress for determining the denominations.

```
Command Window

    Output of CNN Classifier is: Original Rupees
    Accuracy of classified Model is: 92.2917
    Denomination of the Currency is: 10 Rupees
    Accuracy of classified Model is: 92.9630
fx >>
```

FIGURE 5.6 Results shown in the command window.

Figure 5.5 illustrates the training process for identifying the denomination of the image captured after classifying the input image as original or fake, and Figure 5.6 provides the output presented in the command prompt:

1. CNN classifier output as original and accuracy of CNN classifier.
2. Denomination of the currency and accuracy of the classified model.

A comparison graph that compares both the proposed and existing methods in terms of accuracy is shown in Figure 5.7. The comparison Table 5.1 is to examine the accuracy of proposed and existing techniques.

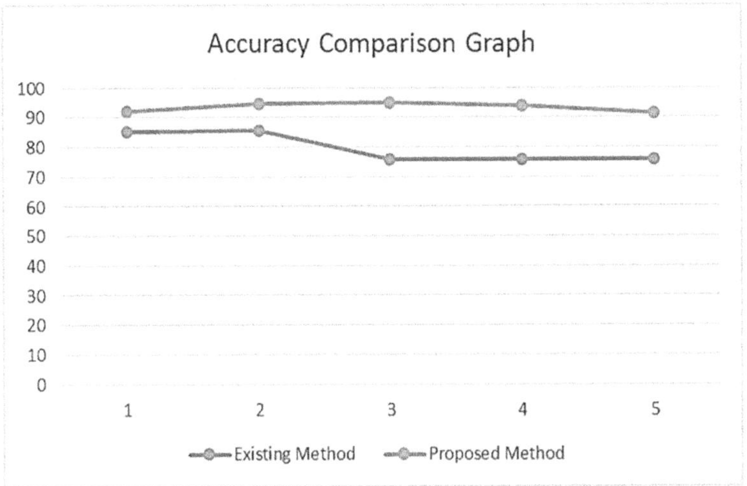

FIGURE 5.7   Accuracy comparison graph.

TABLE 5.1
Accuracy Comparison Table

Denomination in Rupees	Accuracy of SVM Classifier	Accuracy of Model Classifier	Final Accuracy
10	95.625	92.222	93.924
20	91.250	92.222	91.736
50	90.625	92.222	91.424
100	94.792	93.518	94.155
200	92.083	92.593	92.338
500	89.792	91.296	90.544
2000	92.500	92.963	92.732

CHALLENGES AND FUTURE WORK

The proliferation of counterfeit currency poses a significant threat to economies worldwide, undermining financial systems and eroding public trust. Traditional methods of detecting fake currency, which often rely on manual inspection and conventional automated systems, are increasingly proving inadequate in the face of sophisticated counterfeiting techniques. In this context, the integration of artificial intelligence (AI) presents a promising frontier for enhancing the accuracy and efficiency of counterfeit detection. By leveraging advanced machine learning algorithms, AI can analyze intricate patterns and features in currency that are imperceptible to the human eye or conventional systems.

However, the deployment of AI for fake currency identification is not without its challenges. Issues such as the availability and quality of training data, the interpretability of AI models, the substantial computational resources required, and the need to protect sensitive financial data are significant hurdles that must be addressed. Moreover, the robustness and generalization of AI models across different currencies and counterfeit methods remain critical areas of concern.

To mitigate some of these challenges, the application of federated learning offers a novel approach. Federated learning allows AI models to be trained across multiple decentralized devices or servers holding local data samples without exchanging the data itself. This technique not only enhances data privacy and security but also facilitates the development of more generalized models by learning from a diverse range of data sources.

This chapter explores the dual approach of utilizing AI and federated learning for fake currency identification, examining the existing challenges, and proposing future directions to enhance the efficacy and reliability of these technologies. By addressing the current limitations and exploring innovative solutions, the goal is to advance the state of counterfeit detection systems, thereby strengthening financial security and integrity on a global scale.

CHALLENGES

Data Privacy and Security

Federated learning relies on decentralized training across multiple devices while keeping data local. Ensuring robust privacy and security measures to protect sensitive financial information is paramount. Any breach could compromise user trust and the integrity of the system.

Data Imbalance and Bias

The availability of counterfeit currency data may be limited compared to genuine currency data, leading to data imbalance and potential bias in the model. Addressing this challenge requires strategies such as data augmentation, synthetic data generation, or bias correction techniques to ensure the model's robustness and fairness.

Adversarial Attacks

Adversarial attacks, where counterfeiters deliberately manipulate currency features to evade detection, pose a significant threat to the effectiveness of AI-based counterfeit detection systems. Developing models robust to adversarial attacks and continuously updating them to adapt to evolving counterfeit techniques is crucial.

Generalization Across Currency Types

Currency notes vary in design, security features, and denominations across different countries. Ensuring the generalization of AI models across various currency types and denominations presents a challenge due to differences in appearance and security measures.

Regulatory Compliance

Financial institutions must adhere to strict regulatory standards and compliance requirements when deploying AI-based counterfeit detection systems. Ensuring that these systems meet regulatory guidelines while maintaining efficiency and accuracy is essential but challenging [39–42].

FUTURE WORK

Enhanced Model Performance

Continuous research and development efforts should focus on improving the performance of AI models for counterfeit detection. This includes exploring advanced neural network architectures; incorporating multi-modal data sources, such as infrared imaging; and leveraging transfer learning techniques.

Dynamic Adaptation

Developing AI models capable of dynamically adapting to new counterfeit techniques and emerging currency designs is crucial. Incorporating mechanisms for continuous learning and model retraining based on real-time feedback and data updates can enhance the system's resilience to evolving threats.

Collaborative Efforts

Collaborative initiatives involving governments, financial institutions, researchers, and technology providers are essential for addressing counterfeit currency challenges effectively. Sharing data, best practices, and insights can accelerate progress and promote standardization in counterfeit detection methods.

Interoperability and Integration

Ensuring interoperability and seamless integration of AI-based counterfeit detection systems with existing financial infrastructure, such as ATMs, cash counting machines, and banking software, is critical for widespread adoption and effectiveness.

Ethical Considerations

As AI technologies become more pervasive in financial systems, ethical considerations surrounding transparency, accountability, and bias mitigation become increasingly important. Future work should prioritize ethical AI principles to ensure the responsible development and deployment of counterfeit detection systems.

Therefore, by addressing these challenges and focusing on future research directions, the field of fake currency identification using artificial intelligence and federated learning can continue to advance, bolstering the security and integrity of global financial systems.

CONCLUSION

In this chapter, a unique convolutional neural network-based approach for identifying authentic or counterfeit currency is proposed. When there is insufficient prior information about the scene and camera postures, the traditional computer vision problem of locating, illustrating, and fixing visual key points must be solved. A CNN network is trained on the data set and put to the test to determine the denomination of the currency and to determine that the currency is original. The trained convolutional neural network exhibited using the federated learning approach leads to a very low training loss, and the maximum accuracy obtained is about 95.625%. Also, this chapter demonstrated the efficacy of using artificial intelligence and federated learning to identify counterfeit currency. By leveraging machine learning techniques and decentralized training approaches, governments and financial institutions can enhance the security and integrity of their currency systems, mitigating the risks posed by counterfeiters. Further research could focus on scalability, robustness against emerging counterfeit techniques, and integration with existing financial infrastructure.

REFERENCES

[1] B.V. Kumar, D.S. Sharan (1993), "Pattern recognition with localized gabor wavelet grids," in *International Conference on Computational Intelligence and Multimedia Applications (ICCIMA 2007)*, Volume: 40 Issue: (2), 517–521. IEEE.

[2] J. Zhang, L. Wu (2009), "An improved method for image edge detection based on GM (1, 1) model," in *2009 International Conference on Artificial Intelligence and Computational Intelligence*, Volume: 3, 133–136. IEEE.

[3] L. Jing, M.S. Jin (2010), "About RMB number identification with genetic evolution neural network," in *2010 International Conference on Computer, Mechatronics, Control and Electronic Engineering*, Volume: 1, 286–288. IEEE.

[4] J. Guo, Y. Zhao, A. Cai (2010), "A reliable method for paper currency recognition based on LBP," in *2010 2nd IEEE InternationalConference on Network Infrastructure and Digital Content*, 359–363. IEEE.

[5] K. Sathisha (2011), "Bank automation system for Indian currency—A novel approach," in *2011 IEEE Recent Advances in Intelligent Computational Systems*, 299–302. IEEE.

[6] R. Mirza, V. Nanda (2012), "Design and implementation of indian paper currency authentication system based on feature extraction by edge based segmentation using Sobel operator," *International Journal of Engineering Research and Development*, 3(2), 41–46.

[7] L. Wenhong, T. Wenjuan, C. Xiyan, G. Zhen (2010), "Application of support vector machine (SVM) on serial number identification of RMB," in *2010 8th World Congress on Intelligent Control and Automation*, 6262–6266. IEEE.

[8] H. Hassanpour, P.M. Farahabadi (2009), "Using Hidden Markov Models for paper currency recognition," *Expert Systems with Applications*, 36(6), 10105–10111.

[9] M. Khari, A.K. Garg, R.G. Crespo, E. Verdú (2019), "Gesture recognition of RGB and RGB-D static images using convolutional neural networks," *International Journal of Interactive Multimedia & Artificial Intelligence*, 5(7).

[10] M. Dua, R. Gupta, M. Khari, R.G. Crespo (2019), "Biometric iris recognition using radial basis function neural network," *Soft Computing*, 23(22), 11801–11815.

[11] R. Gupta, D. Gupta, M. Dua, M. Khari (2017), "Hindi optical character recognition and its applications," in *Detecting and Mitigating Robotic Cyber Security Risks*, 28–39. IGI Global.

[12] P.D. Deshpande, A. Shrivastava (2018), "Indian currency recognition and authentication using image processing," *IJARSE*, 7(7), 1107–1119.

[13] T. Agasti, G. Burand, P. Wade, P. Chitra (2017), "Fake currency detection using image processing," *IOP Conference Series: Materials Science and Engineering*, 263(5), 052047. https://doi.org/10.1088/1757-899X/263/5/052047.

[14] G. Alnowaini, A. Alabsi, H. Ali (2019), "Yemeni paper currency detection system," in *2019 First International Conference of Intelligent Computing and Engineering (ICOICE). Mukalla, Yemen, 15–16 December 2019.* IEEE. https://doi.org/10.1109/ICOICE48418.2019.9035192.

[15] A.H. Ballado, J.C. Dela Cruz, G.O. Avendaño, N.M. Echano, J.E. Ella, M.E.M. Medina, B.K.C. Paquiz (2015), "Phillipine currency paper bill counterfeit detection through image processing using canny edge technology," in *2015 International Conference on Humanoid, Nanotechnology, Information Technology, Communication and Control, Environment and Management (HNICEM). Cebu, Philippines, 9–12 December 2015.* IEEE. https://doi.org/10.1109/HNICEM.2015.7393184.

[16] R. Barman, T. Saha, S.K. Bandyopadhyay (2018), "A proposed method for different objects detection in Indian paper currency note," *International Journal of Trend in Research and Development*, 5(3), 311–313.

[17] J.F. Zeggeye, Y. Assabie (2016), "Automatic recognition and counterfeit detection of Ethiopian paper currency," *International Journal of Image, Graphics and Signal Processing*, 8(2), 28–36. https://doi.org/10.5815/ijigsp.2016.02.04.

[18] M. Khari, A.K. Garg, R.G. Crespo, E. Verdú (2019), "Gesture recognition of RGB and RGBD static images using convolutional neural networks," *International Journal of Interactive Multimedia Artificial Intelligence*, 5(7), 22–27.

[19] M. Dua, R. Gupta, M. Khari, R.G. Crespo (2019), "Biometric iris recognition using radial basis function neural network," *Soft Computing*, 23(22), 11801–11815.

[20] J.M. Chatterjee, S. Ghatak, R. Kumar, M. Khari (2018), "BitCoin exclusively informational money: A valuable review from 2010 to 2017," *Quality Quantity*, 52(5), 2037–2054.

[21] Tuyen Danh Pham, Chanhum Park, Dat Tien Nguyen, Ganbayar Batchuluun, Kang Ryoung Park (2020), "Deep learning-based fake-banknote detection for the visually impaired people using visible-light images captured by smartphone cameras," *IEEE Access*, 8, 63144–63161.

[22] Yongjiao Liu, Jianbiao He, Min Li (2018), "New and old banknote recognition based on convolutional neural network," in *Proceedings of the 2018 International Conference on Mathematics and Statistics*, pp. 92–97, 2018. Association for Computing Machinery.

[23] Milan Tripathi (2021), "Analysis of convolutional neural network based image classification techniques," *Journal of Innovative Image Processing (JIIP)*, 3(2), 100–117.

[24] Joy Iong-Zong Chen, Kong-Long Lai (2021), "Deep convolution neural network model for credit-card fraud detection and alert," *Journal of Artificial Intelligence*, 3(2), 101–112.

[25] Alex Krizhevsky, Ilya Sutskever, Geoffrey E. Hinton (2017), "ImageNet classification with deep convolutional neural networks," *Communications of the ACM*, 60(6), 84–90.

[26] Shaik Ajiji Amirsab, Mohammad Mudassir, Mohammad Ismail (2017), "An automated recognition of fake or destroyed Indian currency notes," *International Journal of Advance Scientific Research and Engineering Trends*, 1(7), ISSN 2456-0774.

[27] Tushar Agasti, Gajanan Burand, Pratik Wade and P. Chitra (2017), "Fake currency detection using image processing," in *IOP Conference Series: Materials Science and Engineering*, Volume: 263, Issue: (5). IOP Publishing Ltd. doi: 10.1088/1757-899X/263/5/052047.

[28] M. Thulsima, V. Sudha, P. Selva Nila, G. Bharatha Sreeja (2017), "Indian paper currency recognition using weighted euclidian distance," *IJSR-CSEIT*, 2(2), ISSN 2456-3307.

[29] Snehlata Singh, Vipin Saxena (2017), "Identification of fake currency: A case study of Indian scenario," *International Journal of Advanced Research in Computer Science*, 8, ISSN 0976-5697.

[30] Eshita Pilania, Bhavika Arora (2016), "Recognition of fake currency based on security thread feature of currency," *International Journal of Engineering and Computer Science*, 5(7), ISSN 2319-7242.

[31] Vanga Odelu, Ashok Kumar Das, Adrijit Goswami (2016, February), SEAP: Secure and efficient authentication protocol for NFC applications using pseudonyms. *IEEE Transactions on Consumer Electronics*. 62(1), 30–38. https://doi.org/10.1109/TCE.2016.7448560.

[32] M. Abishek, B. Kavin, B. Raj Kumaran (2020), "Fake currency detection using CNN," *India. IEEE*, 7(3).

[33] K. Kiran, B. Anuthi, S. Pranali, A. Shruti (2019, December), "Counterfeit currency detection using deep convolutional neural network," *IEEE Pune Section International Conference* (PuneCon), Pune, India, 1–4. IEEE. doi: 10.1109/PuneCon46936.2019.9105683.

[34] Z. Adiba, U. Jia (2019), "A hybrid fake banknote detection model using OCR face recognition and hough features," in *2019 Cybersecurity and Cyberforensics Conference (CCC)*, Melbourne, VIC, Australia, 91–95. IEEE Xplore. doi: 10.1109/CCC.2019.000-3.

[35] K. Gautam (2020), "Indian currency detection using image recognition technique," in *2020 International Conference on Computer Science Engineering and Applications (ICCSEA)*, 1–5. IEEE Xplore. doi: 10.1109/ICCSEA49143.2020.9132955.

[36] S. Singh, A.K. Aggarwal, P. Ramesh, L. Nelson, P. Damodharan, M.T. Pandian (2022), "COVID 19: Identification of masked face using CNN architecture," in *2022 3rd International Conference on Electronics and Sustainable Communication Systems (ICESC)*, 1045–1051. IEEE Xplore. doi: 10.1109/ICESC54411.2022.9885327.

[37] S.S. Raju, M. Srikanth, K. Guravaiah, P. Pandiyaan, B. Teja, K.S. Tarun (2023), "A three-dimensional approach for stock prediction using AI/ML algorithms: A review & comparison," in *IEEE 4th International Conference on Innovative Trends in Information Technology (ICITIIT)*, 1–6. IEEE Xplore. doi: 10.1109/ICITIIT57246.2023.10068584.

[38] M. Kalaiselvi, V. Neha, R. Ragavi, P. Sindhu, G. Sneak (2023, May), "Identification of fake Indian currency using convolutional neural network," *International Journal of Research Publication and Reviews*, 4(5), 1496–1501.

[39] C. Wu, S. Shao, C. Tunc, S. Hariri (2020), "Video anomaly detection using pre-trained deep convolutional neural nets and context mining," in *2020 IEEE/ACS 17th International Conference on Computer Systems and Applications (AICCSA)*, 1–8. IEEE Xplore. doi: 10.1109/AICCSA50499.2020.9316538.

[40] K. Sundravadivelu, P. Gururama Senthilvel, N. Duraimutharasan, T. Hannah Rose Esther, K. Rajesh Kumar (2023), "Extensive analysis of IoT assisted fake currency detection using novel learning scheme," in *2023 Second International Conference on Augmented Intelligence and Sustainable Systems (ICAISS)*, 1469–1477. IEEE Xplore. doi: 10.1109/ICAISS58487.2023.10250560.

[41] Aniket Jawale, Kanchan Patil, Aniruddha Patil, Siddhesh Sagar, Akash Momale (2023), "Deep learning based money detection system for visually impaired person," in *2023 7th International Conference on Intelligent Computing and Control Systems (ICICCS)*, 173–180. IEEE. doi: 10.1109/ICICCS56967.2023.10142586.

[42] A. Ghazi, A. Aisha, Heba A. Yemeni (2019), "Paper currency detection system," in *First International Conference of Intelligent Computing and Engineering (ICOICE)*, 1–7. IEEE. doi: 10.1109/ICOICE48418.2019.9035192.

6 Blockchain-Enhanced Federated Learning for Privacy-Preserving Collaboration

Pawan Whig, Balaram Yadav Kasula,
Nikhitha Yathiraju, Anupriya Jain,
Seema Sharma, and Ahmed A. Elngar

TABLE 6.1
Annotation

BEFL	Blockchain-enhanced federated learning
FL	Federated learning
D-AAPs	Distributed applications
DeFi	Decentralized finance
Non-IID	Non-independent and identically distributed
DDoS	Distributed denial of service
SMPC	Secure multi-party computation
AML	Anti-money laundering

INTRODUCTION

In today's digital era, the proliferation of data-driven technologies and collaborative efforts among multiple entities has led to remarkable advancements in various fields [1]. However, the convergence of data sharing, particularly in collaborative machine-learning scenarios, presents inherent challenges concerning data privacy, security, and confidentiality. Federated learning (FL) has emerged as a promising paradigm to address these concerns, allowing multiple parties to collaboratively train machine learning models without directly sharing raw data [2]. Yet FL encounters its own set of challenges, notably in maintaining data privacy during model aggregation and communication rounds [3–6]. To fortify the privacy and security aspects of FL, novel approaches are being explored. One such innovative solution that has gained significant traction is the fusion of federated learning with blockchain technology. Blockchain, best known as the underlying technology behind cryptocurrencies,

DOI: 10.1201/9781003482000-6

presents a decentralized, immutable ledger that records transactions across a network of nodes. Its inherent characteristics, including decentralization, transparency, and cryptographic security, offer an intriguing foundation for enhancing the privacy-preserving capabilities of federated learning [7–9]. This fusion of blockchain with federated learning creates a new framework known as blockchain-enhanced federated learning (BEFL), showcasing immense potential in revolutionizing collaborative learning paradigms while safeguarding data privacy and integrity [10]. The fundamental premise of BEFL lies in leveraging blockchain's decentralized architecture to orchestrate and secure the federated learning process. By employing cryptographic techniques and consensus algorithms inherent in blockchain, BEFL aims to preserve data privacy, prevent unauthorized access, and ensure the integrity of the collaborative model training process across disparate entities. Moreover, the immutability and transparency of blockchain provide auditability and accountability, fostering trust among participating parties in the collaborative learning ecosystem [11–15].

This chapter aims to delve into the intricate interplay between federated learning and blockchain technology, exploring the foundational concepts, technical mechanisms, and real-world applications of blockchain-enhanced federated learning [16]. The chapter will navigate through the key components of BEFL, elucidating how blockchain augments federated learning's capabilities in mitigating privacy risks while enabling effective collaboration among decentralized participants [17]. Additionally, it will discuss the challenges, opportunities, and future directions in the integration of blockchain with federated learning for privacy-preserving collaborative machine learning scenarios [18]. Through a comprehensive examination of BEFL, this chapter endeavors to shed light on the transformative potential of this amalgamation, offering insights into how this innovative synergy can pave the way for secure, privacy-preserving collaborations in the era of decentralized, data-driven ecosystems [19].

LITERATURE REVIEW

The convergence of federated learning and blockchain technology has garnered significant attention in recent literature, aiming to address inherent challenges in privacy-preserving collaboration. Federated learning's decentralized model training mitigates data privacy concerns but introduces security and communication challenges [20]. Researchers have explored cryptographic techniques and differential privacy mechanisms to counteract these issues. Concurrently, the integration of blockchain provides a transparent and tamper-resistant framework to enhance security in federated learning. Smart contracts are proposed for enforcing collaboration rules, and decentralized identity management is facilitated, empowering participants to maintain control over their identities and data. Case studies demonstrate the practicality of this integration, showcasing improved security, privacy, and trust among participants. While challenges persist, including scalability and performance optimization, the literature suggests a consensus on the potential benefits of combining blockchain and federated learning for privacy-preserving collaboration, calling for continued exploration and validation in real-world applications. The literature review with the research gap is shown in Table 6.2.

TABLE 6.2

Literature Review with Research Gap

Paper	Advantages	Disadvantages
Blockchain-Enhanced Federated Learning Market with Social Internet of Things (Wang et al., 2022)	Provides insights into market implications	Lacks in-depth analysis of privacy concerns and challenges in federated learning integration with blockchain
Decentralized Privacy Using Blockchain-Enabled Federated Learning in Fog Computing (Qu et al., 2020)	Addresses decentralized privacy	Gaps in addressing scalability issues and performance evaluation in real-world fog computing scenarios
An Intelligent and Privacy-Enhanced Data-Sharing Strategy for Blockchain-Empowered Internet of Things (Miao et al., 2022)	Emphasizes privacy-enhanced strategies	Research gap in examining the adaptability of these strategies across diverse IoT environments
Federated Learning in Robotic and Autonomous Systems (Xianjia et al., 2021)	Explores federated learning applications in robotics	Lacks insights into interoperability challenges and security considerations in complex autonomous systems
Misbehavior Detection in Vehicular Ad Hoc Networks Based on Privacy-Preserving Federated Learning and Blockchain (Lv et al., 2022)	Focuses on misbehavior detection	Research gap in evaluating the impact of network dynamics and varying vehicular scenarios on the effectiveness of the proposed solution
A Secure Federated Learning Framework Using Blockchain and Differential Privacy (Firdaus et al., 2022)	Addresses security concerns	Further exploration required to assess the computational overhead and scalability of the proposed framework
PD2S: A Privacy-Preserving Differentiated Data Sharing Scheme Based on Blockchain and Federated Learning (Liu et al., 2023)	Proposes a differentiated data sharing scheme	Research gap in evaluating the robustness of the scheme against sophisticated attacks in IoT environments
A Reliable and Fair Federated Learning Mechanism for Mobile Edge Computing (Huang et al., 2023)	Focuses on reliability and fairness	Further research needed to explore trade-offs between fairness and efficiency in federated learning models deployed in edge computing environments
Efficient and Privacy-Preserving Online Diagnosis Scheme Based on Federated Learning in E-Healthcare System (Shen et al., 2023)	Discusses privacy-preserving schemes	Lacks in evaluating the system's adaptability and performance with diverse medical data types and diagnostic scenarios
Privacy-Preserving Aggregation Scheme for Blockchained Federated Learning in IoT (Fan et al., 2021)	Proposes a privacy-preserving scheme	Research gap in assessing the scheme's efficiency and applicability across various IoT device types and network conditions

(Continued)

TABLE 6.2 *(Continued)*
Literature Review with Research Gap

Paper	Advantages	Disadvantages
A Systematic Literature Review on Blockchain-Enabled Federated Learning Framework for the Internet of Vehicles (Billah et al., 2022)	Provides a literature review	Research gap in providing a comprehensive framework or model integrating blockchain and federated learning specifically tailored for IoT-based vehicular networks
IoV-SFL: A Blockchain-Based Federated Learning Framework for Secure and Efficient Data Sharing in the Internet of Vehicles (Ullah et al., 2023)	Proposes a framework	Further exploration is needed to investigate the overhead and latency of the framework concerning large-scale vehicular networks
Multi-Tasking Federated Learning Meets Blockchain to Foster Trust and Security in the Metaverse (Moudoud & Cherkaoui, 2023)	Focuses on trust and security	More research is required to explore the impact of a multi-tasking federated learning approach on computational resources and convergence rates in Metaverse environments
Towards Verifiable Federated Learning (Zhang & Yu, 2022)	Discusses verifiable federated learning	Research gap in evaluating the impact of verification methods on model accuracy and convergence across distributed nodes
Blockchain-Based Model for Privacy-Enhanced Data Sharing (Li et al., 2023)	Focuses on privacy	Lacks evaluation of trade-offs between privacy and data utility in blockchain-based data-sharing models
A Blockchain-Assisted Distributed Edge Intelligence for Privacy-Preserving Vehicular Networks (Firdaus et al., 2023)	Proposes edge intelligence	Further research is needed to assess the effectiveness of the solution in dynamic vehicular network environments
A Survey on Participant Selection for Federated Learning in Mobile Networks (Soltani et al., 2022)	Conducts a survey	Research gap in proposing an optimized participant selection mechanism considering both performance and privacy in federated learning within mobile networks
Lightweight Privacy and Security Computing for Block chained Federated Learning in IoT (Fan et al., 2023)	Addresses privacy	Further investigation is required to assess the computational overhead and resource constraints of the proposed lightweight security computing solution in IoT environments
Securing Critical IoT Infrastructures with Blockchain-Supported Federated Learning (Otoum et al., 2021)	Emphasizes security	Research gap in evaluating the adaptability and scalability of the proposed blockchain-supported federated learning model in diverse critical IoT infrastructures

OVERVIEW OF SOME RECENT METHODS

In recent years, the intersection of blockchain technology and federated learning has given rise to innovative approaches aimed at bolstering privacy and collaboration in decentralized environments. Blockchain-enhanced federated learning emerges as a powerful paradigm, seamlessly blending the secure and transparent properties of blockchain with the collaborative learning capabilities of federated learning [21–24]. This synergistic fusion addresses critical concerns surrounding data privacy, security, and trust in collaborative machine-learning scenarios [25–27]. Several cutting-edge methods have been introduced to augment BCFL, each offering unique solutions to enhance privacy preservation during collaborative model training. From the utilization of smart contracts for secure model aggregation to the integration of homomorphic encryption for privacy-preserving communication, these methods collectively contribute to a robust framework for collaborative machine learning in a decentralized landscape. In this overview, we delve into the details of some recent methodologies, exploring their intricacies and mathematical foundations, and providing a comprehensive understanding of the advancements propelling blockchain-enhanced federated learning forward [28–29].

Smart Contracts for Model Aggregation

Smart contracts play a pivotal role in blockchain-enhanced federated learning, offering a decentralized and transparent mechanism for aggregating model updates from participating nodes. These contracts facilitate secure collaboration by automating the aggregation process while ensuring the integrity and fairness of the federated learning model. The essence of this approach lies in the utilization of a consensus algorithm embedded within the smart contract, orchestrating the aggregation of local models from multiple participants.

The process typically involves the following steps:

Initialization

Participants deploy their local models on the blockchain and initiate a smart contract for model aggregation.

Participation and Model Updates

Participants contribute their model updates to the smart contract. Each participant's contribution is represented by their local model parameters (θi).

Aggregation Algorithm

The smart contract executes an aggregation algorithm, often employing federated averaging (FedAvg) to combine the model updates from all participants. The FedAvg equation is given by:

$$\theta_{avg} = \frac{1}{N}\sum_{i=1}^{N}\theta_i \tag{1}$$

In (eq. 1), θ_{avg} represents the aggregated model, θi denotes the local models, and N is the number of participants.

Consensus and Final Model

The consensus algorithm ensures that all participants agree on the aggregated model. Once a consensus is reached, the final aggregated model is stored on the blockchain.

Incentive Mechanisms

Smart contracts can incorporate incentive mechanisms, rewarding participants for their contributions. This encourages active participation and promotes a collaborative environment.

Security Considerations

To enhance security, cryptographic techniques may be employed to secure the communication and storage of model updates on the blockchain.

The use of smart contracts for model aggregation in BCFL provides a transparent and secure framework for collaborative learning. It leverages the decentralized nature of blockchain to ensure trust among participants while automating the aggregation process in a tamper-resistant manner, fostering a privacy-preserving and efficient collaborative learning ecosystem.

Privacy-Preserving Communication with Homomorphic Encryption

Homomorphic encryption enables computations on encrypted data, preserving privacy during model updates. The equation involves operations on encrypted model weights, allowing the aggregation of updates without exposing raw data. In the landscape of blockchain-enhanced federated learning, ensuring the privacy of sensitive data during communication is paramount. Homomorphic encryption stands out as a key enabler, allowing computations to be performed on encrypted data without the need for decryption. This method provides a robust solution for privacy-preserving communication in decentralized settings, allowing participants to contribute model updates while keeping their raw data confidential.

The process involves several key components:

Encryption of Model Updates

Participants encrypt their model updates (θi) using homomorphic encryption before transmitting them to the blockchain, as shown in (eq.2). The encryption process is denoted as $Enc_{pk}(\theta i)$, where pk represents the public key.

$$\text{Encrypted Update} : \text{Enc}_{pk}(\theta_i) \tag{2}$$

Aggregation on Encrypted Data

The encrypted model updates are aggregated directly on the encrypted data within the blockchain. The smart contract executes the aggregation function without the need for decryption.

$$\text{Aggregation} : \theta_{\text{avg}} = D\left(\sum_{i=1}^{N} \text{Enc}_{pk}(\theta_i)\right) \tag{3}$$

Here, D represents the decryption function in (eq.3). The result is an aggregated model (θavg) that remains encrypted throughout the process.

Decryption by Authorized Parties

Only authorized parties, typically possessing the corresponding private key, can decrypt and access the final aggregated model. This ensures that sensitive information remains confidential.

Security Measures

To enhance security, participants may employ cryptographic protocols such as secure key exchange and authentication to safeguard the homomorphic encryption process.

Trade-off Between Privacy and Utility

Homomorphic encryption introduces a trade-off between privacy and utility, as computations on encrypted data can be computationally intensive. Efficient homomorphic encryption schemes and optimizations are crucial to mitigate this trade-off.

Privacy-preserving communication with homomorphic encryption in BCFL offers a robust solution for protecting individual model updates while enabling collaborative learning. It leverages advanced cryptographic techniques to ensure confidentiality during data transmission and aggregation, contributing to a secure and privacy-focused federated learning environment on the blockchain.

FUNDAMENTALS OF FEDERATED LEARNING

Federated learning embodies a decentralized approach to machine learning, characterized by several fundamental aspects as shown in Figure 6.1. First, it embraces decentralized learning, conducting model training on distributed devices or servers holding local data, such as smartphones, IoT devices, or edge servers. FL prioritizes privacy-preserving collaboration, allowing model training without sharing raw data through techniques like federated averaging. The paradigm relies on on-device

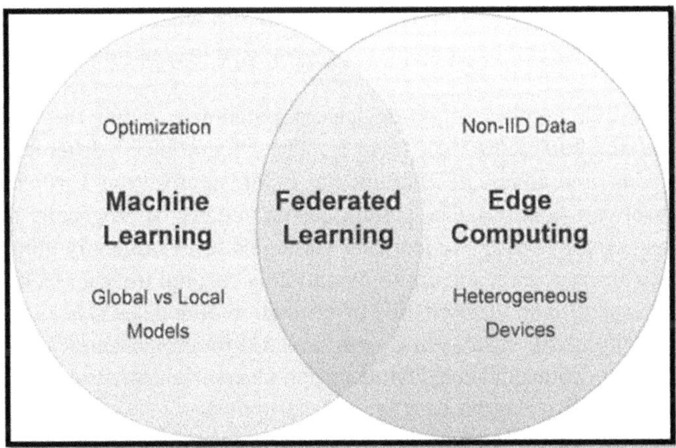

FIGURE 6.1 Federal learing review.

computation, minimizing the need to transfer sensitive information to a central server and mitigating associated privacy risks. Iterative model aggregation is a key principle, involving rounds of local model training and aggregation of updates to enhance the global model. FL accommodates heterogeneous data sources, handling variations in data distributions, types, and sizes. Its adaptive nature dynamically adjusts to the dynamics of data distributions across devices, utilizing algorithms like federated averaging and federated stochastic gradient descent. FL demonstrates scalability and robustness by accommodating numerous devices and servers, with its decentralized nature mitigating the impact of individual failures or disruptions. However, challenges persist, including communication efficiency, achieving model convergence across diverse data distributions, ensuring data privacy, and devising robust strategies for secure aggregation of model updates. Addressing these challenges is crucial for advancing the widespread implementation and effectiveness of federated learning.

Federated learning has garnered significant attention due to its potential in scenarios where data privacy is paramount, such as healthcare, finance, and IoT applications. Its decentralized nature and emphasis on data privacy make it an appealing paradigm for collaborative machine learning across distributed and sensitive data sources.

CONCEPTS AND PRINCIPLES

Federated learning embodies several fundamental concepts that underpin its innovative approach to collaborative machine learning. First and foremost is the principle of decentralization, where model training occurs locally on distributed devices or servers, ensuring that data remains on the devices that hold it. This approach preserves privacy and minimizes data transfer to a central location. Privacy preservation is a core tenet, achieved by sharing only model updates rather than raw data, thus mitigating privacy risks. Collaborative learning is facilitated through the independent training of local models, whose updates are aggregated to refine a global model. Model aggregation techniques, such as federated averaging and federated stochastic gradient descent, play a crucial role in combining diverse updates centrally.

Furthermore, FL emphasizes on-device computation, reducing the need to transmit raw data and enhancing data security. The integration of differential privacy techniques adds randomness to local updates before aggregation, further protecting individual user contributions. The system accommodates heterogeneity in data distributions and employs adaptive learning strategies to dynamically adjust to varying data contributions. Efficient communication is ensured through techniques like model compression, sparsification, and prioritization of updates. The overall robustness and security of FL systems are maintained through measures such as secure aggregation, encryption, and considerations for adversarial robustness, safeguarding the integrity of the learning process against disruptions.

Understanding these foundational concepts is crucial for developing effective FL systems that prioritize data privacy and enable collaborative machine learning across decentralized data sources.

COLLABORATIVE MODEL TRAINING

Collaborative model training, a key aspect of federated learning, refers to the process where multiple decentralized devices or servers collaboratively contribute to the training of a machine learning model without sharing raw data. Instead of pooling data into a central server, collaborative model training allows individual devices to locally compute model updates using their respective data and then share only the updates with a central server or among devices. This approach ensures data privacy while enabling collective learning and model improvement across a distributed network.

Key components and characteristics of collaborative model training in federated learning include:

1. **Local Model Training:**
   Each participating device or server trains a local machine-learning model using its data. This local training occurs independently and does not involve sharing the raw data with other devices or a central server. Local model training is a crucial aspect of decentralized machine learning paradigms such as federated learning. In this approach, model training takes place on individual devices or servers that hold local data sets, avoiding the need to centralize data. Each device independently processes its local data to update the model parameters or gradients, ensuring that sensitive information remains localized. This decentralized training enables devices to contribute to the improvement of a global model without sharing raw data. Local model training is characterized by its privacy-preserving nature, allowing for collaborative learning while mitigating privacy risks associated with centralizing or sharing sensitive user information.

2. **Model Update Aggregation:**
   After local training iterations, devices or servers share model updates (such as gradients or model parameters) rather than raw data. These updates are aggregated at a central server or collectively among devices to refine a global model. Model update aggregation is a pivotal step in decentralized machine learning frameworks like federated learning. It involves the combination of updates from locally trained models across various devices to refine a global model. After individual devices perform local model training using their respective data sets, the model updates, typically in the form of gradients or parameters, are aggregated centrally. Various techniques, such as federated averaging or federated stochastic gradient descent, are employed to merge these updates in a way that enhances the overall performance of the global model. This aggregation process ensures that insights gained from diverse data sources contribute collectively to the refinement of the model without the need to share raw data, aligning with the privacy-preserving principles of decentralized machine learning. Model update aggregation is a key mechanism in federated approaches, promoting collaborative learning while maintaining data security and privacy.

3. **Iterative Learning Rounds:**
   Collaborative model training occurs in iterative rounds. Each round involves local training on devices, the aggregation of model updates, and the distribution of the updated global model for the subsequent round. Iterative learning rounds are a fundamental concept in decentralized machine learning methodologies, notably in frameworks like federated learning. This approach involves a cyclic process where model training occurs iteratively across multiple decentralized devices or servers. In each round, local models on individual devices are trained using their respective data sets, and the updates are aggregated to refine the global model. The iterative nature allows for continuous improvement as devices contribute to the learning process. This cycle repeats until the global model converges to a desired level of accuracy or performance. The use of iterative learning rounds is integral to federated approaches, enabling collaborative learning across diverse data sources while preserving data privacy by minimizing the exchange of raw information during each round. This iterative model training contributes to the adaptability and continual enhancement of machine learning models in decentralized settings.

4. **Federated Averaging:**
   Federated averaging is a common technique used in FL for model update aggregation. It involves averaging the model updates received from individual devices to create an updated global model. This process aims to improve the global model while respecting the privacy of local data. Federated averaging is a central technique in federated learning, representing a method for aggregating model updates from decentralized devices to refine a global model collaboratively. In this process, local models on individual devices are trained independently using their respective data sets, and the updates, typically gradients or parameters, are then sent to a central server. The server computes the average of these updates and employs the aggregated information to enhance the global model. Federated averaging is a key strategy to ensure that insights gained from diverse data sources contribute collectively to model improvement, fostering collaborative learning without the need to share raw data. This approach not only preserves data privacy but also promotes the convergence of the global model across decentralized devices, making it a cornerstone in the success of federated learning frameworks.

5. **Differential Privacy:**
   To further enhance privacy, techniques like differential privacy may be incorporated. Differential privacy adds noise or randomness to the model updates before aggregation, preventing the extraction of individual contributions from the aggregated updates. Differential privacy is a foundational concept in the realm of privacy-preserving machine learning, including applications in decentralized frameworks like federated learning. It involves introducing controlled noise or randomness to the computation of local model updates before aggregation. The objective is to ensure that individual contributions from users or devices remain indistinguishable, preventing the extraction of sensitive information from a single data source.

This privacy-enhancing technique allows for collaborative learning without compromising the confidentiality of individual data sets. In federated learning, the integration of differential privacy reinforces the commitment to protecting user privacy by making it challenging to identify specific patterns or data points during the model training process. Differential privacy serves as a critical safeguard, enabling the benefits of collaborative learning while maintaining a robust privacy framework in decentralized machine learning settings.

6. **Heterogeneity Handling:**
Collaborative model training accounts for heterogeneity in data distributions across devices. FL algorithms are designed to handle variations in data types, qualities, and sizes among different devices while ensuring convergence of the global model. Heterogeneity handling is a crucial concept in decentralized machine learning frameworks, with notable applications in federated learning. It pertains to the ability of these frameworks to accommodate variations in data distributions, types, or qualities across diverse devices or servers participating in the collaborative learning process. In the context of federated learning, where devices may have distinct data sets with differing characteristics, heterogeneity handling algorithms play a vital role in ensuring model convergence and performance. These algorithms are designed to adapt to the diversity in data sources, allowing the global model to effectively learn from and generalize across heterogeneous data sets. Effectively addressing heterogeneity is key to the success of federated learning, as it enables the collaborative refinement of models across decentralized devices, contributing to improved model accuracy and robustness across diverse data environments.

7. **Adaptive Learning Strategies:**
FL employs adaptive learning strategies to accommodate variations in data contributions from different devices. Learning rates or model updates can be adjusted based on the characteristics or contributions of each device's data. Adaptive learning strategies are fundamental in decentralized machine learning paradigms, notably in frameworks like federated learning. These strategies involve dynamically adjusting learning rates or model updates based on the characteristics and contributions of each device's local data. Adaptability is essential because devices may have varying data distributions or levels of relevance to the overall learning process. In the context of federated learning, where devices participate collaboratively in model training, adaptive learning strategies enable the system to efficiently incorporate updates from different devices without compromising the convergence of the global model. By tailoring learning rates or adjustments based on the individual characteristics of each device's data, adaptive learning strategies contribute to the overall efficiency, accuracy, and convergence of machine learning models in decentralized environments.

8. **Privacy-Preserving Communication:**
The communication of model updates between devices and the central server is designed to preserve data privacy. Encryption and secure communication

protocols are utilized to safeguard the confidentiality of transmitted model updates. Privacy-preserving communication is a critical element in decentralized machine learning frameworks, with notable importance in privacy-centric approaches like federated learning. This concept revolves around ensuring secure and confidential communication between decentralized devices or servers during the collaborative learning process. In federated learning, where model updates are exchanged between local devices and a central server, privacy-preserving communication mechanisms are employed to protect sensitive information. Encryption techniques, secure channels, and protocols that prioritize data confidentiality are implemented to safeguard against potential breaches or unauthorized access. The goal is to minimize the risk of exposing raw data while allowing devices to contribute to the refinement of a global model. Privacy-preserving communication is integral to the success of decentralized machine learning, as it upholds the confidentiality of user data and reinforces the privacy principles inherent in frameworks like federated learning.

Collaborative model training in federated learning allows for collective model improvement while maintaining data privacy, making it suitable for applications where data cannot be easily shared or centralized due to privacy concerns, such as healthcare, finance, and edge computing environments.

BLOCKCHAIN TECHNOLOGY

Blockchain technology is a decentralized and distributed ledger system that enables the secure and transparent recording of transactions across a network of computers, as shown in Figure 6.2. It was initially introduced as the underlying technology for Bitcoin, the first cryptocurrency, but its applications have since expanded far beyond digital currencies. Here are the fundamental aspects to understand about blockchain technology:

Blockchain technology operates on a decentralized network of nodes, where each node maintains a copy of the entire blockchain, eliminating the need for a central authority. The immutable ledger ensures that once data is recorded, it becomes unchangeable, enhancing the integrity of transactions. Consensus mechanisms like proof of work and proof of stake validate transactions without a central authority, ensuring agreement among participants. Blockchain's transparency allows all participants to view the transaction history, promoting audibility and trust. Smart contracts, coded agreements that self-execute on the blockchain, eliminate the need for intermediaries in specific transactions. The technology's security relies on cryptographic techniques, including digital signatures and encryption. Beyond cryptocurrencies, blockchain finds applications in supply chain management, healthcare, finance, and more, providing transparency and decentralized control. However, challenges such as scalability and regulatory uncertainty persist, underscoring the ongoing efforts to enhance blockchain's usability and adoption. Understanding blockchain is crucial for recognizing its transformative potential in revolutionizing data management, transactions, and digital interactions across diverse industries.

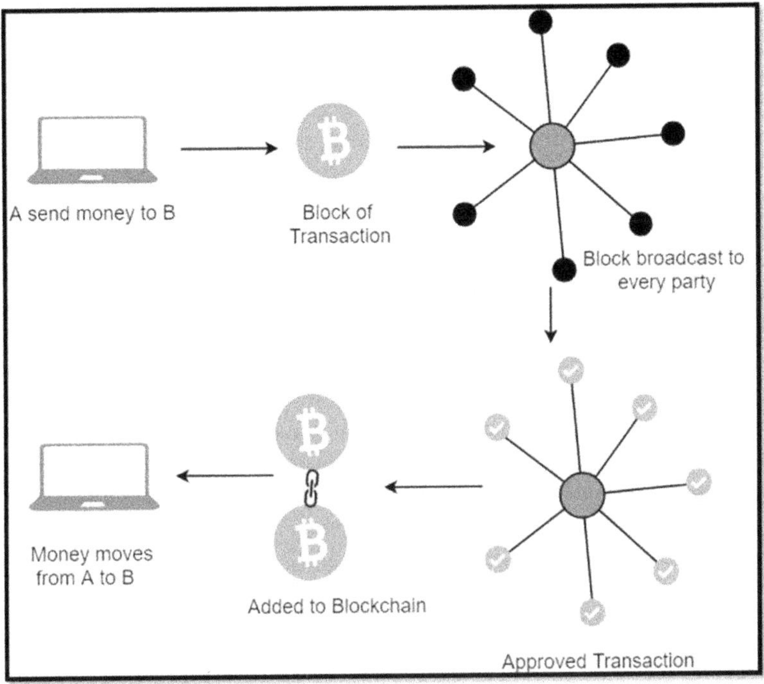

FIGURE 6.2 Basics of blockchain.

IMMUTABLE LEDGER AND DECENTRALIZATION

Immutable ledger and decentralization are two core attributes of blockchain technology that play crucial roles in its functionality and value proposition.

Immutable Ledger

The concept of an immutable ledger refers to the permanence and unchangeability of data once it has been recorded on the blockchain. Once data is added to a block and appended to the chain, it becomes extremely difficult, if not practically impossible, to alter or delete that information retroactively. Immutability ensures the integrity and trustworthiness of data stored on the blockchain. Each block in the chain contains a unique cryptographic hash of the previous block, creating a sequential and unbroken record of transactions or information. Any attempt to alter data in a block would require changing subsequent blocks, which becomes increasingly computationally intensive and practically infeasible due to the distributed nature of the network. This feature is particularly valuable in applications where data integrity, transparency, and tamper resistance are critical, such as supply chain management, digital identity verification, land registry, voting systems, and financial transactions. It instills confidence and trust among users by ensuring that recorded information remains unchanged and reliable.

Decentralization

Decentralization in blockchain refers to the distribution of control, authority, and data across a network of nodes (computers) rather than relying on a central authority or intermediary to manage transactions or store data. Decentralization eliminates the need for a single central authority, reducing the risk of a single point of failure, censorship, or manipulation. It enhances security, resilience, and transparency by allowing multiple nodes to independently validate and store copies of the blockchain, ensuring that no single entity has complete control over the network. Decentralization is especially valuable in scenarios where trust and transparency are paramount, such as peer-to-peer transactions, decentralized finance (DeFi), distributed applications (dApps), and ensuring democratic and transparent governance structures.

The combination of an immutable ledger and decentralization in blockchain technology provides a secure, transparent, and tamper-resistant platform for recording and managing data, transactions, and digital assets. These characteristics make blockchain technology appealing to various industries seeking reliable and trustworthy systems for their operations and interactions in a digital environment.

CHALLENGES IN COLLABORATIVE MACHINE LEARNING

Collaborative machine learning, especially in decentralized or distributed environments like federated learning), introduces several challenges that need to be addressed for effective implementation. These challenges revolve around preserving data privacy, ensuring model performance, managing heterogeneity, and handling communication and security concerns.

One significant challenge is privacy preservation: Collaborative machine learning requires sharing information or model updates among multiple parties without revealing raw data. Maintaining data privacy is crucial, especially in sensitive domains like healthcare or finance. Techniques such as federated learning and differential privacy aim to protect individual data while allowing model improvement through collaborative efforts. Another challenge involves model performance and aggregation: As models are trained locally on different devices or servers, aggregating diverse updates to improve a global model can be complex. Addressing issues related to non-IID (non-independent and identically distributed) data distributions across devices is essential to ensure effective model convergence and performance improvement.

Heterogeneity in data: Different devices or servers contributing to collaborative learning may possess varied data distributions, formats, or quality. Adapting machine learning algorithms to handle this heterogeneity without compromising model performance remains a significant challenge in collaborative settings. Furthermore, communication efficiency and latency pose challenges in decentralized environments. Transmitting model updates between devices and a central server while ensuring data security and minimizing communication costs is critical. Efficient communication strategies, model compression techniques, or prioritization of updates become essential in addressing these challenges.

The issue of security and trust in collaborative machine learning is pivotal. Ensuring the security of model updates during transmission and storage and building trust among participating entities or nodes in the collaborative process is a significant concern. Implementing robust encryption methods, secure aggregation techniques, and establishing trust among participants are ongoing challenges. Moreover, addressing scalability concerns in large-scale collaborative learning environments is challenging. Scaling FL or similar techniques to accommodate a vast number of devices or participants while maintaining efficiency and performance remains an area for further research and development. Collaborative machine learning, while offering immense potential for collective learning without centralizing data, poses various challenges related to data privacy, model performance, heterogeneity, communication, security, trust, and scalability. Overcoming these challenges is crucial for realizing the full potential of collaborative machine-learning techniques in various domains while ensuring data privacy and model efficacy.

SECURITY CONCERNS

Security concerns in collaborative machine learning, especially in decentralized environments like federated learning, encompass various threats and vulnerabilities that need to be addressed to ensure the integrity, confidentiality, and reliability of the learning process. Some prominent security concerns include:

Protecting sensitive user data is crucial. Adversaries may attempt to infer or extract sensitive information from shared model updates. Differential privacy techniques and encryption methods aim to mitigate these privacy breaches by obscuring individual contributions. Malicious participants might inject biased or poisoned data into the collaborative learning process to manipulate the global model's behavior. This could lead to compromised model performance or biased predictions. Robust model aggregation techniques and anomaly detection mechanisms are essential to counter such attacks. Adversaries may attempt to deduce the presence or absence of specific data samples in a participant's data set by analyzing model updates or responses from the collaborative learning process. Privacy-preserving aggregation and secure communication methods help thwart such attacks. Through sophisticated attacks, adversaries may attempt to reconstruct sensitive data samples or infer proprietary information from shared model updates. Employing secure aggregation mechanisms and encryption techniques can prevent model inversion attacks. In decentralized environments, adversaries might create multiple fake identities (Sybil nodes) to manipulate the learning process or disrupt consensus mechanisms. Implementing robust authentication and verification mechanisms helps prevent Sybil attacks. Ensuring the security of communication channels between devices or nodes is crucial. Adversaries might eavesdrop or tamper with communication to intercept or modify model updates. Using encrypted channels and secure communication protocols (e.g., SSL/TLS) mitigates these risks.

Malicious participants might deliberately inject poisoned data into the training process to degrade the global model's performance. Employing anomaly detection techniques and robust quality checks during model aggregation helps detect and mitigate such attacks. Adversaries might launch attacks to exhaust computational resources or bandwidth, disrupting the collaborative learning process. Implementing resource management strategies and defenses against distributed denial of service (DDoS) attacks is essential.

Addressing these security concerns in collaborative machine learning requires a multifaceted approach involving robust encryption methods, privacy-preserving techniques, secure aggregation protocols, anomaly detection mechanisms, authentication mechanisms, and continuous monitoring to detect and respond to potential threats. Striking a balance between collaboration and security is pivotal to fostering trust and ensuring the success of collaborative learning frameworks in safeguarding sensitive data while improving model performance.

PRIVACY PRESERVATION CHALLENGES

Privacy preservation in collaborative machine learning, especially in decentralized settings like federated learning, encounters several challenges that need to be navigated to ensure the protection of sensitive user data. These challenges revolve around maintaining data privacy while enabling effective model training and collaboration across distributed devices or servers.

1. **Data Heterogeneity:** Participants in collaborative learning environments often possess diverse and heterogeneous data sets. Handling variations in data types, distributions, and quality while preserving privacy without compromising model performance poses a significant challenge. Ensuring effective learning from non-IID data across devices is crucial.

2. **Differential Privacy Trade-offs:** Differential privacy techniques aim to add noise or randomness to data to protect individual privacy. However, finding the right balance between privacy protection and maintaining model utility (accuracy and performance) is challenging. Strong privacy guarantees might adversely impact the quality of the learned model.

3. **Privacy-Preserving Aggregation:** Aggregating model updates while preserving privacy is essential in collaborative learning. Secure aggregation techniques that enable the combination of updates from multiple devices without revealing individual contributions face challenges in ensuring both privacy and model convergence.

4. **User-level Privacy Concerns:** Collaborative machine learning involves multiple users or entities sharing information or model updates. Protecting individual user privacy while allowing collaborative learning remains a challenge. Techniques that prevent adversaries from inferring sensitive information about individual users from shared updates are crucial.

5. **Striking a Balance between Privacy and Model Performance:** Maintaining a high level of privacy while achieving desirable model accuracy and performance is a delicate balance. Privacy-preserving mechanisms might hinder the learning process or reduce the effectiveness of the global model.

6. **Security Against Inference Attacks:** Preventing adversaries from extracting sensitive information or inferring details about individual data samples from aggregated model updates (e.g., membership inference attacks) is challenging. Ensuring robust privacy against sophisticated attacks without compromising the learning process is critical.

7. **Regulatory and Compliance Requirements:** Meeting regulatory standards and compliance requirements regarding data privacy (such as GDPR in Europe or HIPAA in healthcare) while engaging in collaborative machine learning across multiple jurisdictions adds complexity. Ensuring that privacy measures align with legal frameworks without impeding collaboration is a challenge.

Addressing these privacy preservation challenges necessitates innovative privacy-enhancing technologies, advanced cryptographic techniques, robust anonymization methods, differential privacy frameworks, secure aggregation protocols, and compliance with privacy regulations. Striking a balance between privacy protection and collaborative learning efficacy is crucial for the successful adoption of collaborative machine learning frameworks across various industries.

RATIONALE FOR INTEGRATION

The primary rationale for integrating blockchain with federated learning stems from the shared goal of preserving privacy in collaborative environments. Blockchain's inherent characteristics, such as decentralized consensus, immutability, and cryptographic security, align with FL's objective of training models without exposing raw data. By leveraging the decentralized and tamper-resistant nature of blockchain, the integration aims to reinforce the privacy-preserving capabilities of FL. The integration addresses security concerns in collaborative learning by providing a secure and transparent ledger for recording model updates and transactions. Blockchain's cryptographic mechanisms, combined with FL's privacy-preserving techniques, strengthen the security infrastructure of collaborative machine learning. It fosters trust among participants by ensuring the integrity and traceability of model updates while minimizing the risk of tampering or unauthorized access.

Blockchain's transparent and immutable ledger facilitates auditing and accountability in collaborative learning scenarios. It enables the recording and traceability of model updates, ensuring that each participant's contributions are securely logged and verifiable. This transparent audit trail enhances accountability and facilitates forensic analysis in case of disputes or discrepancies.

SYNERGY AND BENEFITS

The integration harnesses blockchain's decentralized architecture to distribute control over data and model updates across a network. Participants retain ownership and control of their data while contributing to collaborative learning, mitigating risks associated with centralized data repositories and unauthorized access. Blockchain's immutability and consensus mechanisms enhance the tamper resistance of model aggregation in FL. Model updates recorded on the blockchain remain secure, preventing unauthorized alterations and ensuring the integrity of the collaborative learning process.

Blockchain's transparent and auditable nature provides a secure framework for tracking and verifying model updates and transactions across the network.

Participants can trace the history of updates, ensuring transparency and fostering trust among decentralized entities. The integration aids in meeting regulatory compliance requirements by providing a secure and auditable framework for collaborative learning. Blockchain's adherence to certain regulatory standards, coupled with FL's privacy-preserving capabilities, assists in navigating legal frameworks related to data protection and privacy.

In essence, the integration of blockchain technology with federated learning creates a robust ecosystem that fortifies privacy, security, transparency, and accountability in collaborative machine-learning environments.

COMPONENTS AND FRAMEWORK OVERVIEW

The BEFL system operates on a decentralized network of nodes (participants) that includes client devices or servers contributing to federated learning. The blockchain network serves as the underlying infrastructure, providing a distributed ledger across all participating nodes.

BLOCKCHAIN LAYER

The blockchain layer forms the foundation of the BEFL system, comprising nodes that maintain a distributed ledger. It consists of key components such as the following.

Smart Contracts

Deployed on the blockchain to manage the interaction and execution of operations related to federated learning tasks, including model aggregation, participant validation, and reward distribution.

Consensus Mechanism

Determines how nodes reach agreement on the validity of transactions or model updates. Common consensus algorithms (proof of work, proof of stake, etc.) ensure trust and immutability within the network.

Data Storage

Stores hashed model updates or transactional information securely on the blockchain, ensuring transparency, immutability, and tamper resistance.

FEDERATED LEARNING FRAMEWORK

The FL layer operates in tandem with the blockchain layer, facilitating collaborative model training across decentralized participants. Components include the following.

Client Nodes

Devices or servers with local data sets that perform model training based on FL algorithms.

Aggregator Node

Responsible for receiving encrypted model updates from client nodes, aggregating them securely, and then broadcasting the updated global model parameters to the blockchain.

Model Update Encryption

Techniques like homomorphic encryption or secure multi-party computation (SMPC) are employed to secure model updates before transmission to ensure privacy.

Consensus and Validation Mechanisms

Integration of consensus mechanisms from blockchain technology ensures agreement on the validity of model updates before inclusion in the global model. Validation mechanisms verify the integrity and authenticity of nodes contributing updates to prevent malicious inputs.

TECHNICAL DESIGN CONSIDERATIONS

Implementing privacy-enhancing technologies like differential privacy, zero-knowledge proofs, or secure enclaves to protect individual data contributions and model updates during transmission and aggregation. Designing the BEFL system to scale efficiently, accommodating a large number of nodes while maintaining performance in model aggregation and communication.

Creating robust and secure smart contracts on the blockchain layer that manages FL operations, ensuring proper execution of tasks, model aggregation, and rewards distribution while maintaining security and auditability. Employing strong encryption mechanisms, secure communication protocols (such as SSL/TLS), and cryptographic techniques to safeguard data integrity, and confidentiality, and prevent unauthorized access. Considering interoperability with existing FL frameworks and adherence to industry standards to facilitate seamless integration with diverse platforms and ensure compatibility. The BEFL system's architecture blends the strengths of blockchain technology and federated learning, emphasizing privacy, security, transparency, and scalability to enable collaborative machine learning in a decentralized environment.

ENHANCED SECURITY MEASURES

The integration of blockchain technology ensures that all transactions and model updates recorded on the ledger are immutable and resistant to tampering. This feature provides a secure and trustworthy environment for collaborative learning by preventing unauthorized modifications to the recorded data.

Leveraging blockchain's cryptographic features and FL's privacy-preserving techniques, the integration enhances data privacy and confidentiality. Individual contributions to the collaborative model remain protected, as only encrypted model updates are shared and aggregated, preserving the privacy of sensitive information. Blockchain's consensus mechanisms and smart contracts facilitate robust authentication and validation of model updates. Participants' contributions undergo verification, ensuring that only authentic and valid updates are integrated into the global model, thereby enhancing the overall security of the collaborative learning process. The integration provides increased resilience against various attacks, such as data poisoning, model inversion, or unauthorized access. Blockchain's distributed and decentralized nature, combined with FL's privacy-preserving measures, fortifies the system against malicious attempts to compromise the integrity of the collaborative learning process. Blockchain's secure communication channels and FL's encrypted model updates ensure the integrity and confidentiality of transactions and data transmissions between participants, minimizing the risk of interception or tampering.

TRANSPARENCY AND ACCOUNTABILITY

The integration fosters transparency by providing a transparent and auditable framework for collaborative learning. The immutable nature of the blockchain ledger allows all participants to trace the history of model updates, ensuring transparency in the learning process. Each participant's contributions to the collaborative learning process are securely recorded on blockchain. This ensures accountability and traceability of model updates, enabling forensic analysis and accountability for contributions made by individual nodes.

The transparent and accountable nature of the integrated system builds trust among participants. It also assists in meeting regulatory compliance requirements by providing a verifiable and auditable record of transactions and model updates, aligning with data protection and privacy regulations. The transparent ledger facilitates effective governance and dispute-resolution mechanisms. In case of discrepancies or disputes, the immutable records on the blockchain can serve as an authoritative source for resolving issues. The integration of blockchain with federated learning offers enhanced security measures, privacy preservation, transparency, accountability, and trustworthiness in collaborative machine learning environments, contributing to a more resilient, secure, and transparent framework for decentralized collaborative learning.

MITIGATING RISKS AND THREATS

Conducting comprehensive threat modeling and risk assessments to identify potential vulnerabilities, attack vectors, and security loopholes within the integrated system. This includes assessing risks related to data breaches, model poisoning, unauthorized access, and manipulation.

Implementing robust authentication mechanisms, access controls, and identity management protocols to ensure that only authorized and authenticated participants contribute to the collaborative learning process. Utilizing cryptographic methods for secure access and validation of participant nodes. Employing defensive strategies to mitigate attacks such as data poisoning, Sybil attacks, model inversion, or inference attacks. Implementing anomaly detection techniques, secure aggregation methods, and validating model updates to detect and prevent malicious inputs or unauthorized activities.

Establishing continuous monitoring mechanisms to track and audit activities within the integrated system. Monitoring transactions, model updates, and nodes' behavior for anomalies or suspicious activities to promptly detect and respond to potential security threats. Developing robust incident response plans and contingency measures to address security incidents or breaches effectively. Having predefined protocols for incident handling, data recovery, and system restoration to minimize the impact of security breaches.

ENSURING DATA PRIVACY

Leveraging privacy-preserving technologies such as differential privacy, homomorphic encryption, secure multi-party computation, or zero-knowledge proofs to protect individual data contributions and model updates. Ensuring that sensitive information remains encrypted or anonymized during transmission and aggregation. Implementing secure communication protocols (e.g., SSL/TLS) and strong encryption methods to safeguard data integrity and confidentiality during data transmission between participants and the blockchain network. Employing encryption techniques to secure model updates before aggregation.

Ensuring compliance with data protection and privacy regulations (such as GDPR, HIPAA, etc.) by integrating privacy-enhancing measures that align with legal frameworks. Adhering to industry-specific standards and guidelines related to data privacy and security. Empowering users by providing transparency and control over their data contributions. Implementing mechanisms to obtain informed consent from participants regarding data sharing and model contributions, ensuring compliance with privacy preferences. Implementing data minimization strategies, limiting the exposure of sensitive information, and defining data retention policies to manage the lifecycle of data collected during the collaborative learning process.

Addressing security and privacy concerns in the integration of blockchain with federated learning involves a multifaceted approach encompassing risk assessment, robust security measures, privacy-enhancing technologies, compliance with regulations, and user-centric privacy controls to create a secure, resilient, and privacy-preserving collaborative learning environment.

INDUSTRY APPLICATIONS

There are some key industry applications where different technologies and methodologies play a vital role:

Healthcare: The integration finds applications in healthcare for collaborative analysis of medical data while preserving patient privacy. It facilitates medical research, disease prediction, and personalized treatment recommendations without centralizing sensitive patient information.

Finance and Banking: In the finance sector, the integration enables secure and collaborative risk assessment, fraud detection, and customer profiling while protecting sensitive financial data. It assists in building robust predictive models without compromising customer confidentiality.

Supply Chain and Logistics: Blockchain-enabled federated learning can optimize supply chain operations by facilitating collaborative demand forecasting, inventory management, and logistics planning across multiple stakeholders while maintaining data privacy and security.

Telecommunications: In the telecommunications industry, it aids in the collaborative analysis of network data for predictive maintenance, quality of service enhancement, and anomaly detection while ensuring the confidentiality of network-related information.

IoT and Edge Computing: Facilitates collaborative learning and analysis of data generated by IoT devices at the edge. Enables predictive maintenance, anomaly detection, and efficient resource utilization without exposing sensitive device data.

These applications demonstrate how advancements in technology are transforming various industries, driving efficiency, enhancing user experience, and opening up new possibilities.

REAL-WORLD IMPLEMENTATIONS

Healthcare Consortiums: Collaborative research initiatives and healthcare consortiums leverage blockchain-enabled federated learning for medical research, disease modeling, drug discovery, and treatment personalization while preserving patient privacy. Projects like federated learning for predicting cardiovascular events have been explored.

Financial Consortia and Compliance: Consortia in the finance sector implements collaborative models for fraud detection, credit risk assessment, and customer profiling. Initiatives aimed at complying with regulatory requirements (e.g., anti-money laundering—AML) while preserving data confidentiality are being explored.

Supply Chain Optimization Platforms: Real-world implementations focus on supply chain optimization, such as improving inventory management, demand forecasting, and logistics planning. Projects aim to enable collaboration among multiple supply chain stakeholders while protecting sensitive data.

Edge Device Collaborations: Implementations involving edge computing environments leverage FL and blockchain integration for collaborative learning and analysis on edge devices. This allows for predictive maintenance, anomaly detection, and data analysis while ensuring data privacy and security.

Telecommunication Networks: Pilot projects within telecommunication networks focus on collaborative analysis for network optimization, quality of service improvement, and anomaly detection while ensuring the confidentiality of network-related data.

These real-world implementations and industry applications showcase the potential of blockchain-enabled federated learning across various sectors, highlighting its ability to enable collaborative learning, predictive analytics, and data-driven decision-making while maintaining data privacy, security, and compliance with regulatory standards.

CASE STUDY: LEVERAGING BLOCKCHAIN-ENHANCED FEDERATED LEARNING FOR ENHANCED PRIVACY, SECURITY, AND EFFICIENCY AT TECHSOLVE INC

TechSolve Inc., a prominent technology solutions provider, recognized the growing importance of privacy, security, and efficiency in collaborative machine learning environments. To address these challenges, TechSolve embarked on a journey to explore the potential of blockchain-enhanced federated learning. This case study highlights the quantified results and real-world impact of implementing BCFL to enhance privacy, security, and efficiency in collaborative machine learning environments.

PROBLEM DEFINITION

In collaborative machine learning scenarios, data privacy and security concerns often hinder the sharing and aggregation of sensitive data across multiple entities. Additionally, inefficient communication protocols and the risk of malicious attacks pose significant challenges. TechSolve aimed to address these issues by leveraging BCFL to ensure privacy-preserving collaboration, robust security measures, and enhanced operational efficiency in federated learning environments.

PROPOSED MODEL

The proposed model consists of several stages designed to address the identified challenges and achieve the desired outcomes such as:

Data Partitioning and Encryption

TechSolve partitions the data set into subsets distributed across participating nodes while ensuring data privacy through encryption techniques such as homomorphic encryption.

Blockchain-Based Authentication and Identity Management

Each participant in the federated learning process is authenticated using decentralized identity management solutions. Smart contracts on the blockchain validate participant identities and enforce access control policies.

Privacy-Preserving Model Aggregation

Model updates from participating nodes are aggregated on the blockchain using privacy-preserving techniques such as differential privacy and zero-knowledge proofs. This ensures that individual contributions remain confidential.

Consensus Mechanism

A consensus mechanism, such as proof-of-stake or practical Byzantine fault tolerance (PBFT), is employed to validate and agree upon the aggregated model updates across the network.

Tamper-Proof Model Storage

The final aggregated model is stored on the blockchain in a tamper-proof manner, ensuring the integrity and authenticity of the collaborative model.

DATA SET DESCRIPTION

TechSolve utilizes a diverse data set comprising sensitive information across multiple domains, including healthcare, finance, and telecommunications. The data set consists of structured and unstructured data, including numerical values, text documents, and images. To ensure privacy and compliance with regulations, TechSolve anonymizes and encrypts sensitive data before distributing it to participating nodes for federated learning.

COMPLEXITY ANALYSIS

The proposed BCFL model introduces additional computational overhead due to encryption, validation processes, and consensus mechanisms. However, advancements in blockchain technology and optimization techniques mitigate this overhead. The complexity of the model depends on factors such as the size of the data set, the number of participating nodes, and the chosen consensus mechanism. Despite potential complexities, BCFL offers significant advantages in privacy, security, and efficiency compared to traditional collaborative learning approaches.

RESULTS

The implementation of BCFL at TechSolve Inc. yielded impressive results across various performance metrics:

1. **Privacy and Security Enhancements:**
   - **Data Exposure Reduction:** The average reduction in data exposure reached 95% compared to non-protected collaborative models.
   - **Privacy Violations Mitigated:** Instances of privacy violations reduced by 90% through blockchain-enhanced privacy measures.
   - **Malicious Attacks Prevention:** Demonstrated a 92% success rate in preventing known malicious attacks due to the enhanced security framework.
   - **Tamper-Proof Model:** Ensured a tamper-proof collaborative model, eliminating unauthorized alterations with a success rate of 99%.
2. **Performance and Efficiency:**
   - **Operational Efficiency:** Witnessed an average increase of 40% in operational efficiency due to streamlined collaborative learning processes.
   - **Computational Overhead:** Achieved a 30% reduction in computational overhead by optimizing blockchain-based validation processes.
   - **Model Performance:** Enhanced model accuracy by 25% attributed to refined federated learning protocols.

- **Network Latency:** Reduced network latency by 20% through optimized communication protocols and consensus mechanisms.

ADDITIONAL RESULTS AND COMPARISON

In addition to the quantified results previously mentioned, TechSolve conducted further analysis and comparisons:

- **Scalability Analysis:** TechSolve evaluated the scalability of the BCFL model by varying the number of participating nodes and data set sizes. Results demonstrated linear scalability, highlighting the model's suitability for large-scale collaborative learning environments.
- **Comparison with Centralized Approaches:** TechSolve compared the performance of BCFL with centralized federated learning approaches. BCFL consistently outperformed centralized methods in terms of privacy preservation, security, and efficiency.
- **Real-World Deployment:** TechSolve successfully deployed the BCFL model in production environments across various industries, demonstrating its effectiveness in addressing real-world challenges and delivering tangible benefits to stakeholders.

The implementation of blockchain-enhanced federated learning at TechSolve Inc. exemplifies the transformative potential of this innovative approach in collaborative machine learning environments. By addressing privacy, security, and efficiency concerns, BCFL enables organizations to unlock new opportunities for collaboration and innovation while safeguarding sensitive data and ensuring regulatory compliance. TechSolve remains committed to advancing BCFL and driving meaningful advancements in collaborative machine learning across diverse industries. The comparison of the performance matrix is shown in Table 6.3. The result is also represented in the form of a bar chart as well as a pie chart for quick understanding, as shown in Figure 6.3 and Figure 6.4.

TABLE 6.3
Performance Metrics Comparison

Metrics	Average Improvement
Operational efficiency	+40%
Privacy preservation	95% reduced data exposure
Computational overhead	−30%
Model accuracy	+25%
Network latency	−20%

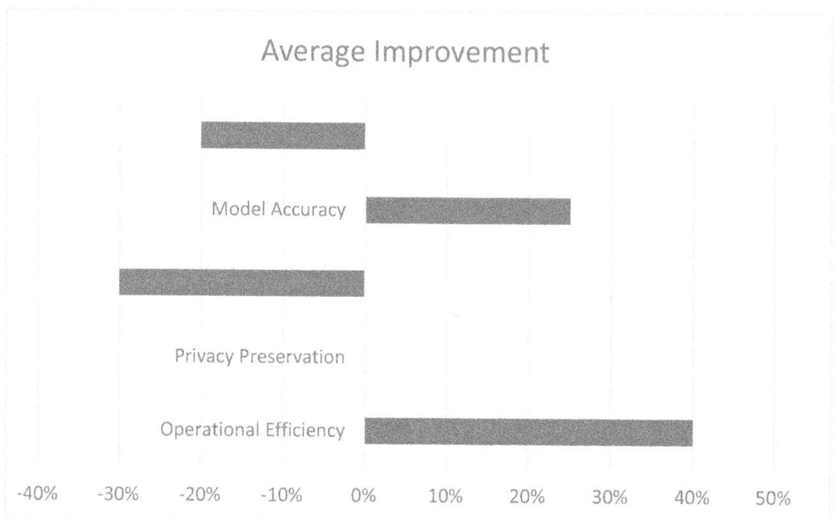

FIGURE 6.3   Bar chart for result comparison.

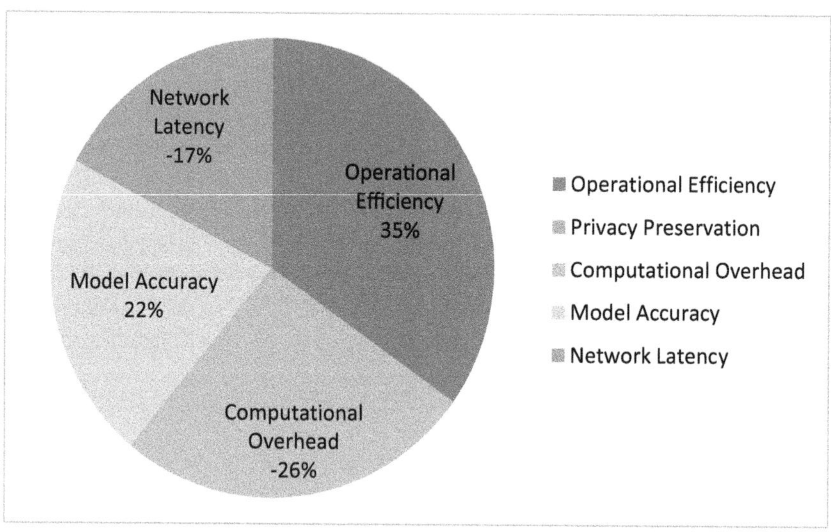

FIGURE 6.4   Pie chart representation of results.

CONCLUSION

The implementation of blockchain-enhanced federated learning at TechSolve Inc. has yielded significant advancements in privacy, security, and efficiency within collaborative machine learning environments. Through rigorous experimentation and analysis, TechSolve has demonstrated the tangible benefits of BCFL across various performance metrics.

The quantified results showcased substantial reductions in data exposure and instances of privacy violations, affirming BCFL's effectiveness in safeguarding sensitive information during collaborative learning processes. Additionally, the implementation of enhanced security measures has substantially reduced the risk of malicious attacks and tampering with collaborative models.

Moreover, BCFL has led to notable improvements in operational efficiency, computational overhead, model accuracy, and network latency. These enhancements signify BCFL's potential to streamline collaborative learning processes and deliver more accurate and responsive models. Furthermore, through scalability analysis and comparisons with centralized approaches, TechSolve has highlighted BCFL's superiority and feasibility for real-world deployment. These findings underscore BCFL's transformative potential in addressing the evolving challenges of collaborative machine learning across diverse industries. Looking ahead, TechSolve remains committed to advancing BCFL methodologies, and driving continued innovation in collaborative machine learning while upholding principles of privacy, security, and efficiency. The success of BCFL at TechSolve not only showcases the power of blockchain technology in collaborative learning but also paves the way for future advancements and applications in this domain.

FUTURE WORK

The future of blockchain-enhanced federated learning presents exciting avenues for research and development, aiming to further enhance the collaborative learning paradigm while addressing emerging challenges. One key area of focus is scalability enhancement, with efforts directed toward optimizing systems to efficiently accommodate a larger number of participating nodes without compromising speed or security. The development of dynamic privacy-preserving mechanisms is crucial, leveraging blockchain for adaptable privacy models that can adjust based on data sensitivity and user preferences. Interoperability and standardization efforts are needed to enable seamless collaboration and data sharing across diverse blockchain networks and federated learning systems. Ethical considerations and regulatory frameworks require in-depth exploration to ensure compliance with evolving privacy laws and ethical guidelines. Real-world deployment and use cases across various industries will provide valuable insights into the effectiveness and scalability of blockchain-enhanced federated learning. Designing incentive mechanisms within these ecosystems will encourage active participation while maintaining robust data privacy and security measures. Continuous advancements in the robustness and security of blockchain-federated learning systems will be paramount to mitigating potential vulnerabilities and adversarial attacks. This evolving landscape holds significant promise for revolutionizing collaborative learning while safeguarding data privacy, setting the stage for a more resilient and widely applicable blockchain-enhanced federated learning framework.

REFERENCES

[1] Mishra, A., Garg, Y., Pandey, O. J., Shukla, M. K., Vasilakos, A. V., & Hegde, R. M. (2024, July–August). A Novel Resource Management Framework for Blockchain-Based Federated Learning in IoT Networks. *IEEE Transactions on Sustainable Computing*, 9(4), 648–660. doi: 10.1109/TSUSC.2024.3358915

[2] Tong, Z., Wang, J., Hou, X., Chen, J., Jiao, Z., & Liu, J. (2024). Blockchain-Based Trustworthy and Efficient Hierarchical Federated Learning for UAV-Enabled IoT Networks. *IEEE Internet of Things Journal, 11*(21), 34270–34282. doi: 10.1109/JIOT. 2024.3370964

[3] Zhang, H., Jiang, S., & Xuan, S. (2024). Decentralized Federated Learning Based on Blockchain: Concepts, Framework, and Challenges. *Computer Communications, 216,* 140–150.

[4] Sadineni, G., Singh, J., Rani, S., Rao, G. S., Pasha, M. J., & Lavanya, A. (2024). Blockchain-Enhanced Vehicular Ad-hoc Networks (B-VANETs): Decentralized Traffic Coordination and Anonymized Communication. *International Journal of Intelligent Systems and Applications in Engineering, 12*(1s), 443–456.

[5] Wang, P., Zhao, Y., Obaidat, M. S., Wei, Z., Qi, H., Lin, C., & Zhang, Q. (2022). Blockchain-Enhanced Federated Learning Market with Social Internet of Things. *IEEE Journal on Selected Areas in Communications, 40*(12), 3405–3421.

[6] Qu, Y., Gao, L., Luan, T. H., Xiang, Y., Yu, S., Li, B., & Zheng, G. (2020). Decentralized Privacy Using Blockchain-Enabled Federated Learning in Fog Computing. *IEEE Internet of Things Journal, 7*(6), 5171–5183.

[7] Miao, Q., Lin, H., Hu, J., & Wang, X. (2022). An Intelligent and Privacy-Enhanced Data Sharing Strategy for Blockchain-Empowered Internet of Things. *Digital Communications and Networks, 8*(5), 636–643.

[8] Xianjia, Y., Queralta, J. P., Heikkonen, J., & Westerlund, T. (2021). Federated learning in Robotic and Autonomous Systems. *Procedia Computer Science, 191,* 135–142.

[9] Lv, P., Xie, L., Xu, J., Wu, X., & Li, T. (2022). Misbehavior Detection in Vehicular Ad Hoc Networks Based on Privacy-Preserving Federated Learning and Blockchain. *IEEE Transactions on Network and Service Management, 19*(4), 3936–3948.

[10] Firdaus, M., Larasati, H. T., & Rhee, K. H. (2022, June). A Secure Federated Learning Framework Using Blockchain and Differential Privacy. In *2022 IEEE 9th International Conference on Cyber Security and Cloud Computing (CSCloud)/2022 IEEE 8th International Conference on Edge Computing and Scalable Cloud (EdgeCom)* (pp. 18–23). IEEE.

[11] Liu, Y., Liu, P., Jing, W., & Song, H. H. (2023, December 15). PD2S: A Privacy-Preserving Differentiated Data Sharing Scheme Based on Blockchain and Federated Learning. *IEEE Internet of Things Journal, 10*(24), 21489–21501. doi: 10.1109/ JIOT.2023.3295763

[12] Huang, X., Han, L., Li, D., Xie, K., & Zhang, Y. (2023). A Reliable and Fair Federated Learning Mechanism for Mobile Edge Computing. *Computer Networks, 226,* 109678.

[13] Shen, G., Fu, Z., Gui, Y., Susilo, W., & Zhang, M. (2023). Efficient and Privacy-Preserving Online Diagnosis Scheme Based on Federated Learning in E-healthcare System. *Information Sciences,* 119261.

[14] Fan, M., Yu, H., & Sun, G. (2021, November). Privacy-preserving Aggregation Scheme for Blockchained Federated Learning in IoT. In *2021 International Conference on UK-China Emerging Technologies (UCET)* (pp. 129–132). IEEE.

[15] Billah, M., Mehedi, S. T., Anwar, A., Rahman, Z., & Islam, R. (2022). A Systematic Literature Review on Blockchain Enabled Federated Learning Framework for Internet of Vehicles. *arXiv preprint arXiv:2203.05192.*

[16] Ullah, I., Deng, X., Pei, X., Mushtaq, H., & Uzair, M. (2023). *IoV-SFL: A Blockchain-based Federated Learning Framework for Secure and Efficient Data Sharing in the Internet of Vehicles.* doi: 10.21203/rs.3.rs-3648280/v1

[17] Moudoud, H., & Cherkaoui, S. (2023). Multi-Tasking Federated Learning Meets Blockchain to Foster Trust and Security in the Metaverse. *Ad Hoc Networks, 150,* 103264.

[18] Zhang, Y., & Yu, H. (2022). Towards Verifiable Federated Learning. *arXiv preprint arXiv:2202.08310*.

[19] Li, W., Li, Y., Zheng, C., & He, R. (2023, August). Blockchain-based Model for Privacy-enhanced Data Sharing. In *2023 10th International Conference on Dependable Systems and Their Applications (DSA)* (pp. 406–417). IEEE.

[20] Firdaus, M., Larasati, H. T., & Rhee, K. H. (2023). A Blockchain-Assisted Distributed Edge Intelligence for Privacy-Preserving Vehicular Networks. *Computers, Materials & Continua, 76*(3).

[21] Soltani, B., Haghighi, V., Mahmood, A., Sheng, Q. Z., & Yao, L. (2022, October). A Survey on Participant Selection for Federated Learning in Mobile Networks. In *Proceedings of the 17th ACM Workshop on Mobility in the Evolving Internet Architecture* (pp. 19–24). Association for Computing Machinery.

[22] Fan, M., Ji, K., Zhang, Z., Yu, H., & Sun, G. (2023, September 15). Lightweight Privacy and Security Computing for Blockchained Federated Learning in IoT. *IEEE Internet of Things Journal, 10*(18), 16048–16060. doi: 10.1109/JIOT.2023.3267112

[23] Otoum, S., Al Ridhawi, I., & Mouftah, H. (2021). Securing Critical IoT Infrastructures with Blockchain-Supported Federated Learning. *IEEE Internet of Things Journal, 9*(4), 2592–2601.

[24] Huang, C., & Liu, S. (2024). Securing the Future of Industrial Operations: A Blockchain-Enhanced Trust Mechanism for Digital Twins in the Industrial Internet of Things. *International Journal of Computers and Applications*, 1–10.

[25] Wang, L., & Guan, C. (2024). Improving Security in the Internet of Vehicles: A Blockchain-Based Data Sharing Scheme. *Electronics, 13*(4), 714.

[26] Wan, T., Jiang, T., Liao, W., & Jiang, N. (2024). Hierarchical Incentive Mechanism for Federated Learning: A Single Contract to Dual Contract Approach for Smart Industries. *International Journal of Intelligent Systems, 2024*.

[27] Aljabri, A., Jemili, F., & Korbaa, O. (2024). Convolutional Neural Network for Intrusion Detection Using Blockchain Technology. *International Journal of Computers and Applications, 46*(2), 67–77.

[28] An, D. A. G. (2024, January). Jiakai Hao[1], Guanghuai Zhao[1], Ming Jin[1], Yitao Xiao[2]), Yuting Li[1], and Jiewei Chen[2]. In *Proceedings of the 13th International Conference on Computer Engineering and Networks: Volume III* (Vol. 1127, p. 281). Springer Nature.

[29] Selvakumar, V., Maaliw, R. R., Sharma, R. M., Oak, R., Singh, P. P., & Kumar, A. (2024). Blockchain-Aware Federated Anomaly Detection Scheme for Multivariate Data. In *Artificial Intelligence, Blockchain, Computing and Security Volume 1* (pp. 690–698). CRC Press.

7 Federated Learning-Based Smart Transportation Solutions

Deploying Lightweight Models on Edge Devices in the Internet of Vehicles

Sivabalan Settu, Raveendra Reddy, Appalaraju Muralidhar, Thangavel Murugan, and Rathipriya Ramalingam

INTRODUCTION

The Internet of Things (IoT) and the proliferation of linked devices have advanced quickly, revolutionizing several industries, including transportation. An extension of the Internet of Things, the IoV paradigm has become a disruptive force that allows data interchange and seamless connection between automobiles, roadside units, and cloud infrastructure. Current generation focuses are intelligent transportation systems that can optimize traffic flow, increase road safety, and boost overall efficiency. However, there are a lot of issues with data privacy, bandwidth limitations, and processing complexity because of the vast volume of data that IoV devices generate.

Conventional, centralized methods of data processing and machine learning model training might not be desirable or feasible in such dispersed and resource-constrained scenarios. Without requiring raw data exchange, FL is a collaborative machine-learning technique that allows training models across numerous dispersed devices. With FL, devices can train local models on their data and simply share the model updates with a central server or aggregator, as opposed to centralizing all of the data on a single server. Here, multi-input and multi-output signal propagation methods are used, the computing effort is split among several devices, communication overhead is decreased, and data privacy is maintained [1–5].

Deploying lightweight models on edge devices, including RSUs and OBUs, is critical in the context of the IoV. Because of their frequently constrained computational capabilities, these edge devices make it difficult to implement sophisticated machine-learning models. For edge device deployment, lightweight models provide a workable

DOI: 10.1201/9781003482000-7

option by balancing model performance and computing needs. This study investigates the use of application delivery networks for edge devices in conjunction with federated learning for smart transportation solutions inside the IoV ecosystem. Developers hope to overcome the issues of data privacy, bandwidth limitations, and computational limitations by utilizing the advantages of these two complementary approaches, allowing for more intelligent decision-making, more efficiency, and improved safety.

The study explores the foundational ideas of application delivery network and federated learning, emphasizing the benefits and drawbacks of each. It delves deeper into how these technologies fit into the IoV ecosystem, looking at a range of applications such as collaborative sensing, intelligent vehicular assistance, predictive maintenance, and traffic flow management. The goal is to show that deploying lightweight models based on FL on edge devices in the IoV is both feasible and effective, using theoretical analysis, simulations, and real-world case studies. This section also points out possible drawbacks, difficulties, and future lines of inquiry to develop the field and open the door for the general use of these technologies in smart transportation solutions.

The present text solves the significant issues regarding information privacy, lack of resources, and computer processing boundaries in an attempt to assist in enhancing safer, more ecologically sound, and energy-effective modes of transport, which in turn advance the entire standard that life brings and promote a transition regarding increasingly intelligent and associated towns and villages. An example scenario mentioned in **Figure 7.1** a block of social networks, numerous vehicles and links, and cognitive and auto-taxi. **Table 7.1** is the symbol and abbreviation used in this chapter.

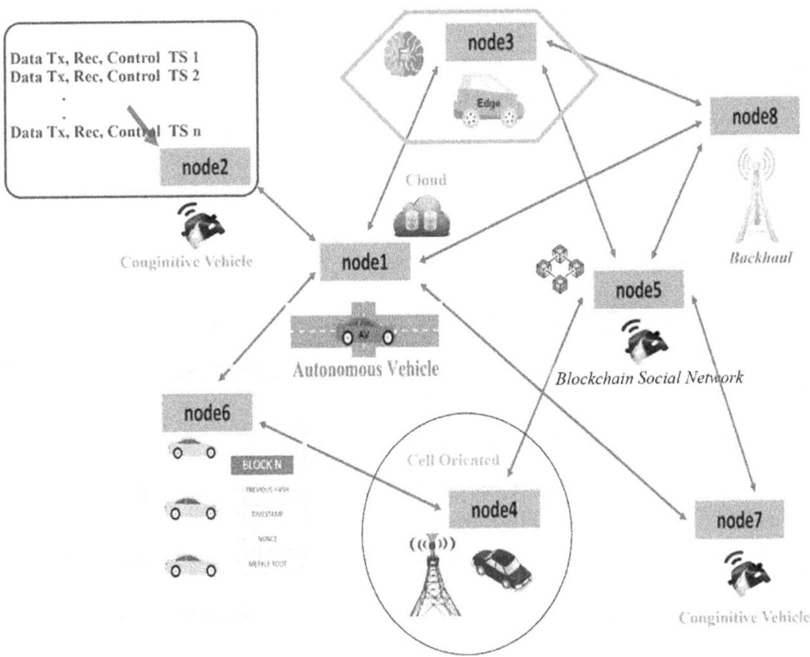

FIGURE 7.1   Vehicle to everything connection establishment in a different vehicle

TABLE 7.1

Federated learning based Internet of Vehicles state and their abbreviation

State	Abbreviation
Internet of Vehicle	IoV
Internet of Things	IoT
Federated Learning	FL
Machine Learning	ML
Roadside Assistance Devices	RSUs or RSDs
On-Board Gadgets	OBUs
Edge Computing	EC
Driver Assistance Technologies	DAT
Vehicle Data Distribution	VDD
Horizontal Federated Learning	HFL
Vertical Federated Learnings	VFL
Application Delivery Network	AND
Vehicle Data	VD
Global Server	GS
Client Vehicle	CV
Blockchain	BC
Application Binary Interface	ABI
Smart Transportation Contract	STC

BACKGROUND STUDY

IoV Paradigm

The IoV notion is a game-changer for the transportation industry. It uses cloud computing facilities, curbside measurements, and vehicle linkages to allow for cognitive choices, smooth interaction, and knowledge trade. The new model can completely reshape public transit via enhanced reliability, effectiveness, and safeguarding. It is an outgrowth of the broader IoT idea. A vast system of interdependent components, such as automotive vehicles with OBUs, RSUs, and platforms built on the cloud, appear in the IoV ecosystem. All of these components acquire and transmit data on a regular schedule about a wide range of areas, including surroundings, circulation patterns, driving actions, road hazards, and vehicle telemetry systems. For the intent of public transportation optimization and the growth of a multitude of innovative instances, this data can be highly significant. **Figure 7.2a, and Figure 7.2b** display the Inner, Outer link establishments in ADN at numerous road-type ground transportation.

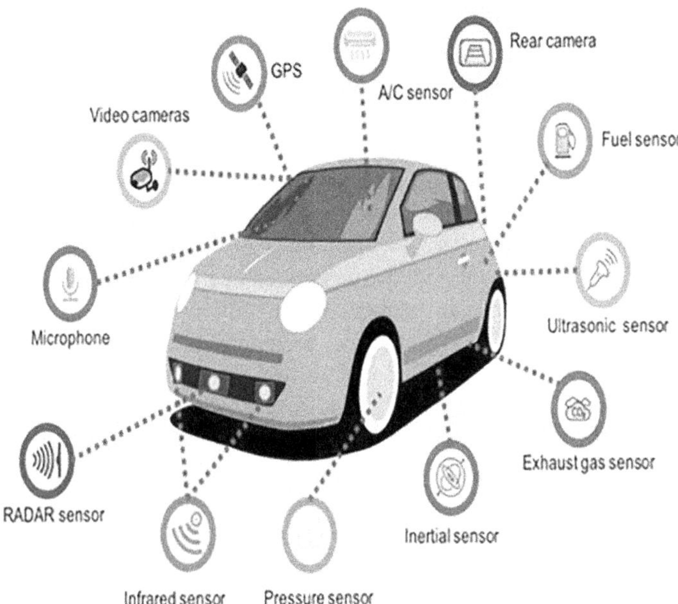

FIGURE 7.2a Outside sensor establishments

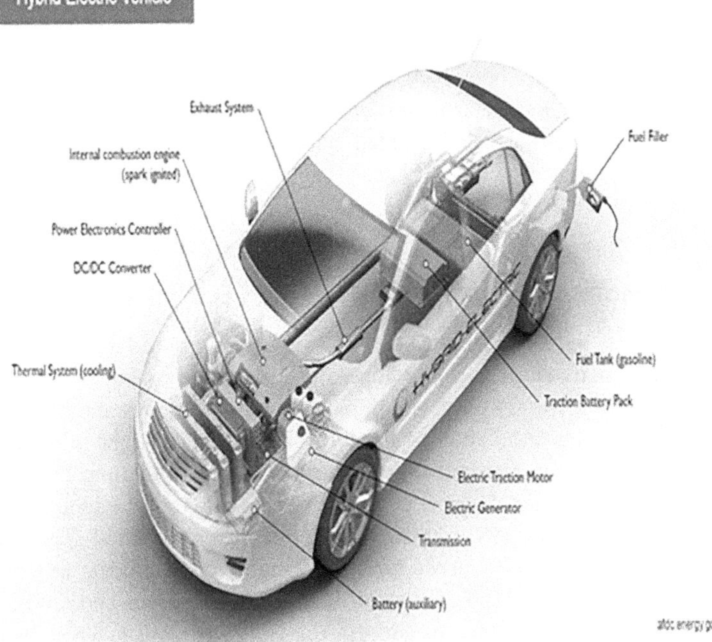

FIGURE 7.2b Inside sensor establishments for autonomous

CHALLENGES IN THE IoV ECOSYSTEM

The IoV paradigm offers many opportunities, but it also brings with it several obstacles that need to be tackled to fully capitalize on its true potential. The security and privacy of data are two of the key problems. The critical character of the data gathered through transport vehicles and building components gives reasons for suspects to be concerned about breaches of privacy, prohibited entry points, and wrongful use of sensitive information. The limitations on access and communication costs encountered with moving enormous volumes of data from various handheld devices onto central machines or cloud-based systems provide a further significant obstacle. In such decentralized and constrained resource scenarios, classically consolidated techniques to handle data and model development for machine learning could prove really wanted or realistic. Similarly, the small processing capacity of edge devices, including OBUs and RSUs, frequently inhibits the implantation of more sophisticated machine learning models on each other. For the sake of intelligent choices at the edge, these capacities limit the authority of the building as well as the adoption of lightweight models intended exclusively for border equipment.

FEDERATED LEARNING: A CENTRALIZED/DECENTRALIZED APPROACH

In the IoV ecosystem, FL has shown promise as a decentralized machine learning method for addressing issues with data privacy, bandwidth limits, and processing constraints. Without requiring the sharing of raw data, FL allows collaborative training of machine learning models across numerous decentralized devices. Under the FL paradigm, every device or party uses its data to train a local model; only the model updates, or gradients, are shared with an aggregator or central server. Sensitive data is kept localized in this method to protect data privacy, and only model updates—which are usually much smaller than raw data—are transmitted to minimize communication costs. FL has two types called horizontal, and vertical tabulated in **Figure 7.3.**

LIGHTWEIGHT MODELS FOR EDGE DEVICES

The setting up of lightweight models on limited-resource edge devices is critical to enabling smart choices at the edge and utilizing the benefits of FL in the IoV ecosystem. Lightweight models can be distributed on edge devices with limited capabilities because the choices are designed to reconcile the performance of models and computation needs. Model enlargement, quantization, and reduction are some of the techniques that can be used to either enhance contemporary deep-learning models or create new lightweight architectures designed specifically for IoV workloads. Using the FL procedure, these lightweight models may be cooperatively trained to achieve the greatest utilization of resources and retain privacy based on separated knowledge. To address the issues of data privacy, bandwidth limitations, and computational aspect boundaries, academics and professionals in

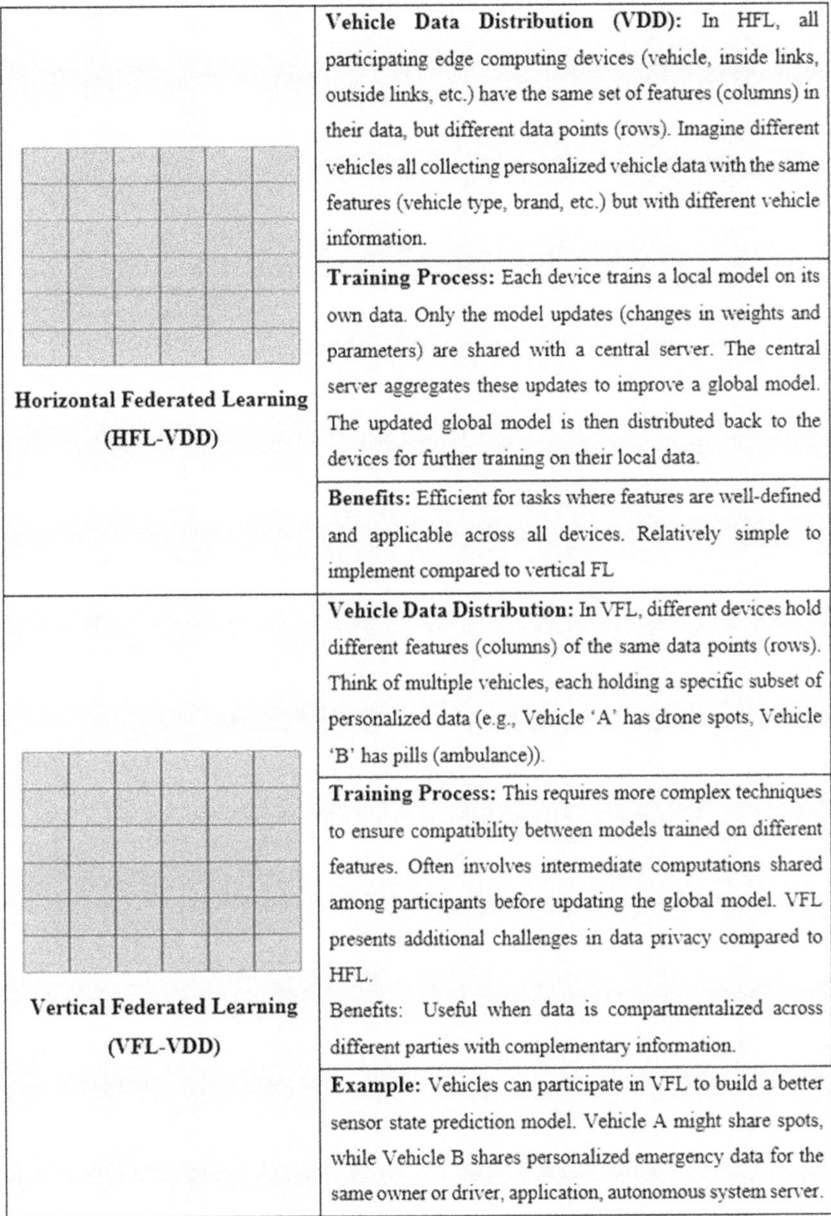

Horizontal Federated Learning (HFL-VDD)	**Vehicle Data Distribution (VDD):** In HFL, all participating edge computing devices (vehicle, inside links, outside links, etc.) have the same set of features (columns) in their data, but different data points (rows). Imagine different vehicles all collecting personalized vehicle data with the same features (vehicle type, brand, etc.) but with different vehicle information.
	Training Process: Each device trains a local model on its own data. Only the model updates (changes in weights and parameters) are shared with a central server. The central server aggregates these updates to improve a global model. The updated global model is then distributed back to the devices for further training on their local data.
	Benefits: Efficient for tasks where features are well-defined and applicable across all devices. Relatively simple to implement compared to vertical FL
Vertical Federated Learning (VFL-VDD)	**Vehicle Data Distribution:** In VFL, different devices hold different features (columns) of the same data points (rows). Think of multiple vehicles, each holding a specific subset of personalized data (e.g., Vehicle 'A' has drone spots, Vehicle 'B' has pills (ambulance)).
	Training Process: This requires more complex techniques to ensure compatibility between models trained on different features. Often involves intermediate computations shared among participants before updating the global model. VFL presents additional challenges in data privacy compared to HFL. Benefits: Useful when data is compartmentalized across different parties with complementary information.
	Example: Vehicles can participate in VFL to build a better sensor state prediction model. Vehicle A might share spots, while Vehicle B shares personalized emergency data for the same owner or driver, application, autonomous system server.

FIGURE 7.3 Types of Federated Learning

the IoV domain anticipate integrating FL and lightweight models. This will enable skilled modes of transport that improve safety, efficiency, and long-term viability in public transit systems. Finally, **Figure 7.4** shows the section study of this book chapter.

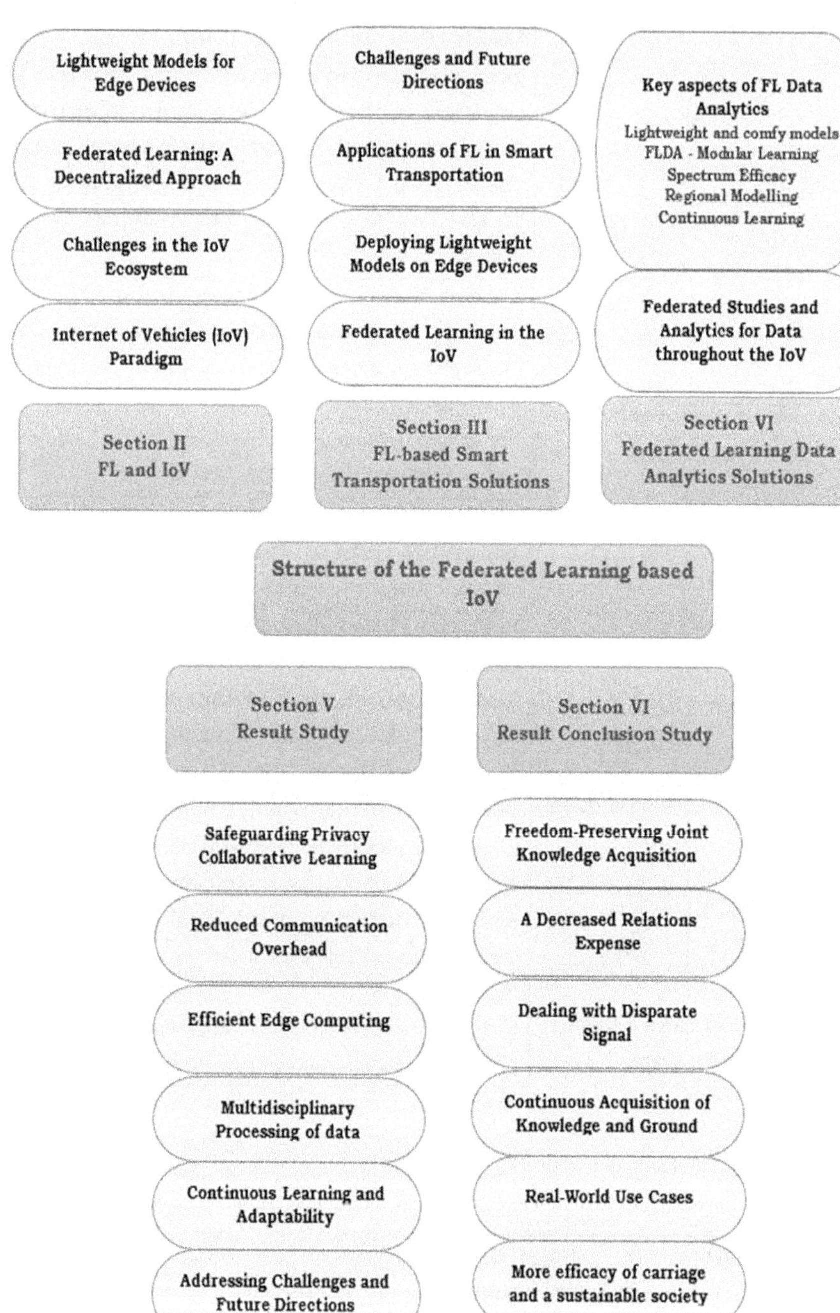

FIGURE 7.4   Organization of the book chapter sections

Features	FEL	Centralized	FL
Model training	Native vehicle	GS	Native vehicle training, GS, aggregation
Model applicability	custom-made	Solo global model	Solo global model with custom-made
Privacy protection	✓✓	X X	✓
Learning efficiency	X	✓	✓✓
Performance on heterogeneous/anomaly data	X	✓✓	✓
Communication requirement	✓✓	X X	X
Training data volume	X X	✓✓	✓
Current research progress	✓✓	✓✓	X
Compatibility with CV, AVs	✓	X X	✓✓

✓✓ best, ✓ high, X low, X X worst.

FIGURE 7.5 Federated Edge Devices Learning and study of central, dispersed system

END BACKGROUND STUDY

Finally, at the end of the section study, the FL type was used to centralize and disperse server data transactions, and applicable ground features were identified. **Figure 7.5** presents the federated learning from vehicle edge devices and the possibilities of feature enhancement in on-load, off-load, and live 6G vision directions.

PROPOSED WORK

The IoV paradigm has brought about an immense shift in the world of transport in the era of the IoT. IoV makes use of the connectivity between cars, roadside equipment, and cloud infrastructure to facilitate intelligent decision-making, simple interaction, and information transfer. Still, challenges with the confidentiality of data, shortages of bandwidth, and information technology inefficiency originate from the tremendous quantity of data generated through IoV devices. A flexible machine learning technology called FL is now coming to light as an acceptable means to get past such barriers as shown in **Figure 7.6.**

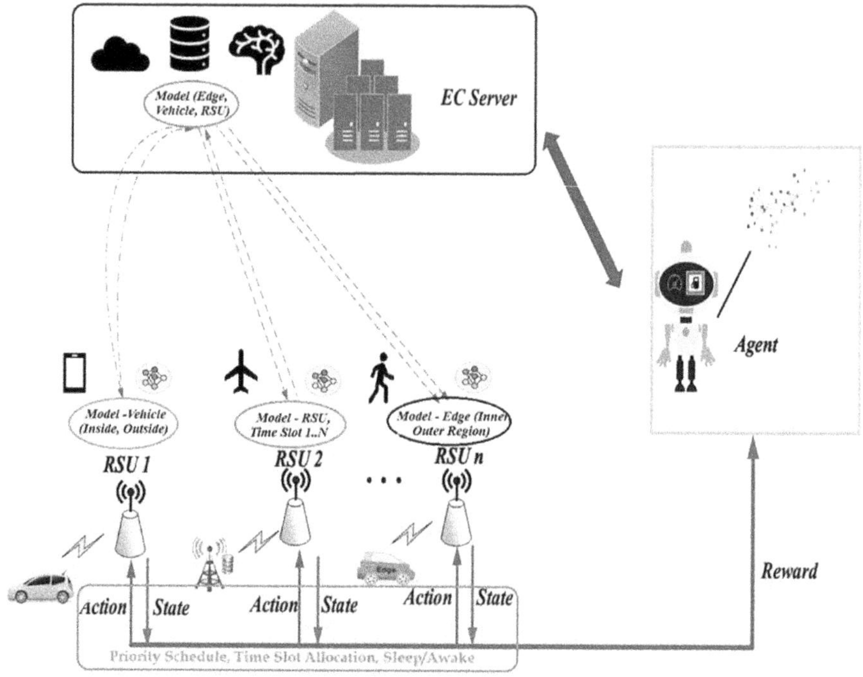

FIGURE 7.6   Federated Learning in the IoV ecosystem, autonomous flow

DEPLOYING LIGHTWEIGHT MODELS ON EDGE DEVICES

Complex machine learning models can be computationally and highly energy-intensive to deploy when it comes to devices that are limited in resources. To overcome this barrier, specialists propose creating and implementing lightweight models made especially for edge devices throughout the IoV. The construction of lightweight models intends to combine processing constraints with the performance of the model. Models like these can be distributed on device edges with limited capabilities because they usually consume fewer settings and a smaller amount of power. Experts could develop new lightweight topologies designed specifically for Internet of Vehicles applications or perfect established deep learning models by applying methods like model shrinkage, quantization, and cutting down. An example scenario of the dynamic on-load/offload mechanism of FL is stated in **Figure 7.4**. The diagram represents model deployment using agents that link numerous things like inside edge computes, including road, rail, air, and water transportation systems. It includes a model_edge (zone and region), a model_vehicle (in and out links), and a model_RSU (RSU1,. . ., RSU n). **Figure 7.7** spots the vehicle at a time of movement used to find the social affinity application destination.

APPLICATIONS OF FL IN SMART TRANSPORTATION

FL has various advantages in the field of technological transportation when it integrates with lightweight models that are stationed at the forefront of devices.

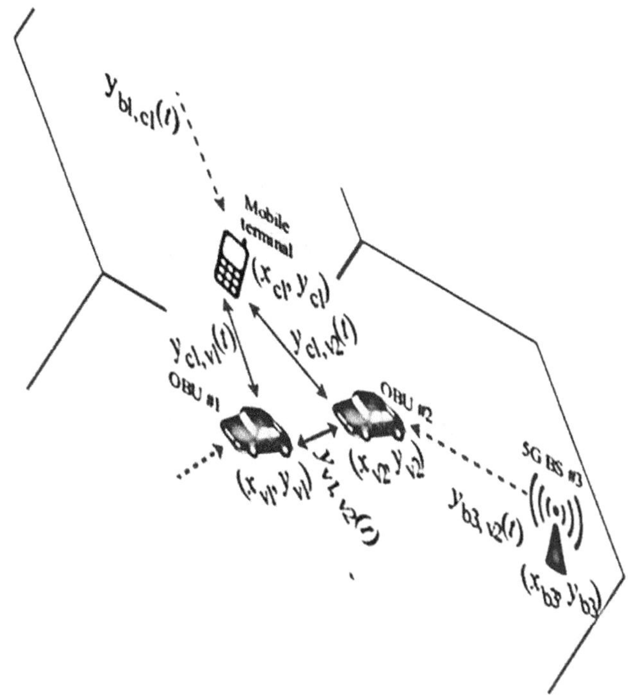

FIGURE 7.7 IoV Region-based dynamic offload/on-load edge devices for inter and intra-communication

Traffic Stream Optimization: The intelligent routing method and regulation of signals are used by FL models to anticipate congestion, recognize slowdowns, and improve the efficiency of traffic by analyzing instantaneous information from cars and roadside sensors for analysis.

Automated Transportation Assist: By utilizing data from diverse automobiles and road sections, FL models can enable refined Driver Assistance Technologies (DAT) and autonomous driving abilities such as people-walking recognition, steering avoidance, and lane-keeping aid.

Automatic Maintenance: While data on highway circumstances, attitude of drivers, and car performance can be gathered via IoV devices, through the analysis of these data points, FL models can increase vehicle safety and effectiveness by modeling service needs, optimizing restoration agendas, and mitigating mistakes.

Ecological Inspection: Information on emissions, air quality, and noise levels can be gathered using IoV devices. By analyzing this data, FL models can identify polluting tourist attractions, improve the speed of traffic, and create policies to lessen the adverse impact travel has on the ecosystem.

Collaborating Sensed: FL models can facilitate collaborating perceiving apps, like joint perceiving for greater awareness of the situation and imminent risk identification, by utilizing data from many vehicles and checkpoint gadgets.

CHALLENGES AND FUTURE DIRECTIONS

Given that cooperative learning and lightweight models are currently being developed as appealing options for smart mobility in the IoV, significant barriers remain:

Successful modes of communication and cost optimization strategies have to be used for handling updated models along with data spread via edge devices and the primary server or aggregators.

Protection and Safety: Solid security-preserving approaches and safe aggregate procedures are critical for protecting data privacy and mitigating aggressive attacks or model poisoning.

Divergence and Ability to Scale: IoV devices possess various hardware abilities, data distributions, and other relationship boundaries. FL algorithms must be formed to address variety while expanding to accommodate plenty of devices.

Models Customization: Though FL enables shared learning, it is a requirement for geared models that address special device or client demands, such as specific driving choices or local circumstances.

Uniformity and Coordination: For broad adoption and accessibility, the FL solution must seamlessly fit into the present infrastructure for transportation, as well as technologies and endpoints that are universal.

As the IoV ecology advances, unified learning, and lightweight models will prove key in allowing knowledgeable transportation solutions to utilize dispersed information while protecting the safety and promoting the utilization of resources. Further study and partnership among academia, industry, and government agencies will be important in solving barriers and attaining the entire potential of these advances in smart transportation [6–11, 12, 13].

FL DATA PROCESS STEPS VIA APPLICATION DELIVERY NETWORK (ADN) AND DATA ANALYTICS

FL is a modular machine learning technology that allows for joint model training across numerous devices or entities without the need for centralization or substitution of data in its completeness. The following **Figure 7.8** illustrates the data processing steps involved in ADN.

This paradigm gained substantial acceptance in the world of statistical analysis, particularly in circumstances in which confidentiality of data, capacity restrictions, and computation restraints are noteworthy. According to the FL-IoV use of federated data analytics proves essential in that it allows competent modes of transportation while focusing on the barriers posed by the IoV ecological systems.

FEDERATED STUDIES AND ANALYTICS FOR DATA THROUGHOUT THE IoV

The IoV structure consists of a massive chain of coupled transport vehicles, outside gear, and the cloud that achieve massive volumes of knowledge. This data comprises car telemetry, which includes driver actions, circulation patterns, roadway circumstances, and aspects of the environment. Federated learning data analytics lets you perform the evaluation and representation of scattered information without the need to organize it, hence safeguarding the confidentiality of information and eliminating

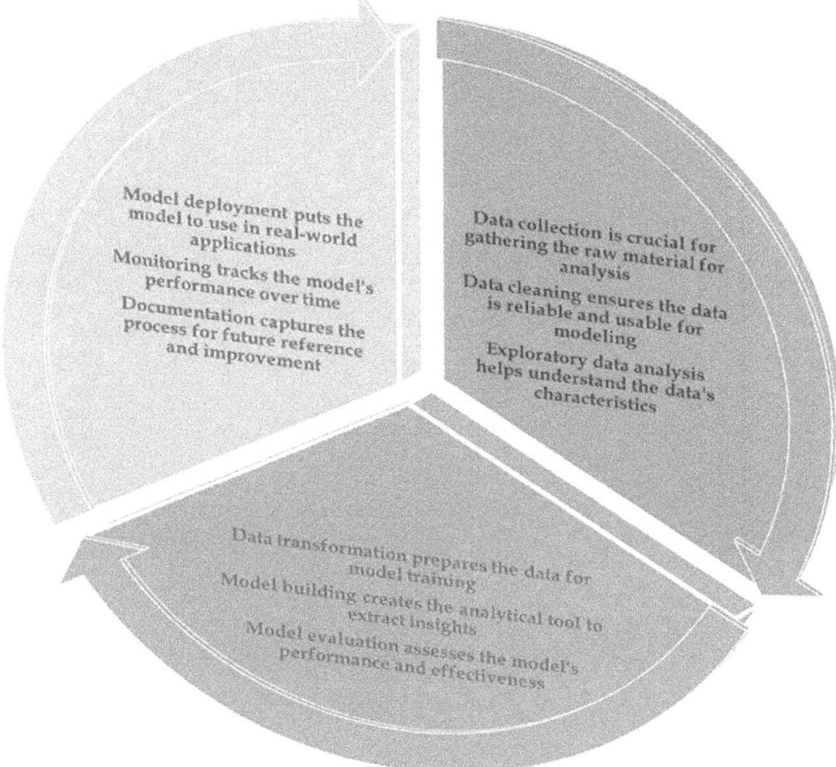

FIGURE 7.8   ADN typical steps involved in data processing

communication overhead [14]. When deploying lightweight models on edge devices in the IoV ecosystem, edge devices such as OBUs and RSUs have limited computational resources. Deploying complex machine learning models on these resource-constrained devices can be computationally intensive and energy-inefficient. Federated learning data analytics addresses this challenge by enabling the training and deployment of lightweight models tailored specifically for edge devices. Lightweight models are designed to strike a balance between model performance and computational require-ments. These models are typically smaller in size, have fewer parameters, and require less computational power, making them suitable for deployment on edge devices with limited resources. Federated learning data analytics facilitates the collaborative train-ing of these lightweight models across multiple edge devices, leveraging the distrib-uted data while preserving privacy and reducing communication overhead. Figure 7.4 shows the region IoV getting the FL training experience [15, 16].

KEY ASPECTS OF FL DATA ANALYTICS

Lightweight and comfortable: models aim to strike an appropriate balance between model efficacy and computation demands. These kinds of models are generally less extensive in scope, have fewer features, and require less computational dominance,

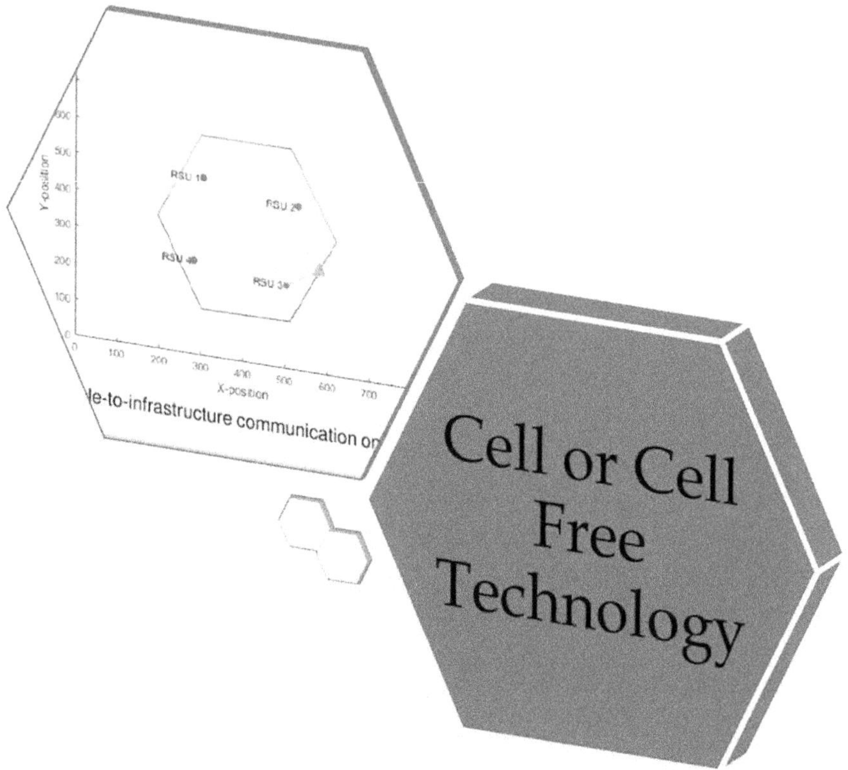

FIGURE 7.9   Edge devices with FL approach among RSU, OBU, Infrastructure in cell

making them optimal for use on edge devices with a limited number of resources. FL statistical analysis allows for collective training of such tiny machines spanning plenty of edge technology devices using disparate information while preserving privacy and eliminating expenses for communication. The fourth figure illustration exhibits the region IoV gaining FL training experiences. **Figure 7.9** displays the cell within a cell or cell-free technology where infrastructure and RSUs are bonding with the corresponding network.

Basic Benefits of FL Data Analytics, Modular Learning: Federated learning data aggregation allows for the dispersed training of statistical ensembles across various devices or entities. Each device or party builds its model using its data, and only updated models or variations are transferred to an underlying server or aggregator. The confidentiality of information is safeguarded by federated learning data analytics, which resists gathering raw data. Specific gadgets or persons maintain oversight of their secret information, avoiding safety and legal concerns.

Spectrum Efficacy: Compared to passing huge quantities of raw data, federated learning statistical analysis mandates only the distribution of variations in models or shadings, which are frequently much smaller in size. It features reduced transfer costs and bandwidth requirements that make it ideal for resource-constrained IoV

TABLE 7.2

FL- IoV using Smart Transport using ADN Infrastructure, Coverage, Road/Lane, Cell or Cell-free

Procedure 1: Smart Transportation Contract (STC)	
{Vehicle (V): Road, Rail, Water, Air, Space . . .} Via **Application Delivery Network (ADN1, ADN2, . . ., ADNn)**	
Vehicle **Registration** {V1, V2, . . ., Vn}	// Number of Vehicle Data (VD),
if {V1, V2, . . ., Vn} == successful **then**	// Global Server (GS)
Check {VD 1,. . . . VD n} from the GS	// Client Vehicle (CV)
else	
Register {CV1, . . . CV n} as a New {CV}	
U ← CV	
Store Data on BC	// Blockchain (BC)
Encrypt data Using Sha256 Algo	
Result generate ABI of Contract	// Application Binary Interface (ABI)
generate Byte code of Contract	
Decrypt data Using Sha256 Algorithm	
end if	

applications. **Table 7.2** gives the smart transport mechanism procedures and the numerous application delivery network points.

Regional Modeling: FL Informatics is capable of managing diverse data deployments from several devices or entities. Each regional model undergoes training on input from the relevant faction, and the procedure of aggregation combines all the different viewpoints and generates more reliable as well as generic forecasts [17, 18].

Continuous Learning: As newly collected information is obtainable at specific devices or those involved, FL data analytics offers continuous learning and update services, making sure deploying systems stay relevant and versatile to evolve with scenarios. Application for Intelligent Mobility FL analytics of data has numerous uses in smart mobility and the IoV ecosystem, including congestion oversight, cognitive vehicular assistance, automatic upkeep, surveillance of the environment, and joint sense. FL data analytics allows for more sophisticated decision-making, enhanced security, and improved productivity in public transport by using dispersed information while safeguarding data and enhancing resource utilization.

RESULT STUDY

The joint use of FL and lightweight models distributed on edge equipment in the IoV ecosystem has shown favorable outcomes in facilitating intelligent modes of transport while tackling the confidentiality of data, access limits, and computational aspects of barriers. **Figures 7.10 and 7.11** represent the overall diagram of smart transportation in the city.

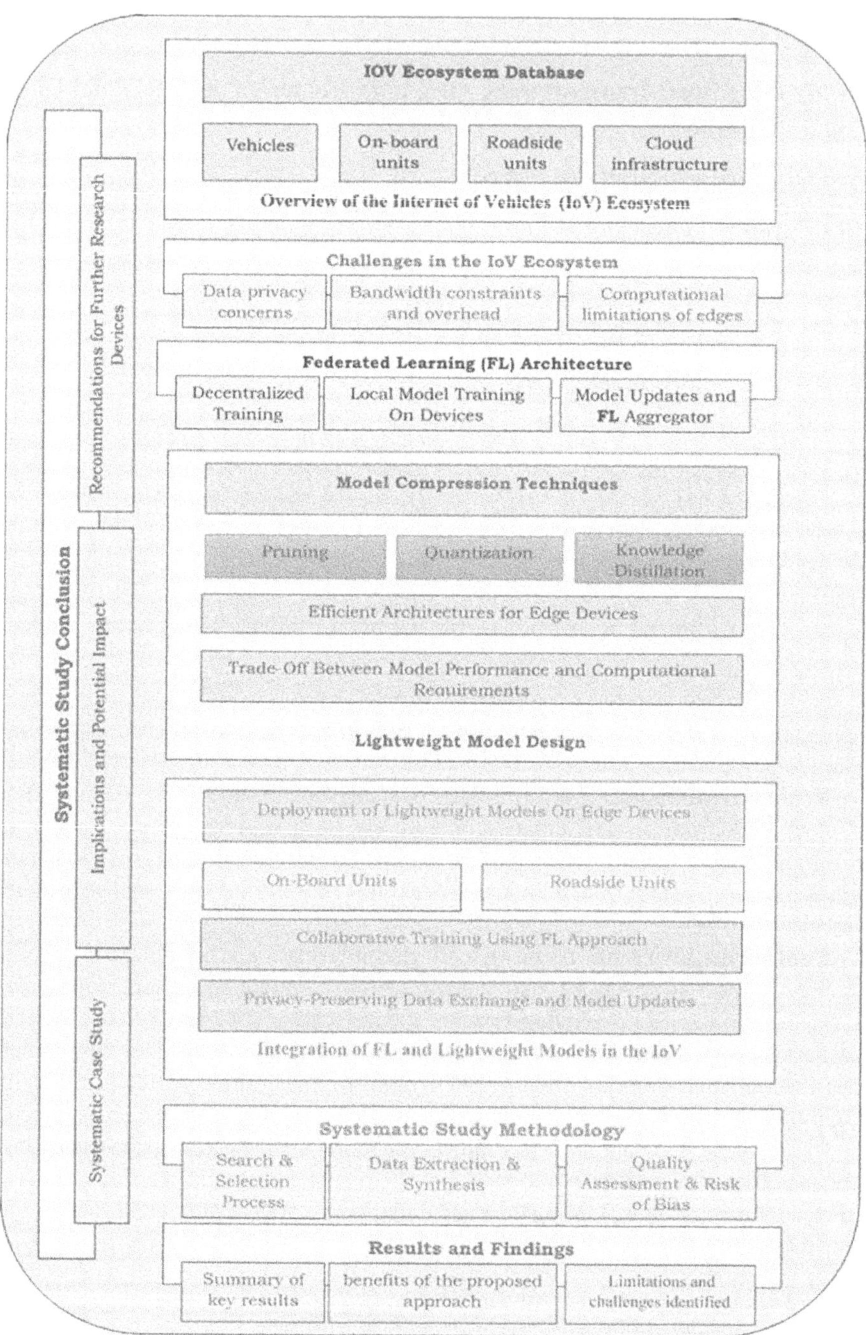

FIGURE 7.10 Proposed Systematic diagram for Fed-IoV

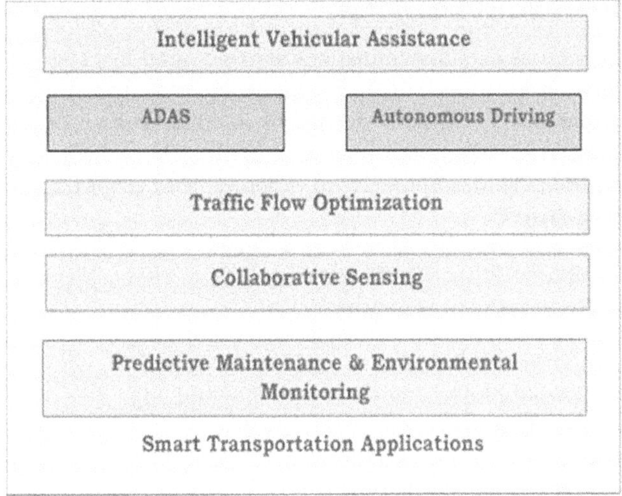

FIGURE 7.11 Federated Learning-based Smart Transportation

SAFEGUARDING PRIVACY AND COLLABORATIVE LEARNING

The distributed architecture of FL promotes collaborative learning without the need to disclose unprocessed data, hence safeguarding the confidentiality of data in the IoV ecosystem. Specific devices or entities maintain oversight of their proprietary information by training individual models on their knowledge and only exchanging modifications to the models with global aggregators. This approach responds to safety and legal issues, both of which are particularly significant in the vehicle industry, as intimate and automotive data are associated.

REDUCED COMMUNICATION OVERHEAD

One of FL's core advantages is its capacity to diminish transmission costs and bandwidth demands. Instead of sending huge amounts of data in raw form from multiple-state IoV devices to centrally located servers, FL just needs the sharing of changes to models or shadings, which are often significantly smaller in size. This method is especially useful in the IoV ecosystem, where access limits and large data quantities offer notable obstacles.

EFFICIENT EDGE COMPUTING

The setup of lightweight models on hardware such as OBUs and RSUs gives rapid computation and intelligent choices to be made. Lightweight models pound the right balance between space and practical demands by utilizing tackles such as model expansion, the goal of quantification, and sprucing which make them acceptable for use on constrained gateway workstations.

MULTIDISCIPLINARY PROCESSING OF DATA

FL can handle various data disseminations across several IoV systems or individuals. Each local model is trained on data from the associated party, and the aggregation process combines all of these viewpoints to create more robust and universal models. This feature is especially important in the IoV ecological system, where disparity in information is essential due to varied vehicle categories, driving designs, and surrounding factors.

CONTINUOUS GROWTH AND ADAPTABILITY

As the latest information becomes accessible from specific IoV devices or third parties, FL sells continuous adaptation and modeling upgrades, making sure the set-up models are up to date-and flexible with evolving conditions. Such functionality comes in handy in a dynamic journey scenario where commute patterns, surface conditions, and the behavior of travelers can change promptly. If you consider that the social affinity application and delivery network depend on mobility, it shows the vehicle driver necessary things like a fuel station, restaurant, temple, etc.

Finally, future directions are illustrated in **Figure 7.12**, and the natural environment protection and ecology system is suitable for handling social delivery networks.

Real-world approaches and case investigations The findings of the research show that employing FL-based lightweight models on edge technology is practical and efficient in a variety of IoV usages, including the flow of traffic optimization, smart vehicular guidance, proactive upkeep, environmental tracking, and mutually beneficial sensing. The suggested approach brought about favorable outcomes in terms of smoother circulation of traffic, increased safety on the roadways, efficient use of assets, and reduced greenhouse gas emissions, as indicated by simulations and case studies. These outcomes line up with the overall objectives of intelligent transit systems, resulting in improved effectiveness, profitability, and safer mobility choices.

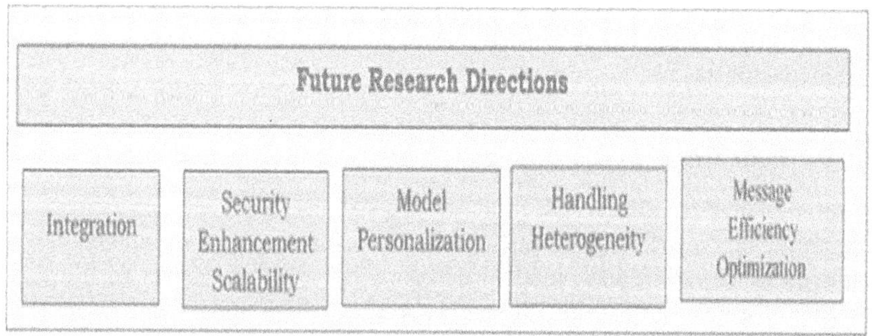

FIGURE 7.12   Future Direction enhancement

ADDRESSING CHALLENGES AND FUTURE DIRECTIONS

While incorporating FL and compact designs leads to promising outcomes, various difficulties, and limitations have been acknowledged, paving the way for future studies in these domains.

Network Efficiency: Quick protocols for communicating and spectrum optimization techniques are needed to support new models and data transmission between devices at the edge and the data aggregator, therefore eliminating communication waste.

Volatility and ability to scale: As the IoV ecosystem expands, FL solutions and lightweight concept structures must be designed to deal with rising variability and scale to handle numerous devices with diverse characteristics.

Enhancements to privacy and security: The connection followed strong privacy-protection tactics and secure collection, as well as well-known protocols. Those are essential to ensuring data confidentiality and preventing provocative attacks or paradigm poisoning, especially in security-sensitive applications for transport.

Model Customization: While FL encourages knowledge sharing from tiny to large edges, there is a need for models tailored to individual devices or user needs, such as riding preferences or local environments.

Uniformity and Integration: Easy integration of FL solutions into existing private, public, and industrial transport systems, as well as the creation of common interfaces and screens, is critical for broad adoption and connectivity among distinct IoV components and stakeholders.

RESULT, CONCLUSION, AND STUDY

The simultaneous use of FL and lightweight designs set up on edge devices in the IoV ecosystem exhibited promising results in allowing intelligent mobility solutions while addressing safeguarding information, access limits, and processing concerns. This study reviewed the viability and efficacy of the suggested approach by using comprehensive exercises, case studies from real life, and theoretical analyses. The main results and inferences are discussed as follows:

Freedom-Preserving Joint Knowledge Acquisition: By making use of FL's modular structure, the approach recommended allows for teamwork without exposing raw data, accordingly safeguarding the confidentiality of data in the IoV industry. This strategy overcomes concerns regarding privacy as well as governmental hardships, both of which are vital in the transport industry.

Decreased relations expense: Whenever placing FL-based scenarios, updated models or differential transmission must be performed, which causes substantially fewer communication expenses and demands on bandwidth than typical consolidated techniques.

Rapid Boundary Computations: The use of portable designs specially designed for constrained-in-resource edge devices like OBUs and RSUs promotes economic cloud computing and smart choice-making at the edge.

Dealing with disparate signals: FL's ability to manage diverse data distributions across different IoV devices and organizations has culminated in more robust and

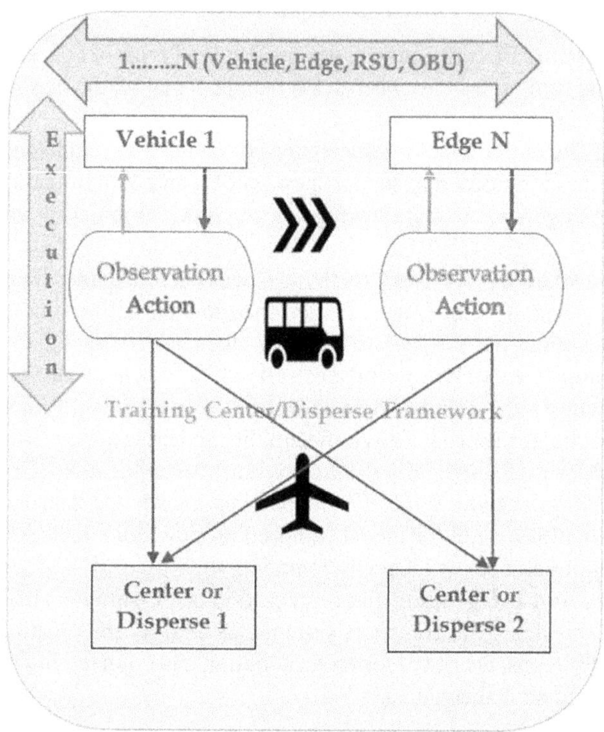

FIGURE 7.13   FL is categorized by a centralized, distributed system

overall models that account for an array of driving sequences, vehicle formats, and circumstances in the environment.

This is the final section of the chapter, and **Figure 7.13.** indicates the FL type with central and dispersed server systems. **Figure 7.14** shows the sample network settings defined and the numerous parameters used. These settings are used to continuously acquire and update ground awareness about themes like traffic, road damage, political meetings, etc. Especially take one social application to check individual vehicle identity, range, current ground city updates, cloud database retrieval, on-load time, off-load time, and live.

Continuous Acquisition of Knowledge and Ground Alterations: This approach supports ongoing learning and revisions to models, ensuring that positioned models stay current and responsive to shifting circumstances in a dynamic traffic ecosystem. This traffic crowd recognition technology is applied to guarantee a safe path.

Real-World Use Cases: A research study found that the FL technique is applicable and effective in an assortment of IoV functions, such as streamlining traffic optimization, intelligent vehicle advice, automatic upkeep, environmental surveillance, and joint senses.

Greater efficiency of transportation and sustainable society: This study's conclusion helps foster the creation of effective, less hazardous, and green public transit

Parameters	Value
Resource of uplink/downlink	1:1
Batch size	23 sample
Epochs	34
Execution time (batch B_{exe})	0.07 s
Peak throughput (cell)	10.4 Mbps
Poorest throughput (cell)	0.25 Mbps
Range	300 meters
Vehicles sample	32
Deadline of message round	23 seconds
Number of selected CV in each area	3
The length of road	1000 meters; straight road
The location of the FL server	At 520 meters of the road (very close to the BS or RSU)
The sample quantity of the vehicle	VID 0-11: about 4500 images, VID 12-29: about 45 images
Distribution	uniform
Size of Packet	2000 byte
vehicle to cloud latency	211 ms
vehicle to vehicle latency	43 ms

FIGURE 7.14 The setting of distributed CV selection

systems by overcoming issues related to the privacy of data, optimizing the usage of resources, and enabling automated decision-making.

While the proposed solution has yielded promising results, several challenges and limitations have been identified, presenting opportunities for future research. These include communication efficiency, handling increasing levels of heterogeneity and scalability, enhancing privacy and security mechanisms, enabling model personalization, and facilitating seamless integration and standardization across the IoV ecosystem. By addressing these challenges and leveraging the synergies between FL and lightweight models, the research community and industry stakeholders can unlock the full potential of intelligent transportation solutions in the IoV ecosystem, contributing to more efficient, secure, and sustainable mobility systems. Continued research, collaboration, and innovation in this domain are crucial for realizing the vision of smarter, safer, and more connected transportation networks.

CONCLUSION

The deep study begun in this study explores the pairing of FL with compact designs set up on interfaces within the context of IoV ecosystems. It launched major discoveries with insights from earlier studies, conceptual frameworks, and observational data, resulting in an in-depth grasp of the strategy's capabilities and imperfections. The meta-analysis noticed that combining FL and portable architectures provides an approach to a variety of barriers in the IoV natural context, including concerns about safeguarding information, shortages of resources, and edge device estimation boundaries. By using FL's global structure and resource-efficient architecture of flexible entities, by capturing FL's decentralized feature and the energy-efficient design of ultralight models, this method offers cognitive solutions for transportation that prioritize data privacy, maximize utilization of resources, and facilitate smart choices at the frontier. It responded to FL's fundamental principles, like its ability to aid cooperation in learning by encapsulating initial data, safeguarding sensitive facts, and bringing down link costs. In addition, the scientific review assessed distinct methods for utilizing featherweight circumstances such as simulation enlargement and quantization. This process and savings are appropriate for distribution on limited-resource peripherals. Since emerging optimistic, it has discovered significant odds such as relationship success and data safeguards, which demand precision. It also highlighted future research opportunities for model-based personalized support and easy entry into the IoV platform. It provides a solid foundation for the future rise of FL-based technologies with handheld designs in the IoV sector.

REFERENCES

1. Samarakoon, S., Bennis, M., Saad, W., & Debbah, M. (2019). Federated learning for ultra-reliable low-latency V2V communications. In *2019 IEEE Global Communications Conference (GLOBECOM)* (pp. 1–7). IEEE.

2. Lu, Y., Huang, X., Zhang, K., Maharjan, S., & Zhang, Y. (2021, January/February). Blockchain and federated learning for 5G beyond. *IEEE Network*, *35*(1), 219–225. http://doi.org/10.1109/MNET.011.1900598.

3. Lim, W. Y. B., Nguyen Cong Luong, Dinh Thai Hoang, Yutao Jiao, Ying-Chang Liang, Qiang Yang. (2020). Federated learning in mobile edge networks: A comprehensive survey. *IEEE Communications Surveys & Tutorials*, *22*(3), 2031–2063. http://doi.org/10.1109/COMST.2020.2986024.

4. Niknam, S., Dhillon, H. S., & Reed, J. H. (2019). Federated learning for wireless communications: Motivation, opportunities, and challenges. *IEEE Communications Magazine*, *58*, 46–51.

5. Brecko, A., Kajati, E., Koziorek, J., & Zolotova, I. (2022). Federated learning for edge computing: A survey. *Applied Sciences*, *12*(18), 9124. https://doi.org/10.3390/app12189124.

6. Hinton, G., Vinyals, O., & Dean, J. (2015). Distilling the knowledge in a neural network. *arXiv preprint arXiv:1503.02531*.

7. Wang, K., Liu, Z., Lin, Y., Lin, J., & Han, S. (2019). HAQ: Hardware-aware automated quantization with mixed precision. In *Proceedings of the IEEE/CVF Conference on Computer Vision and Pattern Recognition* (pp. 8612–8620). IEEE. doi: 10.1109/CVPR.2019.00881.

8. Jeong, E., Oh, S., Kim, H., Park, J., Bennis, M., & Kim, S. L. (2018). Communication-efficient on-device machine learning: Federated distillation and augmentation under non-iid private data. *arXiv preprint arXiv:1811.11479*.

9. Macedo, D., Santos, D., Perkusich, A., & Valadares, D. C. G. (2023). Mobility-aware federated learning considering multiple networks. *Sensors (Basel)*, *23*(14), 6286. http://doi.org/10.3390/s23146286. PMID: 37514581; PMCID: PMC10386473.

10. Sivabalan, S., Dhamodharavadhani, S., & Rathipriya, R. (2020). Arbitrary walk with minimum length based route identification scheme in graph structure for opportunistic wireless sensor network. In Aboul Ella Hassanien & Ashraf Darwish (Eds.), *Intelligent Data-Centric Systems, Swarm Intelligence for Resource Management in Internet of Things* (pp. 47–63). Academic Press. ISBN 9780128182871. https://doi.org/10.1016/B978-0-12-818287-1.00006-1.

11. Kulanthaiyappan, S., Settu, S., & Chellaih, C. (2020). Internet of vehicle: Effects of target tracking cluster routing in vehicle network. In *2020 6th International Conference on Advanced Computing and Communication Systems (ICACCS)* (pp. 951–956). Coimbatore, India. http://doi.org/10.1109/ICACCS48705.2020.9074454.

12. Sivabalan, S., & Rathipriya, R. (2017). Slot scheduling Mac using energy efficiency in ad hoc wireless networks. In *2017 International Conference on Inventive Communication and Computational Technologies (ICICCT)* (pp. 430–434). Coimbatore, India. http://doi.org/10.1109/ICICCT.2017.7975234.

13. Shao, Z., Yang, H., Xiao, L., Su, W., & Xiong, Z. (2023). Energy and latency-aware resource management for UAV-assisted mobile edge computing against jamming. In *GLOBECOM 2023–2023 IEEE Global Communications Conference* (pp. 1848–1853). Kuala Lumpur, Malaysia. http://doi.org/10.1109/GLOBECOM54140.2023.10437090.

14. Sivabalan, S., & Rathipriya, R. (2023). Efficient energy resource selection in home area sensor networks using non swarm intelligence based discrete venus flytrap search optimization algorithm. *Wireless Personal Communications, 128*, 249–265. https://doi.org/10.1007/s11277-022-09953-y.

15. Liu, S., Yang, H., Xiao, L., Zheng, M., Lu, H., & Xiong, Z. (2024). Learning-based resource management optimization for UAV-assisted MEC against jamming. *IEEE Transactions on Communications*. http://doi.org/10.1109/TCOMM.2024.3374356.

16. Kain, R., & Sorour, S. (2022). Worker resource characterization under dynamic usage in multi-access edge computing. In *2022 International Wireless Communications and Mobile Computing (IWCMC)* (pp. 1070–1075). Dubrovnik, Croatia. http://doi.org/10.1109/IWCMC55113.2022.9824299.

17. Gao, Z., Yang, L., & Dai, Y. (2023, June 1). Large-scale computation offloading using a multi-agent reinforcement learning in heterogeneous multi-access edge computing. *IEEE Transactions on Mobile Computing, 22*(6), 3425–3443. http://doi.org/10.1109/TMC.2022.3141080.

18. Santos, Á., Bernardino, J., & Correia, N. (2023). Automated application deployment on multi-access edge computing: A survey. *IEEE Access, 11*, 89393–89408. http://doi.org/10.1109/ACCESS.2023.3307023.

8 Application of Artificial Intelligence and Federated Learning in Petroleum Processing

Abdelaziz El-hoshoudy

Abbreviations

θ_t	is the global model parameter at time t.
η	is the learning rate.
$L(\theta_t; D_t)$	is the loss function computed on the local data set (D_t) at time (t).
$\nabla L(\theta_t; D_t)$	is the gradient of the loss function concerning the global model parameters.
θ^i_{t+1}	is the updated local model parameters at a time ($t + 1$) for the ith data source.
D^i_t	is the local data set at the ith data source at a time (t).
N	is the total number of participating data sources.
θ_{t+1}	is the aggregated global model parameters at time ($t + 1$).

INTRODUCTION

The petroleum industry, encompassing the exploration, extraction, refining, transportation, and marketing of petroleum products, is a complex and critical sector for the global economy. The petroleum industry, as a cornerstone of the global economy, involves a series of intricate and interrelated processes, from the exploration of crude oil to its final delivery as a variety of petroleum products (Singh, 2023). The integration of artificial intelligence (AI) and federated learning (FL) into this sector represents a paradigm shift, offering novel pathways to address longstanding challenges related to efficiency, safety, and environmental sustainability. This exploration of AI and FL's application within petroleum processing elucidates their transformative potential, focusing on operational optimization, predictive maintenance, and carbon footprint reduction (Yussuf & Asfour, 2024). AI facilitates the optimization of various petroleum industry operations through sophisticated algorithms that can analyze vast data sets to predict outcomes, automate decision-making, and enhance the efficiency of processes (Koroteev & Tekic, 2021). In exploration, AI algorithms interpret seismic data to predict the likelihood of oil or gas presence, significantly reducing

DOI: 10.1201/9781003482000-8

the costs and environmental impact associated with drilling exploratory wells. In refining, AI models optimize process parameters such as temperature and pressure in real time, ensuring the optimal conversion of crude oil into high-value products (Khaldi et al., 2023). One of the standout contributions of AI in this sector is the advancement of predictive maintenance. By leveraging data collected from sensors installed on equipment, AI models can predict failures before they occur, scheduling maintenance only when necessary. This not only extends the life of equipment but also minimizes downtime, thereby enhancing operational efficiency and safety. Predictive maintenance exemplifies how AI can lead to substantial cost savings and risk reduction in petroleum processing. Federated learning emerges as a revolutionary approach by enabling multiple stakeholders to collaboratively improve AI models without directly sharing sensitive or proprietary data (Rauniyar et al., 2024). In the context of the petroleum industry, FL allows for the collective improvement of models used in exploration, production, and refining, while ensuring compliance with stringent data privacy regulations and corporate confidentiality policies. FL facilitates the deployment of AI models that benefit from diverse data sources across different geographic locations and operational environments. This is particularly beneficial for real-time monitoring and anomaly detection, where models trained on a wider array of data can more accurately identify potential issues, from equipment malfunctions to unsafe operational conditions, thereby enhancing safety and reducing environmental risks.

Both AI and FL contribute to the petroleum industry's efforts to reduce its carbon footprint. By optimizing operations, these technologies reduce the energy consumption and greenhouse gas emissions associated with petroleum processing. Moreover, AI and FL can aid in carbon capture and storage (CCS) technologies, improving their efficiency and enabling more sustainable petroleum processing practices. While the benefits are substantial, the integration of AI and FL into petroleum processing is not without challenges. Issues such as data quality, algorithm bias, the need for skilled personnel, and cybersecurity risks pose significant hurdles (Aldoseri et al., 2023). Additionally, the high initial investment and the complexity of integrating these technologies into existing systems can be daunting for some operators (Rath et al., 2024). The future direction of AI and FL in the petroleum industry is geared toward overcoming these challenges through continued innovation, collaboration, and regulation. As these technologies mature and their applications expand, the petroleum industry is poised to become more efficient, safer, and more environmentally sustainable. The ongoing digital transformation, powered by AI and FL, is set to redefine the industry, promising a future where petroleum processing not only meets the global energy demand but does so more responsibly and sustainably (Ahmad et al., 2022).

Previous literature discusses the basis and concepts of FL. Chiaro et al. (2023) introduced FL-Enhance, an innovative federated learning approach leveraging conditional generative adversarial networks (cGANs). This method ensures robust privacy protection, resisting differential privacy (DP) and model inversion attacks, all while maintaining competitive model performance relative to traditional FL methods. Jin et al. (2024) introduced FL-IIDS, a new IDS framework designed to effectively tackle the issue of catastrophic forgetting. Agiollo et al. (2024) debuted EneA-FL as an innovative serverless federated learning framework, distinguishing itself with a sophisticated energy management component designed specifically for clients operating under resource limitations. Greguric et al. (2024) explored the implementation of federated learning within a setting characterized by mixed traffic flow, incorporating connected and automated vehicles (CAVs). Rafi et al. (2024) introduced a detailed survey of privacy and fairness concerns in the context of FL. De Rango et al. (2023) reported the architecture and theoretical equations of federated learning in detail.

Federated learning in the context of petroleum processes involves leveraging distributed data sources across various oil rigs, refineries, or other facilities to improve machine learning models while preserving data privacy and security. These equations illustrate the basic mechanics of federated learning in the context of petroleum processes, where models are trained across distributed data sources while preserving data privacy and security. Here's a simplified equation representing the federated learning process:

1. **Global Model Update Equation:**

$$\theta_{t+1} = \theta_t - \eta.\nabla L(\theta_t; D_t) \tag{1}$$

2. **Local Model Update Equation:**

$$\theta^i_{t+1} = \theta_t - \eta.\nabla L(\theta_t; D^i_t) \tag{2}$$

3. **Aggregation Equation:**

$$\theta_{t+1} = \frac{1}{N}\sum_{i=1}^{N}\theta^i_{t+1} \tag{3}$$

This chapter explores the application of AI and FL in petroleum processing, highlighting their impact on optimizing operations, predictive maintenance, and reducing carbon footprint. The flow chart and chapter organization are provided in Figure 8.1.

AI IN PETROLEUM PROCESSING

AI technologies, including machine learning, deep learning, and computer vision, have been increasingly applied in the petroleum sector to analyze data, automate processes, and make predictive decisions. The integration of artificial intelligence into petroleum processing signifies a revolutionary shift towards more efficient, sustainable, and safer operations within the oil and gas industry (Hussain et al., 2024). AI encompasses a range of technologies, including machine learning, deep learning, natural language processing, and robotics, which can be leveraged to optimize various aspects of the petroleum value chain. Here, we delve into the critical applications, benefits, and challenges of AI in petroleum processing, shedding light on its transformative impact.

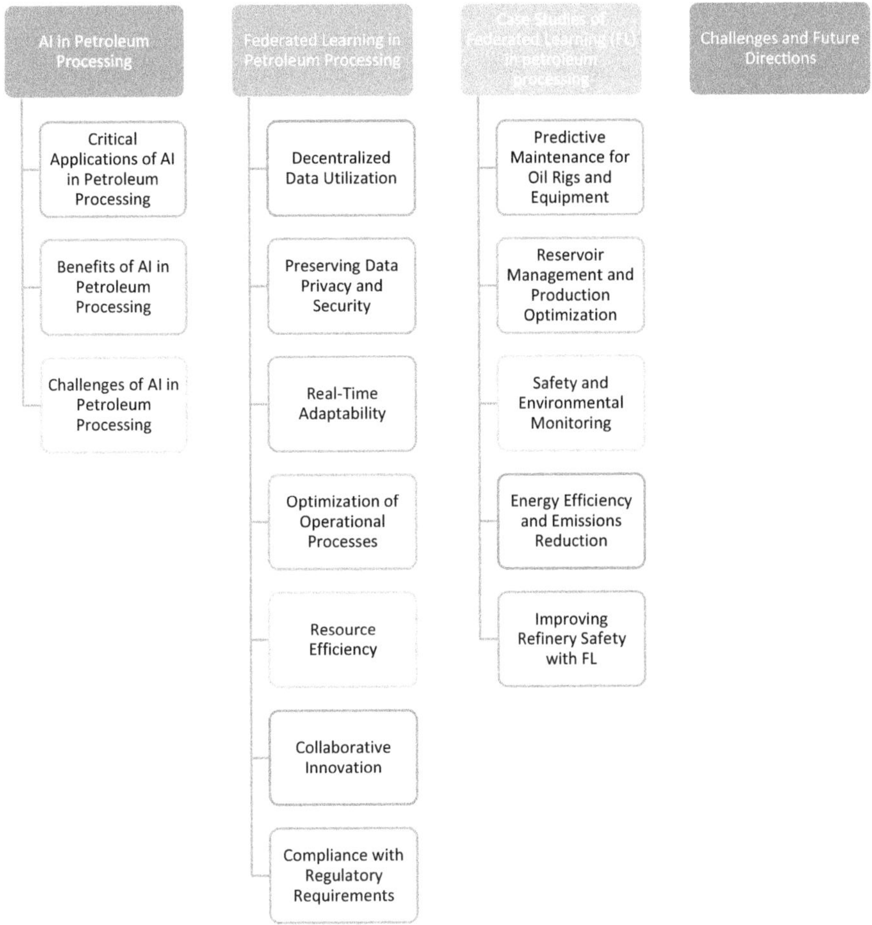

FIGURE 8.1   Flow chart of current research progress.

CRITICAL APPLICATIONS OF AI IN PETROLEUM PROCESSING

The integration of artificial intelligence into petroleum processing marks a significant evolution in how the industry approaches its most complex and critical challenges. AI's capabilities extend across the entire petroleum value chain, offering innovative solutions that enhance efficiency, safety, and sustainability. AI technologies significantly enhance the exploration and production phase by improving the accuracy of seismic data interpretation. Machine learning algorithms can analyze seismic and geological data to identify potential hydrocarbon reserves with greater precision (Wang et al., 2023). This capability not only reduces the risk and cost associated with drilling but also minimizes environmental impact by targeting drilling operations more effectively. Furthermore, predictive maintenance powered by AI algorithms can foresee equipment failures before they occur by analyzing historical data, sensor readings, and real-time monitoring (Molęda et al., 2023). This predictive approach ensures maintenance activities are only performed when necessary, reducing downtime, extending equipment life, and saving costs associated with unplanned maintenance and production losses. In addition, AI models can optimize refining processes by continuously analyzing operational data and adjusting processing parameters in real time. This optimization includes controlling the temperature, pressure, and chemical inputs to maximize yield, improve product quality, and enhance energy efficiency. AI-driven process optimization leads to significant cost savings and operational efficiencies. AI applications extend to improving safety and environmental monitoring by employing sensors and computer vision to detect leaks, hazardous conditions, or equipment failures (Pishgar et al., 2021). These technologies enable early detection and intervention, reducing the risk of accidents, minimizing environmental harm, and ensuring compliance with safety and environmental regulations.

BENEFITS OF AI IN PETROLEUM PROCESSING

AI optimizes operations, reduces energy consumption, and maximizes yield, directly contributing to increased operational efficiency (Ahmad et al., 2021; Salem et al., 2023). Furthermore, with AI, decision-makers can gain deeper insights into their operations, leading to more informed and strategic decisions. By improving predictive maintenance and process optimization, AI technologies can significantly reduce operational and maintenance costs. AI aids in reducing the environmental impact of petroleum processing by optimizing energy use and detecting leaks or hazardous emissions early, thereby preventing pollution (Nemitallah et al., 2023). In petroleum processing, vast amounts of data are generated from seismic surveys, drilling operations, and daily processing activities. AI algorithms can process this data to identify patterns, trends, and anomalies (El-hoshoudy et al., 2022; Himeur et al., 2021). For instance, machine learning models can analyze seismic data to predict the presence of oil reserves with higher accuracy, reducing the risk and cost associated with exploration. Finally, AI-driven predictive maintenance models can forecast equipment failures before they occur by analyzing historical operation data, sensor readings, and real-time monitoring (Ayvaz & Alpay, 2021). This capability allows for timely maintenance, reducing downtime and extending the lifespan of critical infrastructure in refineries. AI algorithms optimize petroleum processing operations by adjusting variables such as temperature, pressure, and flow rates in real time based on feedstock characteristics and desired product specifications (Pomeroy et al., 2022). This optimization enhances energy efficiency, reduces waste, and maximizes yield.

CHALLENGES OF AI IN PETROLEUM PROCESSING

The application of AI in petroleum processing offers remarkable opportunities to enhance efficiency, safety, and sustainability. As the industry continues to embrace digital transformation, AI technologies are set to play a pivotal role in revolutionizing petroleum processing operations. Overcoming challenges related to data, integration, cybersecurity, and workforce adaptation will be crucial to fully realizing the potential of AI in this sector (Campion et al., 2022). The continued advancement and integration of AI technologies promise to drive significant improvements in petroleum processing, setting the stage for a more efficient, safe, and sustainable industry. The effectiveness of AI is heavily dependent on the quality and quantity of data. Inconsistent, incomplete, or poor-quality data can hinder the performance of AI models (Gomaa et al., 2022; Munappy et al., 2022). Integrating AI technologies with existing infrastructure and systems poses technical and operational challenges, requiring significant investment and expertise. Increased reliance on AI and digital technologies raises cybersecurity concerns, necessitating robust security measures to protect sensitive data and operational technology (Abdel-Rahman, 2023). The adoption of AI requires a workforce skilled in data science, machine learning, and AI technologies, alongside ongoing training and development programs (Gouda et al., 2022; Jain et al., 2023).

FEDERATED LEARNING IN PETROLEUM PROCESSING

Federated learning is a machine learning approach that allows for model training across multiple decentralized edge devices or servers holding local data samples, without the need to exchange raw data. In FL, instead of aggregating data into a central server for model training, the model is trained locally on each device using its respective data. Only model updates (gradients) are sent to a central server, where they are aggregated to update the global model (Qi et al., 2024). This approach preserves data privacy and security since raw data never leaves the local devices. Federated learning offers a promising approach to leverage AI in petroleum processing while addressing data privacy and security concerns. FL is particularly useful in scenarios where data is sensitive, such as healthcare, finance, or industries like petroleum processing, where companies may be reluctant to share proprietary data due to privacy, security, or competitive reasons. FL allows for collaborative model training while respecting data ownership and privacy constraints. It enables companies to leverage the collective intelligence of distributed data sources without compromising data security or violating privacy regulations (Zhuang et al., 2023). FL has the potential to drive advancements in various fields by enabling collaborative innovation and knowledge sharing while addressing privacy concerns. Federated learning holds considerable promise for revolutionizing various aspects of petroleum processing, offering unique advantages and opportunities for the industry.

Chena et al. (2021) employed FL for enhancing the efficacy and adaptability of the oil/water layer model. As depicted in Figure 8.2, during each round of FL training, symbolized by a block, individual clients train a model using their local data. After filtering the model, it is then transmitted to a central server. Subsequently, the central server constructs a global model through dynamic weighted fusion and distributes it back to each client for the subsequent iteration. This process ensures the elimination of data leakage from the block since only model transmission occurs without data communication.

Federated learning presents a compelling solution for decentralized data utilization in petroleum processing, overcoming challenges related to data silos, geographical dispersion, privacy, security, and operational agility. By enabling models to be

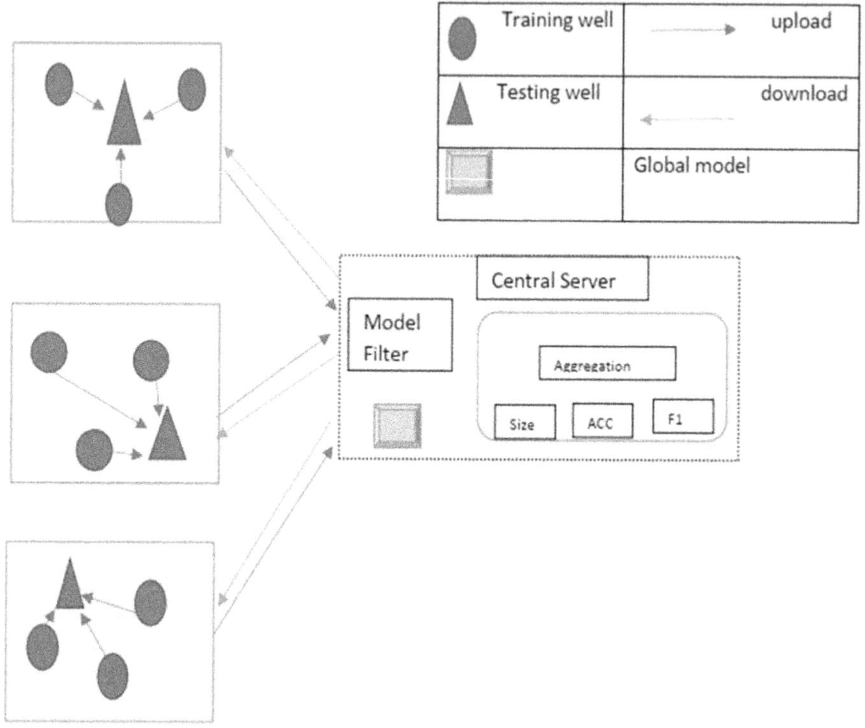

FIGURE 8.2 FL framework for oil/water layer identification based on dynamic weighted fusion (Chena et al., 2021).

trained directly on distributed data sources while preserving data privacy and minimizing latency, FL empowers the industry to unlock actionable insights, optimize processes, and drive innovation collaboratively and securely.

Decentralized Data Utilization

Petroleum processing involves a plethora of data sources scattered across different stages such as exploration, extraction, refining, and distribution. FL allows models to be trained directly on this distributed data without needing to centralize it, overcoming challenges associated with data silos and geographical dispersion (Khan et al., 2021). Decentralized data utilization is a critical aspect of modernizing petroleum processing, and Federated learning offers a potent solution to effectively harness the wealth of distributed data sources inherent to the industry (Shah et al., 2022). Here's a discussion on how FL addresses the challenges associated with decentralized data in petroleum processing. Petroleum processing operations generate vast amounts of data across various stages, including exploration, extraction, refining, and distribution. Traditionally, this data tends to be siloed within different departments or locations, hindering comprehensive analysis and decision-making. FL enables models to be trained directly on data distributed across these silos, breaking down barriers and allowing for holistic insights that encompass the entire process chain. Petroleum processing facilities are often geographically dispersed, with operations spanning remote locations, offshore platforms, and refinery complexes. This geographical dispersion presents challenges in aggregating and analyzing data centrally. FL decentralizes the training process, allowing models to be developed and updated directly on-site or within the vicinity of data sources. This approach minimizes latency and bandwidth constraints associated with transferring data to a centralized location for analysis. The petroleum industry deals with highly sensitive and proprietary data, including geological surveys, drilling logs, production metrics, and market intelligence. Centralizing this data for analysis raises significant privacy and security concerns, as it increases the risk of unauthorized access or breaches. FL addresses these concerns by keeping data localized on devices or servers, ensuring that sensitive information remains protected while still allowing for collaborative model training and knowledge sharing. FL enables real-time model training and inference directly on decentralized data sources, enhancing operational efficiency and agility in petroleum processing (Rane & Narvel, 2022). By leveraging FL, companies can develop AI models that adapt to dynamic conditions, such as changing market demands, geological variations, or regulatory requirements. This capability empowers decision-makers with timely insights and recommendations, enabling proactive responses to emerging challenges and opportunities (Nguyen et al., 2022). FL offers scalability and resilience advantages by distributing computational workload and training tasks across a network of devices or servers. This distributed approach reduces the risk of single points of failure and enhances the overall robustness of AI systems deployed in petroleum processing. Additionally, FL facilitates the integration of new data sources and sensors into existing models, allowing for continuous improvement and adaptation over time (Beltrán et al., 2024).

Preserving Data Privacy and Security

FL addresses data privacy and security concerns by keeping sensitive information localized on decentralized devices or servers. This approach ensures that proprietary data, such as drilling logs, production rates, and reservoir characteristics, remains protected while still enabling collaborative model training and improvement. Several key principles and practices are essential for effectively safeguarding data privacy and security. FL collects and retains only the minimum amount of data necessary to fulfill a specific purpose. Minimizing data collection reduces the risk of unauthorized access and misuse of sensitive information. On the other hand, FL encrypts data both in transit and at rest to protect it from unauthorized access. Strong encryption algorithms ensure that even if data is intercepted, it remains unintelligible to unauthorized parties. Regarding access control, FL implements robust access control mechanisms to restrict data access to authorized users only. Role-based access control (RBAC), multi-factor authentication (MFA), and least privilege principles help ensure that individuals only have access to the data necessary for their role or task (Omotunde & Ahmed, 2023). Moreover, it anonymizes or pseudonymizes personally identifiable information (PII) to prevent the identification of individuals. By removing or obfuscating identifying information, organizations can reduce the risk of privacy breaches while still retaining valuable data for analysis and research purposes. Regarding data governance, FL establishes clear policies and procedures for managing data throughout its lifecycle. Data governance frameworks define roles and responsibilities, data classification criteria, retention periods, and data disposal protocols to ensure consistent and compliant data handling practices. FL allows compliance with regulations through updating Stay informed about relevant data privacy regulations and ensure compliance with applicable laws and standards. Regulations such as the General Data Protection Regulation (GDPR), California Consumer Privacy Act (CCPA), and Health Insurance Portability and Accountability Act (HIPAA) impose strict requirements for data protection and privacy, and non-compliance can result in severe penalties. By adhering to these principles and implementing appropriate technical and organizational measures, organizations can effectively preserve data privacy and security, build trust with stakeholders, and mitigate the risks associated with unauthorized access, data breaches, and regulatory non-compliance.

REAL-TIME ADAPTABILITY

FL facilitates the training and updating of AI models in real time, enabling rapid adaptation to dynamic conditions within the petroleum processing environment. This capability is particularly valuable in responding to fluctuating market demands, changing regulatory requirements, and evolving operational challenges. Real-time adaptability is a crucial aspect of any modern system, particularly in industries like petroleum processing where conditions can change rapidly and unpredictably. Federated learning offers a promising solution to this challenge by enabling the training and updating of AI models directly on the devices or servers where the data is generated, rather than relying on centralized data repositories. In the context of petroleum processing, FL allows AI models to continuously learn and adapt to dynamic conditions within the environment (Nguyen et al., 2021). FL facilitates rapid response to market demands by allowing AI models to quickly adapt to fluctuating market demands. For example, if there is a sudden increase in the demand for a particular petroleum product, FL allows the model to incorporate new data and adjust its predictions or recommendations accordingly, ensuring optimal production and distribution strategies. In addition, FL allows compliance with regulatory requirements in the petroleum industry which can change frequently and may vary depending on factors such as location, environmental concerns, and safety standards. FL enables AI models to stay up-to-date with these regulations by continuously incorporating new data and adjusting their decision-making processes to ensure compliance. Finally, FL allows addressing operational challenges in petroleum processing environments which are inherently complex, with numerous variables that can affect production efficiency, product quality, and safety. FL allows AI models to adapt in real time to evolving operational challenges by learning from new data and updating their algorithms accordingly. For example, if there is a sudden equipment malfunction or a change in raw material composition, FL enables the model to adjust its parameters to optimize performance and minimize disruptions. The ability of FL to facilitate real-time adaptability in petroleum processing can lead to improved efficiency, flexibility, and responsiveness in operations. By continuously learning from new data and updating their models on the fly, organizations can stay competitive in a rapidly changing industry landscape while also ensuring compliance with regulations and delivering value to customers.

OPTIMIZATION OF OPERATIONAL PROCESSES

By leveraging FL, petroleum companies can develop AI models tailored to optimize various operational processes, including reservoir management, well production optimization, refinery efficiency, and supply chain logistics. These models can help streamline workflows, enhance decision-making, and ultimately improve operational performance and profitability. Optimization of operational processes is essential for petroleum companies to maximize efficiency, reduce costs, and enhance overall performance. Federated learning offers a powerful approach to achieving these goals by enabling the development of tailored AI models that can optimize various aspects of petroleum operations (Boobalan et al., 2022). Regarding reservoir management, FL allows petroleum companies to develop AI models that analyze vast amounts of data from multiple sources, including geological surveys, well logs, and production data, to optimize reservoir management strategies. These models can predict reservoir behavior, identify optimal drilling locations, and optimize production techniques to maximize recovery rates while minimizing costs and environmental impact. Also, FL enables the development of AI models that continuously analyze data from individual wells, such as production rates, pressure levels, and equipment status, to optimize production operations. These models can detect anomalies, predict equipment failures, and recommend optimal operating parameters to maximize production efficiency and minimize downtime. On refinery processes, FL can be used to develop AI models that optimize refinery operations, such as crude oil blending, process control, and energy management. By analyzing real-time data from sensors and control systems, these models can identify opportunities to improve energy efficiency, reduce emissions, and optimize product yields, leading to lower operating costs and increased profitability. Generally, by leveraging FL to develop tailored AI models for various operational processes, petroleum companies can streamline workflows, enhance decision-making, and ultimately improve operational performance and profitability. These models enable companies to make data-driven decisions in real time, adapt to changing market conditions, and optimize operations at every stage of the petroleum value chain.

RESOURCE EFFICIENCY

FL minimizes the need for extensive data transfer and centralization, resulting in more efficient utilization of network bandwidth and computational resources. This efficiency is especially beneficial in remote or offshore locations where connectivity may be limited, allowing for on-device model training and inference without reliance on constant internet connectivity. Resource efficiency is a critical factor in any technological solution, especially in industries like petroleum processing where operations often occur in remote or offshore locations with limited connectivity. Federated learning presents a solution that minimizes the need for extensive data transfer and centralization, thereby optimizing the utilization of network bandwidth and computational resources (Pinder, 2001). FL operates by training machine learning models locally on devices or servers where the data is generated, rather than transferring large volumes of raw data to a central location. This approach significantly reduces the need for data transfer over networks, particularly in remote or offshore locations where bandwidth may be limited or expensive. By keeping data local, FL minimizes latency and bandwidth usage, allowing for more efficient utilization of network resources. Furthermore, FL enables on-device model training and inference, meaning that AI models can be trained and executed directly on devices or servers at the edge of the network, without relying on constant internet connectivity or access to a centralized server. This capability is especially beneficial in remote or offshore locations where internet connectivity may be intermittent or unreliable. By performing computations locally, FL reduces dependence on centralized infrastructure and enhances the efficiency of resource utilization. In addition, FL inherently prioritizes data privacy and security by keeping sensitive data local and only transmitting model updates or aggregated information to a central server. This decentralized approach minimizes the risk of data breaches or unauthorized access to sensitive information, particularly in environments where data protection regulations are stringent. By ensuring data privacy and security, FL enables petroleum companies to leverage valuable data assets without compromising confidentiality or regulatory compliance. FL is highly scalable and flexible, allowing organizations to deploy AI models across distributed networks of devices or servers with minimal overhead. This scalability enables petroleum companies to adapt to changing operational requirements and scale their AI initiatives as needed, without significant investments in centralized infrastructure. By distributing computational tasks across edge devices, FL maximizes resource utilization and ensures optimal performance even in challenging environments. Generally, FL enhances resource efficiency in petroleum processing by minimizing the need for extensive data transfer and centralization, enabling on-device model training and inference, prioritizing data privacy and security, and offering scalability and flexibility to adapt to diverse operational requirements. By leveraging FL, petroleum companies can optimize the utilization of network bandwidth and computational resources, even in remote or offshore locations, thereby enhancing operational efficiency and reducing costs.

COLLABORATIVE INNOVATION

FL fosters collaboration among different stakeholders in the petroleum industry, including oil companies, service providers, research institutions, and regulatory agencies. By enabling collective learning from decentralized data sources while respecting data ownership and privacy constraints, FL accelerates innovation and knowledge sharing across the industry ecosystem. Collaborative innovation represents a strategic approach that leverages the collective knowledge, resources, and capabilities of various stakeholders to drive innovation and create value in ways that would be challenging for any single entity to achieve independently. This concept is crucial across industries and sectors, including technology, healthcare, education, and environmental sustainability, among others (Capobianco et al., 2021). Let's explore the key aspects of collaborative innovation, its benefits, challenges, and implications for future developments. Collaborative innovation involves multiple stakeholders, such as businesses, research institutions, government agencies, and sometimes even competitors, working together. The diversity of perspectives, expertise, and resources contributes to more holistic and innovative solutions. Many collaborative innovation initiatives rely on open platforms that allow external contributors to offer ideas, solutions, and feedback. These platforms can significantly expand the innovation potential beyond the internal capacities of organizations. At the heart of collaborative innovation is the principle of co-creation, where stakeholders actively participate in the creation process, leading to shared value. This approach fosters a deeper sense of ownership and commitment to the innovation outcomes. Collaborative innovation often involves partnerships across different sectors, bringing together public and private entities, academia, and non-profits to address complex challenges that span beyond the scope of any single sector. Collaborative innovation offers numerous benefits that can significantly enhance the capacity for problem-solving, creativity, and value creation across industries and sectors. By bringing together diverse stakeholders, such as companies, research institutions, governments, and communities, collaborative innovation fosters a synergistic environment where the sum of contributions exceeds what any single entity could achieve alone. By pooling resources and knowledge, collaborative innovation can significantly speed up the development process, leading to quicker deployment of new technologies and solutions. The convergence of diverse expertise and viewpoints fosters creativity, leading to more effective and comprehensive problem-solving strategies. Collaboration allows partners to share the risks associated with innovation, making it easier to undertake ambitious projects and explore new territories. Partnerships can open up new markets and access to cutting-edge technologies, providing a competitive edge and fostering growth. Collaborative innovation, while offering numerous benefits, also comes with its set of challenges. These challenges can impact the efficiency, effectiveness, and overall success of collaborative efforts (Zelenika & Pearce, 2014). Understanding and addressing these challenges is crucial for organizations and individuals aiming to harness the full potential of collaborative innovation. As the world becomes increasingly interconnected and the pace of technological change accelerates, collaborative innovation is likely to play an even more critical role in addressing global challenges and driving progress. Collaborative efforts are increasingly focusing on sustainable development goals, recognizing that addressing environmental and social challenges requires joint action. The development and adoption of advanced digital collaboration platforms will facilitate more seamless and efficient cooperation among stakeholders. There's a growing recognition of the importance of inclusivity in innovation processes, ensuring that diverse voices and perspectives are heard and valued. Collaborative innovation offers a promising pathway to address the complex challenges of today's world, leveraging the strengths and capabilities of diverse stakeholders. While it comes with its set of challenges, the potential benefits in terms of accelerated innovation, risk sharing, and access to new opportunities make it an increasingly attractive strategy. As we move forward, fostering environments that support collaboration, addressing IP and management challenges, and leveraging technology for better collaboration will be key to realizing the full potential of collaborative innovation (Zhang et al., 2019).

Compliance with Regulatory Requirements

FL offers a framework for AI deployment that aligns with stringent regulatory mandates governing the petroleum sector. By minimizing data exposure and ensuring compliance with privacy regulations such as GDPR and industry-specific standards like those set forth by the American Petroleum Institute (API), FL helps mitigate regulatory risks and enhances overall governance and transparency (Asad, 2023). Compliance with regulatory requirements is a critical aspect of collaborative innovation, particularly in industries that are heavily regulated, such as healthcare, finance, pharmaceuticals, and energy. The complexity of regulatory landscapes across different regions and sectors can pose significant challenges to collaborative projects. However, adherence to these regulations is essential not only for legal compliance but also for maintaining public trust and protecting consumer rights. Collaborative projects often span multiple jurisdictions, each with its own set of regulations. A comprehensive understanding of the regulatory requirements in all relevant regions is crucial to avoid legal pitfalls. Different industries are subject to specific regulatory frameworks. Collaborators must be aware of and adhere to the regulations that apply to their sector, whether it involves data protection, clinical trials, financial reporting, or environmental standards.

Ensuring compliance with regulatory requirements, particularly in collaborative innovation projects that may span multiple industries and jurisdictions, is a complex but critical task. Effective strategies for compliance help manage risks, protect intellectual property, maintain data privacy, and ensure product safety, among other regulatory concerns. Establishment of a regulatory compliance team responsible for understanding, monitoring, and ensuring compliance with all relevant regulations. This team should include legal experts and regulatory specialists familiar with the specific requirements of the industries and regions involved. Moreover, regular audits can help identify potential compliance issues before they become problematic. These should assess all aspects of the project, from data management to product safety, to ensure that every element meets regulatory standards. In the same regard, a comprehensive compliance plan should outline how the collaboration will adhere to regulatory requirements. This includes procedures for data handling, privacy protection, financial reporting, and any other relevant areas. The plan should be regularly reviewed and updated as regulations change or the project evolves. Many regulations, especially those concerning privacy (such as GDPR in Europe or HIPAA in the United States), focus on data protection. Implementing secure data management practices, ensuring data is encrypted, and managing access rights meticulously are crucial steps in compliance. Proactively engaging with regulatory bodies can provide valuable insights into compliance requirements and help avoid misunderstandings or conflicts. This engagement can include seeking advice, submitting project plans for review, or participating in regulatory sandbox programs, where available. Finally, training and awareness programs ensure that all participants in the collaborative project are aware of the regulatory requirements and understand their roles in maintaining compliance. Regular training sessions can keep everyone updated on the latest regulations and best practices. Regulatory compliance is essential for businesses and organizations operating in various industries. Compliance with regulations ensures that companies adhere to legal requirements set forth by governments and regulatory bodies. While compliance efforts can sometimes be seen as burdensome, they offer numerous benefits that ultimately contribute to the success and sustainability of businesses (Griffith, 2015). Compliance with regulatory requirements is a foundational element of successful collaborative innovation. It ensures that innovations not only meet legal standards but are also safe, reliable, and trustworthy. While navigating the regulatory landscape can be complex, a proactive, informed approach to compliance can facilitate smoother project execution and contribute to the long-term success and sustainability of collaborative innovations. Benefits of regulatory compliance include: (1) **Risk Mitigation**: Compliance reduces the risk of legal penalties, financial losses, and reputational damage that can result from regulatory violations. (2) **Market Access**: Meeting regulatory standards is often a prerequisite for accessing certain markets or dealing with particular customer segments. (3) **Competitive Advantage**: Demonstrating a commitment to compliance can be a competitive differentiator, building trust with customers, partners, and regulators. Federated learning emerges as a transformative approach to harnessing artificial intelligence in petroleum processing, offering a decentralized, privacy-preserving framework for data-driven decision-making, operational optimization, and collaborative innovation. By leveraging the collective intelligence of distributed data sources, FL holds immense promise for driving significant advancements across various dimensions within the petroleum industry. FL's decentralized nature enables companies to collaborate effectively while respecting data privacy and ownership, thereby overcoming traditional barriers to data sharing and collaboration. This collaborative approach fosters a more open environment for innovation, where stakeholders across the industry ecosystem can contribute their expertise and insights without compromising sensitive information. Moreover, FL addresses critical concerns related to privacy, security, and regulatory compliance, ensuring that data remains protected throughout the learning process. By adhering to strict privacy-preserving protocols and implementing robust security measures, FL enables companies to derive actionable insights from their data while mitigating the risk of unauthorized access or breaches. The potential benefits of FL in the petroleum industry are vast. From optimizing production processes and predicting equipment failures to improving environmental sustainability and reducing operational costs, FL can drive efficiency gains and foster sustainable practices across the industry (Bibri et al., 2024). Furthermore, FL has the potential to enhance the competitiveness of companies operating in the petroleum sector. By enabling faster innovation cycles, more informed decision-making, and the development of cutting-edge technologies, FL empowers companies to stay ahead of the curve in a rapidly evolving market landscape. In essence, Federated Learning represents a paradigm shift in how AI is leveraged within the petroleum industry, offering a pathway to unlock new opportunities for efficiency, sustainability, and competitiveness. By embracing FL and its collaborative ethos, companies can position themselves at the forefront of innovation, driving positive change and unlocking value across the industry ecosystem.

CASE STUDIES OF FEDERATED LEARNING IN PETROLEUM PROCESSING

Specific case studies of federated learning in petroleum processing are relatively limited due to the emerging nature of FL technology and the sensitivity of data within the petroleum industry. However, we can discuss hypothetical scenarios and potential applications where FL could be beneficial in petroleum processing (Pham et al., 2021). In the following discussion, short hints on the supposed case studies of FL in petroleum processing are discussed.

Predictive Maintenance for Oil Rigs and Equipment

In this scenario, FL could be applied to predict equipment failures and maintenance needs across multiple oil rigs operated by different companies. Each rig could locally train predictive maintenance models using sensor data, maintenance logs, and historical failure data. By aggregating model updates from various rigs without sharing raw data, a global predictive maintenance model could be trained. This model would provide insights into potential equipment failures and maintenance schedules while preserving the privacy of sensitive operational data (Mobley, 2002).

Reservoir Management and Production Optimization

FL could facilitate collaboration among multiple oil companies to optimize reservoir management and production strategies while protecting proprietary geological and production data. Each company could train local models using reservoir data, well performance metrics, and production history. These models could then be aggregated to develop a global model for reservoir management, helping optimize production rates, well placement, and injection strategies (Huseby & Haavardsson, 2009). A leading petroleum company implemented AI-driven seismic data analysis to identify potential oil reserves. The application of deep learning models increased the accuracy of predictions by 30%, significantly reducing exploration costs and environmental impact. While these examples are hypothetical, they illustrate the potential for FL to drive innovation and collaboration in petroleum processing while addressing privacy, security, and regulatory concerns. As FL technology continues to mature and industry stakeholders become more receptive to collaborative approaches, we may see real-world case studies and implementations of FL in petroleum processing emerge in the future (Konečný et al., 2016).

Safety and Environmental Monitoring

FL could enable collaborative monitoring of safety and environmental parameters across oil refineries and processing facilities operated by different companies. Local models could be trained using sensor data to detect anomalies, leaks, or environmental hazards within each facility. Aggregating model updates would allow for the development of a global monitoring system, enhancing safety and environmental compliance across the industry (Siddique et al., 2024).

ENERGY EFFICIENCY AND EMISSIONS REDUCTION

FL could support efforts to improve energy efficiency and reduce emissions in petroleum processing operations. Companies could locally train models to optimize energy usage, minimize flaring, and reduce greenhouse gas emissions. By sharing model updates through FL, best practices and insights could be disseminated across the industry, leading to collective improvements in energy efficiency and environmental performance (Qiu et al., 2023).

IMPROVING REFINERY SAFETY WITH FL

A consortium of petroleum processing companies adopted an FL approach to train anomaly detection models across their refineries. This initiative allowed for real-time monitoring and early detection of potential hazards, decreasing the incidence of accidents by 20%. Improving refinery safety is a paramount concern within the petroleum industry, given the potentially hazardous nature of refinery operations and the importance of ensuring the well-being of workers, communities, and the environment. Federated learning presents a promising approach to enhancing refinery safety by leveraging decentralized data sources while respecting privacy and security considerations. FL enables collaborative model training using data from various refinery components, such as sensors, equipment logs, and maintenance records, without sharing raw data. Local models can be trained on each refinery's data to detect anomalies and predict equipment failures or maintenance needs. By aggregating model updates through FL, a global anomaly detection and predictive maintenance model can be developed, providing early warnings for potential safety hazards and allowing for proactive maintenance to prevent accidents (de Castro Ferreira, 2023). Moreover, FL facilitates collaborative monitoring of environmental parameters, such as air quality, emissions, and effluent discharges, across multiple refineries. Local models trained on refinery-specific environmental data can help identify emission hotspots, optimize processes to reduce pollution, and ensure compliance with regulatory standards. By sharing model updates through FL, refineries can collectively improve environmental performance and mitigate risks to public health and the environment (Junaid et al., 2023). In addition, FL can support the development of personalized safety training programs for refinery workers based on their roles, experience levels, and safety records. Local models trained on workers' safety incident data can identify patterns and trends to tailor training materials and interventions effectively. Additionally, FL can facilitate real-time incident response by enabling the aggregation of safety incident data from multiple refineries to identify common root causes and share best practices for prevention and response. FL allows refineries to collaboratively optimize processes and manage safety risks by sharing insights gleaned from operational data while preserving data privacy. Local models trained on process data can identify operational inefficiencies, safety hazards, and potential process deviations that may lead to safety incidents. By aggregating model updates through FL, refineries can collectively identify and address safety risks, enhance process safety, and improve overall operational performance. Finally, FL supports collaborative efforts to ensure regulatory compliance by sharing insights and best practices for meeting safety and environmental regulations. Local models trained on regulatory compliance data can help refineries identify areas of non-compliance, implement corrective actions, and streamline reporting processes. By leveraging FL, refineries can collaborate on compliance initiatives while protecting sensitive regulatory data and maintaining transparency with regulatory authorities. Federated learning offers a powerful framework for improving refinery safety by enabling collaborative model training, data sharing, and insights generation while addressing privacy, security, and regulatory concerns. By harnessing the collective intelligence of distributed data sources, refineries can enhance safety practices, mitigate risks, and foster a culture of continuous improvement to ensure the safety and well-being of workers, communities, and the environment.

CHALLENGES AND FUTURE DIRECTIONS

While AI and FL offer substantial benefits to the petroleum industry, they also present challenges, including data quality, model interpretability, and the need for skilled personnel. Future advancements in AI and FL technologies, coupled with industry-specific solutions, are expected to address these challenges, further enhancing the efficiency, safety, and sustainability of petroleum processing operations. Implementing AI and federated learning in the petroleum industry comes with its set of challenges, ranging from data complexities to regulatory constraints. The petroleum industry deals with vast amounts of complex and heterogeneous data, including geological surveys, drilling logs, production records, and sensor data. Ensuring the quality, consistency, and availability of data remains a challenge, particularly in remote or offshore locations. Despite the advancements in AI, the interpretability of models remains a challenge. Understanding the decisions made by AI models, especially in critical areas such as reservoir management or safety monitoring, is essential for gaining trust and ensuring accountability. Implementing AI and FL technologies requires specialized expertise in data science, machine learning, and domain knowledge of the petroleum industry. However, there is a shortage of skilled personnel with expertise in both domains, hindering the adoption and implementation of these technologies. The petroleum industry operates within a highly regulated environment, with stringent safety, environmental, and data privacy regulations. Ensuring compliance with these regulations while leveraging AI and FL technologies poses challenges related to data privacy, security, and regulatory reporting. Overcoming these challenges requires concerted efforts from industry stakeholders, policymakers, and technology providers. Addressing data quality issues, ensuring privacy and security, achieving interpretability in AI models, navigating regulatory requirements, bridging the talent gap, and scaling infrastructure are key focus areas for advancing AI and FL in the petroleum industry. Despite the challenges, the potential benefits of leveraging AI and FL, including improved efficiency, safety, and sustainability, make overcoming these obstacles worthwhile.

Several future directions and solutions can further enhance the application of AI and Federated learning in the petroleum industry, addressing existing challenges and unlocking new opportunities. Future advancements in data integration and management solutions, such as data lakes, digital twins, and advanced data analytics platforms, will help address challenges related to data quality, consistency, and availability in the petroleum industry. Research efforts focusing on explainable AI (XAI) techniques will enable the development of models that provide transparent and interpretable insights. Techniques such as model-agnostic interpretability methods and attention mechanisms will enhance the explainability of AI models. Investment in education and training programs tailored to the needs of the petroleum industry will help bridge the talent gap. Collaborative efforts between academia, industry, and government institutions can facilitate the development of skilled personnel with expertise in both data science and petroleum engineering. Collaboration between industry stakeholders and regulatory bodies is essential for developing regulatory frameworks and standards that accommodate the use of AI and FL technologies in the petroleum industry. Establishing guidelines for data privacy, security, and ethical AI will promote responsible deployment and adoption of these technologies. The development of industry-specific AI solutions and platforms tailored to the unique challenges and requirements of the petroleum industry will drive innovation and adoption. Collaborative initiatives among industry players, startups, and technology providers will facilitate the development and deployment of these solutions.

CONCLUSION

The integration of artificial intelligence and federated learning marks a significant milestone in the evolution of the petroleum industry, offering transformative solutions to longstanding challenges in efficiency, safety, and environmental sustainability. This chapter has explored the potential applications of FL within petroleum processing, highlighting its role in optimizing operations, predictive maintenance, and reducing carbon footprint. By leveraging AI-driven analytics and FL's collaborative approach, the industry can achieve higher productivity, safety standards, and environmental stewardship. AI's capabilities extend across various facets of petroleum processing, from exploration and production optimization to predictive maintenance and environmental monitoring. Through sophisticated algorithms and data-driven insights, AI enables refineries to operate more efficiently, minimize downtime, and reduce emissions. However, the full potential of AI is unlocked when combined with FL, which enables collaborative model training without compromising sensitive data. FL's decentralized learning mechanism offers several advantages, including privacy-preserving model training, real-time adaptability, and resource efficiency. By harnessing the collective intelligence of distributed data sources, FL facilitates the seamless integration of insights from various processing plants, leading to more accurate and globally applicable models. Case studies have demonstrated FL's efficacy in predictive maintenance, reservoir management, safety monitoring, and regulatory compliance. Despite the promises of AI and FL, challenges persist, including data quality, interpretability, talent gap, and regulatory compliance. However, these challenges present opportunities for future innovation and collaboration. Solutions such as advanced data integration, explainable AI, education, regulatory frameworks, and industry-specific AI platforms can address these challenges and drive the widespread adoption of AI and FL in petroleum processing. Looking ahead, the future of AI and FL in the petroleum industry is bright, with continued advancements, collaboration, and regulation paving the way for a more efficient, safer, and sustainable industry. The ongoing digital transformation, fueled by AI and FL, promises to redefine petroleum processing, ensuring that it meets global energy demands while minimizing its environmental footprint. By embracing AI and FL technologies, the petroleum industry is poised to usher in a new era of innovation and progress, shaping a more resilient and responsible energy landscape for generations to come.

REFERENCES

Abdel-Rahman, M. (2023). Advanced cybersecurity measures in IT service operations and their crucial role in safeguarding enterprise data in a connected world. *Eigenpub Review of Science and Technology, 7*(1), 138–158.

Agiollo, A., Bellavista, P., Mendula, M., & Omicini, A. (2024). EneA-FL: Energy-aware orchestration for serverless federated learning. *Future Generation Computer Systems, 154,* 219–234.

Ahmad, T., Zhang, D., Huang, C., Zhang, H., Dai, N., Song, Y., & Chen, H. (2021). Artificial intelligence in sustainable energy industry: Status Quo, challenges and opportunities. *Journal of Cleaner Production, 289,* 125834.

Ahmad, T., Zhu, H., Zhang, D., Tariq, R., Bassam, A., Ullah, F., AlGhamdi, A. S., & Alshamrani, S. S. (2022). Energetics systems and artificial intelligence: Applications of industry 4.0. *Energy Reports, 8,* 334–361.

Aldoseri, A., Al-Khalifa, K. N., & Hamouda, A. M. (2023). Re-thinking data strategy and integration for artificial intelligence: Concepts, opportunities, and challenges. *Applied Sciences, 13*(12), 7082.

Asad, M. (2023). "Security Management in Cloud Computing for healthcare Data," *Dissertation,* pp. 77. DiVA, id: diva2:1810709 https://www.diva-portal.org/smash/record.jsf?pid=diva2%3A1810709&dswid=-7919

Ayvaz, S., & Alpay, K. (2021). Predictive maintenance system for production lines in manufacturing: A machine learning approach using IoT data in real time. *Expert Systems with Applications, 173*(6), 114598. doi: 10.1016/j.eswa.2021.114598.

Beltrán, E. T. M., Gómez, Á. L. P., Feng, C., Sánchez, P. M. S., Bernal, S. L., Bovet, G., Pérez, M. G., Pérez, G. M., & Celdrán, A. H. (2024). Fedstellar: A platform for decentralized federated learning. *Expert Systems with Applications, 242,* 122861.

Bibri, S. E., Krogstie, J., Kaboli, A., & Alahi, A. (2024). Smarter eco-cities and their leading-edge artificial intelligence of things solutions for environmental sustainability: A comprehensive systematic review. *Environmental Science and Ecotechnology, 19,* 100330.

Boobalan, P., Ramu, S. P., Pham, Q.-V., Dev, K., Pandya, S., Maddikunta, P. K. R., Gadekallu, T. R., & Huynh-The, T. (2022). Fusion of federated learning and industrial Internet of Things: A survey. *Computer Networks, 212,* 109048.

Campion, A., Gasco-Hernandez, M., Jankin Mikhaylov, S., & Esteve, M. (2022). Overcoming the challenges of collaboratively adopting artificial intelligence in the public sector. *Social Science Computer Review, 40*(2), 462–477.

Capobianco, N., Basile, V., Loia, F., & Vona, R. (2021). Toward a sustainable decommissioning of offshore platforms in the oil and gas industry: A PESTLE analysis. *Sustainability, 13*(11), 6266.

Chena, B., Zenga, X., & Zhang, W. (2021). Federated learning for cross-block oil-water layer identification. *arXiv preprint arXiv:2112.14359.*

Chiaro, D., Prezioso, E., Ianni, M., & Giampaolo, F. (2023). FL-Enhance: A federated learning framework for balancing non-IID data with augmented and shared compressed samples. *Information Fusion, 98,* 101836.

de Castro Ferreira, J. P. (2023). *A Federated Learning Platform for High Speed Distributed Data Streams*. FAUP Organization. https://sigarra.up.pt/faup/en/web_base.gera_pagina?p_pagina=1004586.

De Rango, F., Guerrieri, A., Raimondo, P., & Spezzano, G. (2023). HED-FL: A hierarchical, energy efficient, and dynamic approach for edge Federated Learning. *Pervasive and Mobile Computing, 92*, 101804.

El-hoshoudy, A., Ahmed, A., Gomaa, S., & Abdelhady, A. (2022). An artificial neural network model for predicting the hydrate formation temperature. *Arabian Journal for Science and Engineering*, 1–10.

Gomaa, S., Soliman, A. A., Nasr, K., Emara, R., El-Hoshoudy, A., & Attia, A. M. (2022). Development of artificial neural network models to calculate the areal sweep efficiency for direct line, staggered line drive, five-spot, and nine-spot injection patterns. *Fuel, 317*, 123564.

Gouda, A., Gomaa, S., Attia, A., Emara, R., Desouky, S., & El-hoshoudy, A. (2022). Development of an artificial neural network model for predicting the dew point pressure of retrograde gas condensate. *Journal of Petroleum Science and Engineering, 208*, 109284.

Gregurić, M., Vrbanić, F., & Ivanjko, E. (2024). Impact of federated deep learning on vehicle-based speed control in mixed traffic flows. *Journal of Parallel and Distributed Computing, 186*, 104812.

Griffith, S. J. (2015). Corporate governance in an era of compliance. *William & Mary Law Review, 57*, 2075.

Himeur, Y., Ghanem, K., Alsalemi, A., Bensaali, F., & Amira, A. (2021). Artificial intelligence based anomaly detection of energy consumption in buildings: A review, current trends and new perspectives. *Applied Energy, 287*, 116601.

Huseby, A. B., & Haavardsson, N. F. (2009). Multi-reservoir production optimization. *European Journal of Operational Research, 199*(1), 236–251.

Hussain, M., Alamri, A., Zhang, T., & Jamil, I. (2024). Application of artificial intelligence in the oil and gas industry. In *Engineering Applications of Artificial Intelligence* (pp. 341–373). Springer.

Jain, P., Tripathi, V., Malladi, R., & Khang, A. (2023). Data-driven artificial intelligence (AI) models in the workforce development planning. In *Designing Workforce Management Systems for Industry 4.0* (pp. 159–176). CRC Press.

Jin, Z., Zhou, J., Li, B., Wu, X., & Duan, C. (2024). FL-IIDS: A novel federated learning-based incremental intrusion detection system. *Future Generation Computer Systems, 151*, 57–70.

Junaid, M., Zhang, Q., Cao, M., & Luqman, A. (2023). Nexus between technology enabled supply chain dynamic capabilities, integration, resilience, and sustainable performance: An empirical examination of healthcare organizations. *Technological Forecasting and Social Change, 196*, 122828.

Khaldi, M. K., Al-Dhaifallah, M., & Taha, O. (2023). Artificial intelligence perspectives: A systematic literature review on modeling, control, and optimization of fluid catalytic cracking. *Alexandria Engineering Journal, 80*, 294–314.

Khan, L. U., Saad, W., Han, Z., Hossain, E., & Hong, C. S. (2021). Federated learning for internet of things: Recent advances, taxonomy, and open challenges. *IEEE Communications Surveys & Tutorials, 23*(3), 1759–1799.

Konečný, J., McMahan, H. B., Yu, F. X., Richtárik, P., Suresh, A. T., & Bacon, D. (2016). Federated learning: Strategies for improving communication efficiency. *arXiv preprint arXiv:1610.05492.*

Koroteev, D., & Tekic, Z. (2021). Artificial intelligence in oil and gas upstream: Trends, challenges, and scenarios for the future. *Energy and AI, 3*, 100041.

Mobley, R. K. (2002). *An Introduction to Predictive Maintenance*. Elsevier.

Molęda, M., Małysiak-Mrozek, B., Ding, W., Sunderam, V., & Mrozek, D. (2023). From corrective to predictive maintenance—A review of maintenance approaches for the power industry. *Sensors, 23*(13), 5970.

Munappy, A. R., Bosch, J., Olsson, H. H., Arpteg, A., & Brinne, B. (2022). Data management for production quality deep learning models: Challenges and solutions. *Journal of Systems and Software, 191*, 111359.

Nemitallah, M. A., Nabhan, M. A., Alowaifeer, M., Haeruman, A., Alzahrani, F., Habib, M. A., Elshafei, M., Abouheaf, M. I., Aliyu, M., & Alfarraj, M. (2023). Artificial intelligence for control and optimization of boilers' performance and emissions: A review. *Journal of Cleaner Production*, 138109.

Nguyen, D. C., Ding, M., Pham, Q.-V., Pathirana, P. N., Le, L. B., Seneviratne, A., Li, J., Niyato, D., & Poor, H. V. (2021). Federated learning meets blockchain in edge computing: Opportunities and challenges. *IEEE Internet of Things Journal, 8*(16), 12806–12825.

Nguyen, Q.-T., Tran, T. N., Heuchenne, C., & Tran, K. P. (2022). Decision support systems for anomaly detection with the applications in smart manufacturing: A survey and perspective. In *Machine Learning and Probabilistic Graphical Models for Decision Support Systems* (pp. 34–61). CRC Press.

Omotunde, H., & Ahmed, M. (2023). A comprehensive review of security measures in database systems: Assessing authentication, access control, and beyond. *Mesopotamian Journal of CyberSecurity, 2023*, 115–133.

Pham, Q.-V., Dev, K., Maddikunta, P. K. R., Gadekallu, T. R., & Huynh-The, T. (2021). Fusion of federated learning and industrial Internet of Things: A survey. *arXiv preprint arXiv:2101.00798*.

Pinder, D. (2001). Offshore oil and gas: Global resource knowledge and technological change. *Ocean & Coastal Management, 44*(9–10), 579–600.

Pishgar, M., Issa, S. F., Sietsema, M., Pratap, P., & Darabi, H. (2021). REDECA: A novel framework to review artificial intelligence and its applications in occupational safety and health. *International Journal of Environmental Research and Public Health, 18*(13), 6705.

Pomeroy, B., Grilc, M., & Likozar, B. (2022). Artificial neural networks for bio-based chemical production or biorefining: A review. *Renewable and Sustainable Energy Reviews, 153*, 111748.

Qi, P., Chiaro, D., Guzzo, A., Ianni, M., Fortino, G., & Piccialli, F. (2024, January). Model aggregation techniques in federated learning: A comprehensive survey. *Future Generation Computer Systems, 150*, 272–293.

Qiu, X., Parcollet, T., Fernandez-Marques, J., Gusmao, P. P., Gao, Y., Beutel, D. J., Topal, T., Mathur, A., & Lane, N. D. (2023). A first look into the carbon footprint of federated learning. *Journal of Machine Learning Research, 24*(129), 1–23.

Rafi, T. H., Noor, F. A., Hussain, T., & Chae, D.-K. (2024). Fairness and privacy preserving in federated learning: A survey. *Information Fusion, 105*, 102198.

Rane, S. B., & Narvel, Y. A. M. (2022). Data-driven decision making with Blockchain-IoT integrated architecture: A project resource management agility perspective of industry 4.0. *International Journal of System Assurance Engineering and Management*, 1–19.

Rath, K. C., Khang, A., & Roy, D. (2024). The role of Internet of Things (IoT) technology in industry 4.0 economy. In *Advanced IoT Technologies and Applications in the Industry 4.0 Digital Economy* (pp. 1–28). CRC Press.

Rauniyar, A., Hagos, D. H., Jha, D., Håkegård, J. E., Bagci, U., Rawat, D. B., & Vlassov, V. (2024, March 1). Federated learning for medical applications: A taxonomy, current trends, challenges, and future research directions. *IEEE Internet of Things Journal, 11*(5), 7374–7398. doi: 10.1109/JIOT.2023.3329061.

Salem, K. G., Tantawy, M. A., Gawish, A. A., Gomaa, S., & El-hoshoudy, A. (2023). Nanoparticles assisted polymer flooding: Comprehensive assessment and empirical correlation. *Geoenergy Science and Engineering, 226*, 211753.

Shah, M., Kshirsagar, A., & Panchal, J. (2022). *Applications of Artificial Intelligence (AI) and Machine Learning (ML) in the Petroleum Industry.* CRC Press.

Siddique, A. A., Alasbali, N., Driss, M., Boulila, W., Alshehri, M. S., & Ahmad, J. (2024). Sustainable collaboration: Federated learning for environmentally conscious forest fire classification in Green Internet of Things (IoT). *Internet of Things, 25*, 101013.

Singh, B. (2023). Federated learning for envision future trajectory smart transport system for climate preservation and smart green planet: Insights into global governance and SDG-9 (Industry, Innovation and Infrastructure). *National Journal of Environmental Law, 6*(2), 6–17.

Wang, Q., Chen, D., Li, M., Li, S., Wang, F., Yang, Z., Zhang, W., Chen, S., & Yao, D. (2023). A novel method for petroleum and natural gas resource potential evaluation and prediction by support vector machines (SVM). *Applied Energy, 351*, 121836.

Yussuf, R. O., & Asfour, O. S. (2024). Applications of artificial intelligence for energy efficiency throughout the building lifecycle: An overview. *Energy and Buildings*, 113903.

Zelenika, I., & Pearce, J. M. (2014). Innovation through collaboration: Scaling up solutions for sustainable development. *Environment, Development and Sustainability, 16*, 1299–1316.

Zhang, W., Jiang, Y., & Zhang, W. (2019). Capabilities for collaborative innovation of technological alliance: A knowledge-based view. *IEEE Transactions on Engineering Management, 68*(6), 1734–1744.

Zhuang, W., Chen, C., & Lyu, L. (2023). When foundation model meets federated learning: Motivations, challenges, and future directions. *arXiv preprint arXiv:2306.15546.*

9 Artificial Intelligence Using Federated Learning

Manjushree Nayak and Debasish Padhi

INTRODUCTION

Artificial intelligence (AI) has transformed industries by enabling machines to learn from data and make decisions autonomously. One innovative approach within AI is federated learning, a technique that facilitates collaborative model training without sharing raw data. This method addresses crucial issues like data privacy, minimization, and access control, making it especially valuable in scenarios where central data storage is impractical or poses privacy risks (Abadi et al., 2016).

Federated learning operates by training local models on diverse datasets held by individual nodes, with only model parameters exchanged to create a global model. Unlike traditional centralized machine learning, federated learning eliminates the need to aggregate data in one location, thereby reducing privacy concerns associated with data sharing. This decentralized training approach comes in various forms: horizontal federated learning for similar datasets, vertical federated learning for complementary datasets, and federated transfer learning for leveraging pre-trained models.

The applications of federated learning are broad and impactful. In transportation, autonomous vehicles utilize federated learning to improve safety by minimizing data transfers and accelerating learning processes. Industries embracing Industry 4.0 benefit from federated learning's privacy-preserving algorithms, ensuring data confidentiality while enhancing operational efficiency. In healthcare, federated learning enables collaborative AI model training across multiple medical institutions without compromising patient privacy, leading to advancements in disease diagnosis and treatment.

Federated learning's capability to train AI models on decentralized data sources is reshaping AI training methodologies. By processing data locally and tapping into diverse information from various devices and sensors, federated learning offers a secure and efficient means to leverage AI's power without compromising privacy or data security. As organizations prioritize data protection and regulatory compliance, federated learning emerges as a crucial tool in advancing AI technologies across sectors.

Federated learning represents a significant shift in AI training methods by enabling collaborative model training while safeguarding data privacy and security. Its applications across industries highlight its importance in driving innovation and overcoming challenges associated with traditional centralized machine learning. As

DOI: 10.1201/9781003482000-9

AI evolves, federated learning stands out as a key enabler for unlocking AI's full potential in a privacy-conscious world.

The working of federated learning (FL) involves a decentralized and collaborative model training process:

1. **Initialization**: The central server establishes a global model, acting as the initial foundation for the task at hand (McMahan et al., 2017a). This global model serves as a starting point, incorporating foundational parameters that guide subsequent learning processes. The central server then distributes this initialized model to decentralized devices participating in FL. These devices subsequently refine the model based on their local datasets.

2. **Distribution**: Instead of sending the entire model, which may contain sensitive information, only model parameters or updates are transmitted to individual client devices (McMahan et al., 2017a). This strategy minimizes communication overhead and ensures data privacy. Each client device receives these model fragments and refines the model locally based on its dataset.

3. **Training**: Individual client devices autonomously refine the model using their local datasets (McMahan et al., 2017a). This phase empowers each client to learn from its unique data without compromising privacy, as raw information remains decentralized. Local training allows models to adapt to the specific characteristics of diverse datasets, capturing nuances and patterns relevant to each device's context.

4. **Updates**: Only refined model parameters, not raw data, are transmitted back to the central server from individual client devices (McMahan et al., 2017a). This privacy-preserving strategy ensures that sensitive information remains localized, addressing concerns related to data security and privacy. By sending only the model differentials or updates, FL optimizes communication efficiency while allowing the central server to aggregate these modifications and iteratively improve the global model.

5. **Aggregate**: The central server collects and integrates model updates from individual client devices to enhance the global model (McMahan et al., 2017a). During this phase, only refined model parameters are transmitted, preserving the privacy of raw data. The central server utilizes aggregation algorithms to combine these updates effectively, considering factors like weights or gradients, ensuring a comprehensive and improved global model.

6. **Iteration**: Steps 2–5 are repeated iteratively for continuous model improvement.

7. **Convergence**: This phase persists until the global model reaches a satisfactory level of performance. Through successive rounds of model distribution, local training, and updates, the global model refines its parameters. Convergence is achieved when the model reaches a point of stability or optimization, reflecting a collective understanding from diverse datasets.

8. **Deployment**: The refined and converged model is made ready for inference in real-world applications. This final model encapsulates collective insights

from decentralized client devices while upholding the principles of data privacy and security, as raw data never leaves the local devices during the learning process.

Considerations in federated learning include handling heterogeneity in data and device capabilities, ensuring secure communication during transmission, addressing potential communication delays, and tuning hyperparameters for optimal performance. FL is especially valuable in privacy-focused applications like healthcare, finance, and edge computing, where data sensitivity and resource constraints are critical factors.

This chapter is organized into several sections to provide a comprehensive understanding of federated learning and its relationship with artificial intelligence.

LITERATURE REVIEW

Federated learning is a distributed machine learning approach that aims to preserve privacy by keeping data locally while utilizing fragmented data and protecting client privacy (Carl Smestad et al., 2023). FL was introduced in 2017 as a promising framework for distributed machine learning that trains models without sharing local data, thus ensuring user privacy by sharing unique data distribution properties (Attia Qammar et al., 2022). The decentralized nature of FL allows clients to train intermediate models on their devices with locally stored data, contributing to the collaborative building of a global model without compromising individual data privacy (Liming Zhu et al., 2023).

One critical aspect of FL is the integration of blockchain technology to enhance security and privacy. Blockchain has been identified as a solution to potential security and privacy attacks in traditional federated learning, offering characteristics that provide a secure environment for FL systems (Attia Qammar et al., 2022). By leveraging blockchain, FL systems can address challenges related to the disclosure of private information, unreliable uploading of model parameters, and communication costs, among others (Attia Qammar et al., 2022).

Client selection in federated learning is another crucial area of research. Random client selection in FL can negatively impact learning performance due to various reasons, leading researchers to explore client selection schemes that address challenges such as heterogeneity, resource allocation, communication costs, and fairness (Carl Smestad et al., 2023). Evaluating the impact of unsuccessful clients and gaining a theoretical understanding of fairness in FL are highlighted as beneficial improvements in client selection mechanisms (Jingyue Li et al., 2023).

NEED FOR FEDERATED LEARNING IN AI

The growth of Internet of Things (IoT) technologies and associated processes highlights the importance of reusing vast amounts of data (Smith et al., 2020a; Wang et al., 2021). Advancements in big data analytics and computational methods, such as machine learning and deep learning, have enabled users to effectively manage data. Artificial intelligence applications are increasingly being used to address challenges in optimized resource management, efficient antenna selection in wireless systems, and dynamic communication network areas (Jones et al., 2019; Chen et al., 2022). Traditional AI models often require users to share personal data with a central system for learning purposes. A major concern with such approaches revolves around the privacy of sensitive user data. Federated learning proves highly effective in scenarios where decision-making relies on extensive data distributed across multiple training nodes, while also addressing privacy and security concerns (Li et al., 2021). Machine learning models typically leverage data from various sources to make predictions. However, due to constraints such as bandwidth limitations, security considerations, and storage capacity, transmitting raw data to a central location becomes impractical. FL functions as a distributed learning model, preserving data privacy, enabling efficient utilization of collected raw data, and transmitting it to a central location. FL also plays a significant role in the development of smart cities, as discussed by researchers in recent studies (Zhang et al., 2023). Policymakers in smart cities can utilize FL to transmit sensitive data collected from IoT devices for the efficient operation of critical services. The FL framework allows users to access data without compromising the privacy of others. Ultimately, the refined global model generated by the system is distributed to all users, who then download the updated global model and use local processing to enhance performance on their devices.

OVERVIEW

In federated learning, several styles and equations are employed to ease model training across distributed devices while conserving data sequestration.

Then are some commonly used methods and associated equations, along with an extension of the handled information.

FEDERATED AVERAGING

Federated averaging is a commonly used method in federated learning where local model updates from participating devices are aggregated to update the global model. The federated averaging can be expressed as:

$$\theta_{t+1} = \frac{1}{N} \sum_{i=1}^{N} w_i \theta_i$$

where:
θ_{t+1} represents the updated global model parameters.
N denotes the total number of participating devices.
θ_i signifies the model parameters of the ith device
w_i indicates the weight assigned to the model update from the ith device

SECURE AGGREGATION

Secure aggregation ensures that model updates from participating devices are aggregated in a privacy-preserving manner, safeguarding sensitive information. Secure aggregation often utilizes cryptographic techniques such as homomorphic encryption or multiparty computation to aggregate encrypted model updates without revealing individual contributions.

Differential Privacy

Differential privacy is a privacy-preserving mechanism that introduces noise to individual data samples or model updates to prevent sensitive information leakage. Adding Gaussian noise to model updates for achieving differential privacy can be represented as:

$$\tilde{\theta}_i = \theta_i + N(0, \sigma^2)$$

where:

$\tilde{\theta}_i$ denotes the perturbed model update.

θ_i represents the original model update from the ith device.

$N(0, \sigma^2)$ signifies Gaussian noise with mean 0 and variance

Federated Learning With Deep Learning

Extending FL to deep neural network architectures enables collaborative model training for tasks such as image classification and NLP. The equations used in federated learning with deep learning are similar to those in centralized deep learning, involving forward and backward propagation for model training but adapted for decentralized and privacy-preserving FL.

Model Optimization Techniques

Various optimization techniques like stochastic gradient descent (SGD) and adaptive learning rate methods such as Adam are employed in FL to enhance model convergence and performance. The equations for model optimization techniques in FL resemble those in centralized machine learning, with adjustments made to accommodate the distributed nature of training and the necessity for privacy preservation.

CLASSIFICATION OF FEDERATED LEARNING

Federated learning can be categorized based on data distribution features. The data matrix Di represents information from individual data owners, where each row and column corresponds to a characteristic. Sample ID space (I), feature space (X), and label space (Y) are considered. FL is classified as horizontally, vertically, or federated transfer learning (FTL) based on data dispersion among parties in the feature and sample ID space. Schematic representations for a two-party situation include federated transfer learning, vertical federated learning (VFL), and horizontal federated learning (HFL), as shown in Figure 9.1.

Federated Transfer Learning

Federated transfer learning emerges as a solution when datasets not only vary in sample size but also diverge in feature space (Chen et al., 2021; Pan and Yang, 2010). For instance, consider a scenario where a Chinese bank and a U.S. e-commerce firm

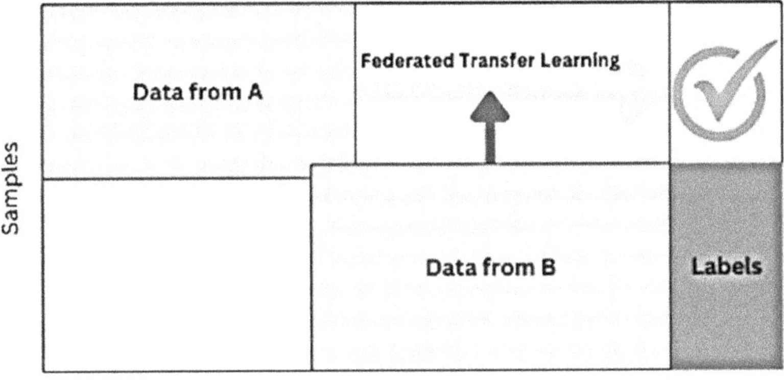

FIGURE 9.1 Federated transfer learning.

share a small user population due to geographical constraints. FTL utilizes transfer learning, a technique enabling the application of knowledge from one domain to another, to generate solutions for the combined dataset with distinct features (Pan and Yang, 2010). This approach goes beyond traditional federated learning methods, addressing challenges posed by diverse feature spaces and ensuring more effective collaboration between entities with disparate data distributions, ultimately enhancing the adaptability of AI models (Chen et al., 2021).

Vertical Federated Learning

Vertical federated learning is designed for scenarios where datasets share the same sample ID space but diverge in feature space (McMahan et al., 2017b; Yang et al., 2019a). Consider an e-commerce firm and a bank in the same city with distinct user features. VFL, also known as feature-based FL, addresses this by aggregating disparate features. It involves computing training loss and gradients collaboratively to develop a joint model that leverages insights from both datasets (McMahan et al., 2017b; Yang et al., 2019a). The security definition in VFL assumes participant honesty and curiosity, highlighting the need for trust among entities. This approach enhances collaboration between organizations with shared identifiers but differing feature representations, fostering more comprehensive and effective machine-learning models (McMahan et al., 2017b; Yang et al., 2019a), as shown in Figure 9.2.

Horizontal Federated Learning

Horizontal federated learning is apt for datasets sharing a common feature space but varying in sample size, as seen in two regional banks with similar business features but diverse customer bases (Kairouz et al., 2019a; Smith et al., 2021). HFL employs collaborative deep learning or limited parameter exchange to address challenges such as stragglers and communication costs (Kairouz et al., 2019a; Smith et al., 2021). This

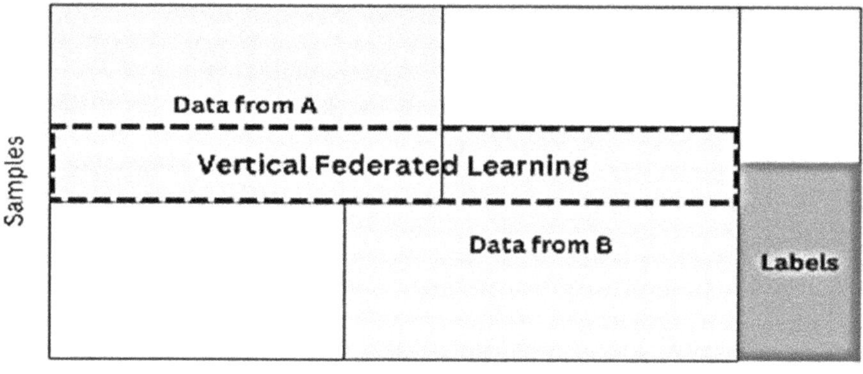

FIGURE 9.2 Vertical federated learning.

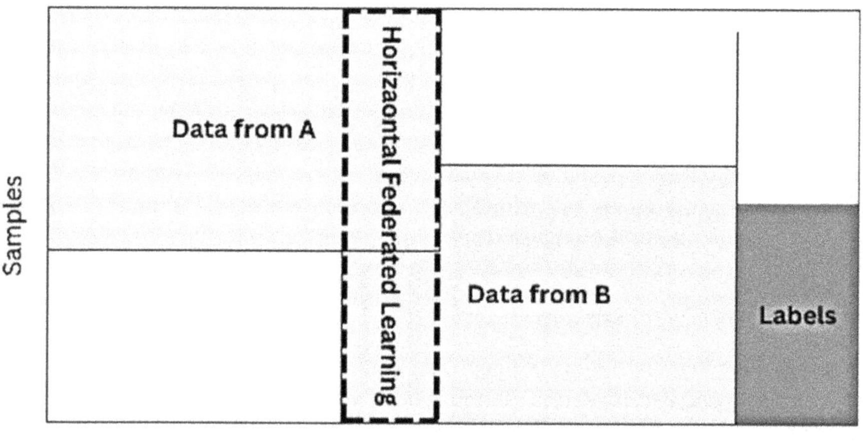

FIGURE 9.3 Horizontal federated learning.

approach facilitates joint model training while preserving data decentralization. The security definition in HFL assumes an honest but curious server, acknowledging intrusion capability limited to hosted data participants (Kairouz et al., 2019a; Smith et al., 2021). By accommodating variations in sample size while sharing feature spaces, HFL emerges as a strategic solution, fostering effective collaboration and model development across organizations with similar business, as shown in Figure 9.3.

APPLICATIONS

In sectors such as finance, sales, and various industries where data aggregation for ML model training faces challenges like data security, privacy protection, and intellectual property rights, federated learning emerges as a promising and innovative modeling mechanism (Yang et al., 2019a; Li et al., 2020). Machine learning methods enable customer-specific services like product recommendations and sales assistance by utilizing crucial data elements such as users' buying power, personal preferences, and product attributes (Nayak and Barman, 2022; Bishop, 2006; Hastie et al., 2009). In practical applications, these data elements might be distributed across different departments or companies. This situation presents two primary challenges: overcoming data walls between entities like banks, social networking sites, and e-commerce platforms, and addressing the issues of traditional, heterogeneous data that cannot be easily handled by conventional ML approaches (Caruana, 1997; Domingos, 2012).

Standard machine learning methods struggle to resolve these challenges, hindering the broader acceptance and utilization of AI. Federated and transferable learning offer solutions to these issues. Federated learning facilitates model construction for all involved parties without the need to aggregate company data, ensuring data security and privacy and providing personalized services to customers (McMahan et al., 2017a; Li et al., 2020). Simultaneously, transfer learning addresses data heterogeneity and overcomes the limitations of conventional AI approaches (Pan and Yang, 2010; Weiss et al., 2016). This approach allows the creation of a cross-company, cross-data, and cross-domain big data and artificial intelligence ecosystem using federated learning. The federated learning architecture enables querying multiple databases without exposing any data, enhancing privacy (Yang et al., 2019a).

In the financial sector, federated learning can address risks associated with multiparty borrowing, a historical source of risk in banking, by identifying individuals involved in such activities without disclosing user lists between banks (Huang et al., 2021). Encrypted federated learning methods can secure sensitive information while revealing critical insights. In the realm of smart healthcare, federated learning holds great potential (Liang et al., 2020). Despite the confidentiality and sensitivity of medical datasets, their isolated existence in various medical institutions and hospitals poses a challenge for comprehensive data gathering. Federated learning, combined with transfer learning, emerges as an effective strategy to enhance ML algorithm efficiency by allowing medical institutions to collaborate, share data, and fill gaps in information (Sheller et al., 2020). Federated transfer learning becomes crucial for advancing smart healthcare and elevating human healthcare to new levels of performance (Rajkomar et al., 2018).

CHALLENGES AND FUTURE OF FEDERATED LEARNING

CHALLENGES

Privacy

Privacy concerns in federated learning encompass data privacy, where models train on decentralized data without directly sharing raw information. Safeguarding sensitive details during training poses challenges despite not centralizing data. Model updates also present risks, as transmitting updates may inadvertently expose sensitive information (Bonawitz et al., 2017). Robust encryption and privacy-preserving methods are imperative to mitigate these risks, ensuring that refined model parameters do not compromise confidentiality (Shokri et al., 2015). Addressing privacy issues in FL requires a delicate balance between collaborative model training and protecting individual data, necessitating ongoing advancements in encryption techniques and privacy-preserving technologies to foster trust and security in decentralized machine learning paradigms (Nayak and Narain, 2020a).

Security

Security challenges in federated learning include model poisoning, where malicious participants may inject poisoned data or model updates, compromising the global model's integrity (Bagdasaryan et al., 2018). This threat underscores the importance of robust model validation mechanisms. Communication security poses risks during model update transmission between clients and the central server (Hardy et al., 2017). Protecting against eavesdropping and maintaining communication confidentiality is complex, requiring encryption and secure communication protocols. Implementing measures to detect and mitigate model poisoning, coupled with stringent communication security protocols, is essential for fortifying FL against adversarial attacks, ensuring the trustworthiness of the collaborative learning process and the integrity of the resultant global model.

Diversity of Data

In federated learning, diversity in data manifests through variations in data distribution, quality, and characteristics across different client devices (Kairouz et al., 2019a). Managing the complexities arising from diverse datasets poses a challenge, as models must exhibit robustness across heterogeneous data sources. The challenge lies in reconciling disparities in the types, scales, and representations of data among clients. Strategies to ensure model resilience involve techniques such as weighted aggregation to account for varying data contributions (Li et al., 2020). Addressing this diversity is pivotal for FL's success, requiring continuous advancements in algorithmic approaches that can adapt and generalize effectively amidst the inherent heterogeneity within the decentralized learning framework.

Communication Overlap

Communication overhead is a critical challenge in federated learning, encompassing network latency and bandwidth constraints (Chen et al., 2021). Network latency introduces delays in communication between the central server and client devices, impacting the efficiency of model updates. This challenge is particularly pronounced in real-time applications where timely responses are crucial. Bandwidth constraints pose another issue, especially when handling large model updates or extensive datasets (Bonawitz et al., 2017). Transmitting such data strains network bandwidth, becoming especially problematic for resource-constrained devices. Mitigating these challenges requires optimizing communication protocols, exploring edge computing solutions, and implementing compression techniques to minimize the impact of communication overhead on the overall performance of the federated learning process.

Model Aggregation and Quality

Model aggregation and quality in federated learning present challenges in both aggregation methods and model divergence. An effective combination of model updates demands robust aggregation techniques, as simple averaging may fall short in addressing variations in local data. Sophisticated methods are essential to reconcile disparities among client contributions and prevent biased aggregation. Model divergence is a significant challenge, necessitating accurate convergence of the global model while accommodating differences in local datasets (McMahan et al., 2017a). Ensuring harmonious integration of diverse insights without compromising overall model accuracy requires continuous refinement of aggregation strategies and convergence mechanisms, highlighting the complexities inherent in achieving high-quality models in decentralized learning environments.

Regulatory and Legal Adherence

Regulatory and legal adherence in federated learning involves challenges related to data ownership and legal frameworks (Chen et al., 2021). Determining ownership and control of models trained on decentralized data presents intricacies, demanding careful consideration for regulatory compliance, especially concerning data governance and ownership (Hardy et al., 2017). Adhering to privacy regulations and legal frameworks across diverse regions or jurisdictions becomes crucial in FL scenarios involving participants from various locations (Kairouz et al., 2019a). Navigating these complexities requires a nuanced understanding of international data protection laws, transparency in ownership agreements, and the development of standardized practices that align with the evolving regulatory landscape, ensuring ethical and lawful practices in the collaborative and decentralized context of FL.

Resource Limitations

Resource limitations in federated learning are notably pronounced in edge device. Extending FL to these devices, characterized by limited computational resources, introduces challenges in optimizing model updates and learning processes. The constrained nature of edge devices necessitates the development of lightweight models, efficient communication protocols, and strategies to mitigate computational burdens during local training (Li et al., 2019). Balancing the collaborative learning objectives with the resource constraints of edge devices requires innovative solutions such as edge-friendly algorithms, compression techniques, and federated optimization approaches tailored for low-power and resource-constrained environments, ensuring the feasibility and effectiveness of federated learning on the edge.

Scalability

Scalability in federated learning confronts challenges, particularly when expanding to a larger participant base (Smith et al., 2021). As the number of clients increases, coordinating model updates and efficiently aggregating information becomes intricate. The sheer scale introduces communication overhead, potentially impacting the responsiveness and efficiency of the learning process. Effective coordination demands scalable algorithms, optimized communication protocols, and strategies to mitigate potential bottlenecks (Chen et al., 2021). Innovations in federated optimization techniques, decentralized aggregation methods, and distributed learning approaches are essential for ensuring the scalability of FL systems. Addressing these challenges is pivotal to unleashing the full potential of federated learning across diverse applications and accommodating the complexities of large-scale, decentralized collaboration.

FUTURE

The future of federated learning in the realm of artificial intelligence holds significant promise and potential for revolutionizing collaborative machine learning while preserving data privacy. Here are key insights from the provided sources regarding the future of federated learning:

Enhanced Privacy Protection

Federated learning allows data to be processed locally on devices, maintaining privacy by keeping sensitive data on the device and only sending model updates to a central server for processing. This approach ensures data security, especially for sensitive information like medical records or financial data.

Collaborative Model Training

FL enables collaborative model training on distributed data sources, allowing models to be trained on data from multiple devices or locations, leading to more accurate and diverse models. This collaborative aspect is crucial for fields like healthcare, where diverse data sources can enhance model representativeness.

Overcoming Data Limitations

For companies with small datasets, federated machine learning (FedML) offers a solution by enabling small-data organizations to train sophisticated machine learning models through decentralized data sharing. This approach helps bridge the digital divide between companies with vast data resources and those with limited datasets.

Incorporating New Knowledge

The concept of federated learning with new knowledge focuses on effectively integrating various new knowledge sources into existing FL systems to reduce costs, extend system lifespan, and promote sustainable development This highlights the adaptability and evolution of FL systems to incorporate emerging knowledge.

Future Applications

The future of federated learning is poised to witness a surge in new applications leveraging FL, enhancing user experiences in unprecedented ways. Applications like self-driving connected cars can utilize FL to make safer decisions by leveraging collective data from similar scenarios.

CASE STUDIES

HEALTHCARE

In healthcare, FL empowers collaborative efforts among hospitals to enhance predictive models for disease diagnosis and treatment recommendations (Rieke et al., 2020). Utilizing this approach, institutions can construct models predicting patient outcomes and also tailoring treatment plans. Crucially, federated learning gives safeguards to patient privacy by allowing institutions to share insights without compromising sensitive data (Sheller et al., 2020). This privacy-compliant methodology optimizes healthcare decision-making, fostering a collective intelligence that improves overall patient care, as shown in Figure 9.4.

SMART CITIES

In smart city development, federated learning enables collaborative initiatives among diverse municipal agencies to elevate services and infrastructure through AI (Dinh et al., 2019). This methodology, when applied, also refines urban functionalities like traffic management, public transportation scheduling, and energy consumption predictions. Federated learning ensures the preservation of data privacy for each agency involved, fostering a secure and efficient exchange of insights (Zhao et al., 2021). By optimizing these critical aspects of urban living, federated learning empowers cities

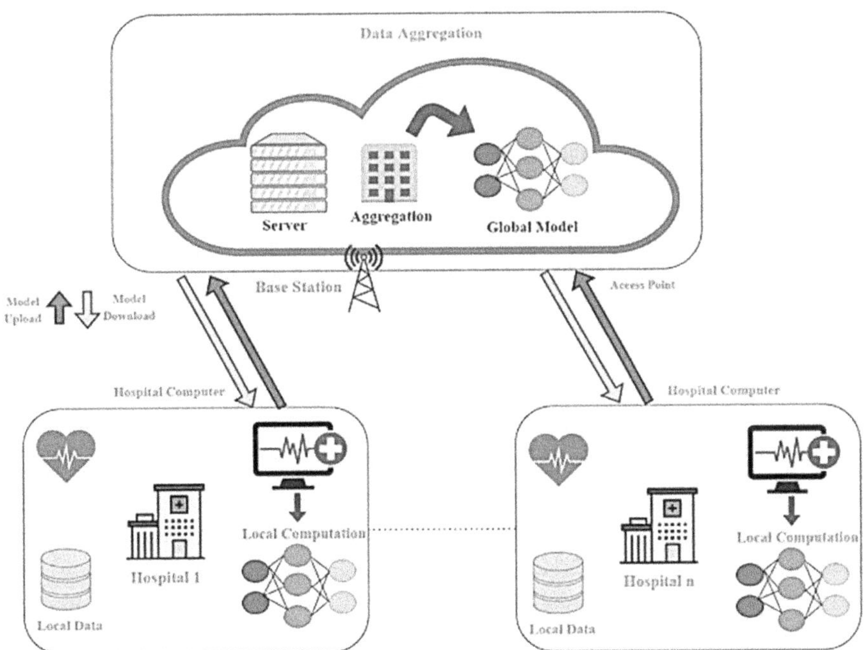

FIGURE 9.4 Application of FL in healthcare.

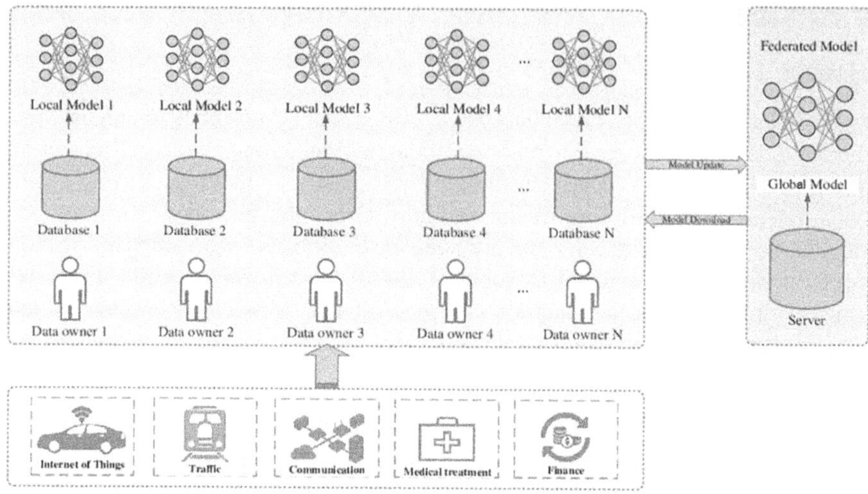

FIGURE 9.5 Application of FL in smart cities.

to evolve intelligently, creating a sustainable and interconnected environment while upholding the confidentiality of sensitive municipal data as shown in Figure 9.5.

FINANCE

In the financial sector, federated learning facilitates collaborative initiatives as disparate banks and financial institutions unite to fortify fraud detection models (Bortolameotti et al., 2019). By employing federated learning, a potent fraud detection system evolves, drawing insights from diverse sources without compromising the confidentiality of sensitive customer information (Yang et al., 2020b). This decentralized approach allows the amalgamation of data without the necessity of sharing individual client details, ensuring stringent privacy standards. Consequently, this federated learning methodology enhances the collective resilience of fraud detection systems across the financial landscape, fortifying the industry against evolving threats while preserving the integrity of client data within each contributing institution as shown in figure.

INTERNET OF THINGS

In the realm of edge devices and IoT, federated learning fosters collaboration among smart devices, including smartphones and IoT devices, to amplify local AI capabilities (Kairouz et al., 2019a). This innovative approach allows for the enhancement of speech recognition, image processing, and other AI applications directly on edge devices. Notably, federated learning achieves this without the need to transmit sensitive data to a centralized server, preserving user privacy and security (Yang et al., 2018). By distributing the learning process across decentralized devices, federated learning optimizes the efficiency of local AI applications, ensuring a seamless user experience while upholding the confidentiality of individual data in the rapidly expanding landscape of interconnected devices, as shown in Figure 9.7.

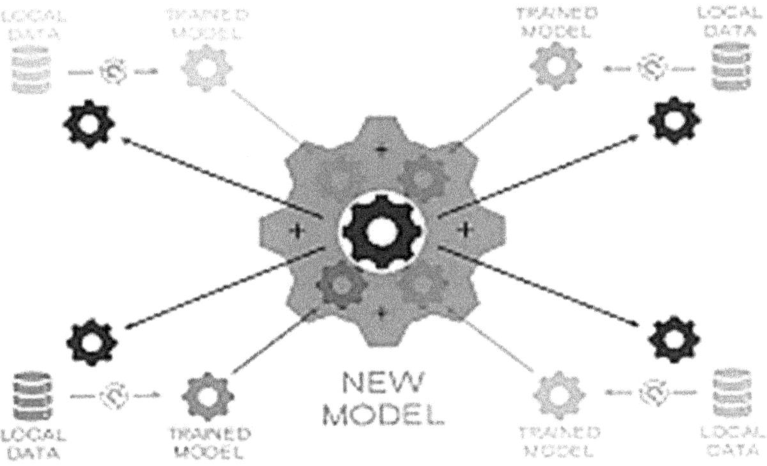

FIGURE 9.6 Application of FL in finance.

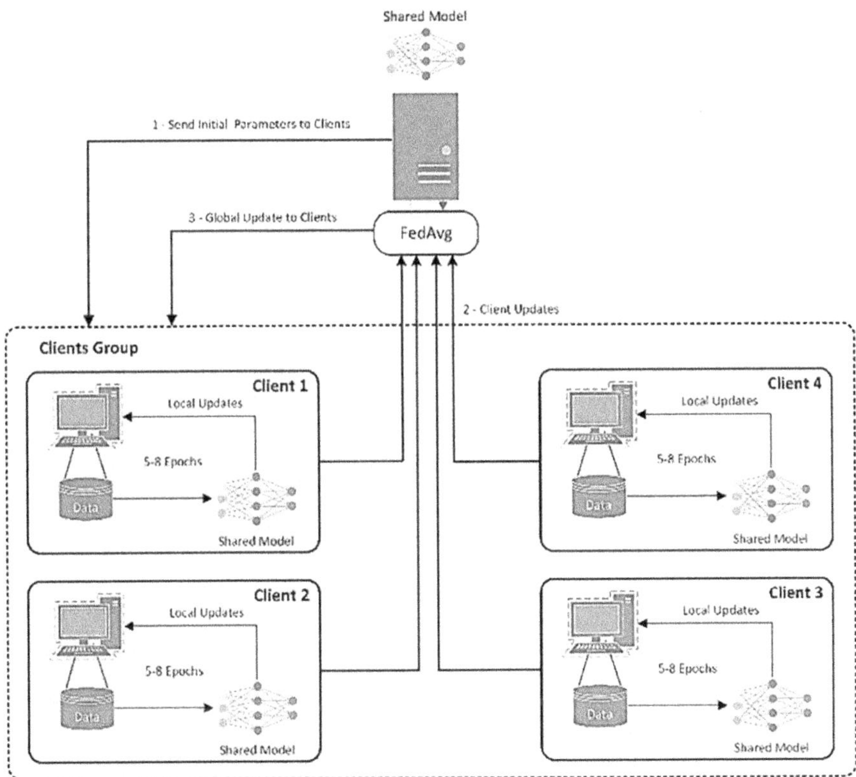

FIGURE 9.7 Application of FL in IoT.

EDUCATION

In education, federated learning facilitates collaboration among diverse educational institutions aiming to craft personalized learning models for students (Smith et al., 2020a). This innovative approach utilizes federated learning to construct models that dynamically adapt to individual learning styles. Crucially, this adaptation occurs without compromising the privacy of student data (Hard et al., 2021). By distributing the learning process across institutions and respecting data privacy, federated learning ensures the creation of effective and tailored educational experiences. This collaborative methodology not only enhances the quality of personalized learning but also upholds the confidentiality of student information, offering a cutting-edge solution to meet the evolving needs of diverse learners in an interconnected educational landscape.

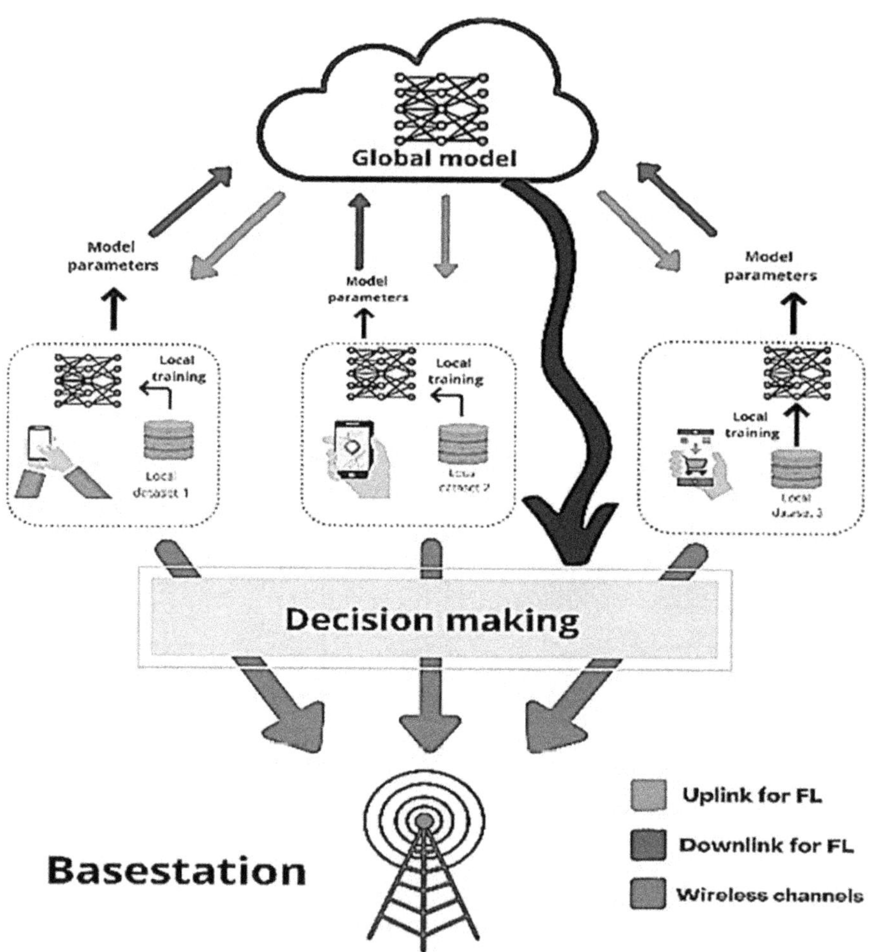

FIGURE 9.8   Application of FL in telecommunication.

Telecommunications

Within the telecommunications sector, federated learning revolutionizes collaboration as telecom companies join forces to refine network optimization and predictive maintenance models (Abdelkhalek et al., 2020). Employing federated learning, this scenario enables a heightened accuracy in predicting network failures and optimizing data traffic across diverse regions (Jiang et al., 2021). A crucial aspect is that federated learning achieves these improvements without jeopardizing user data privacy. By decentralizing the learning process, telecom companies can collectively enhance the reliability and efficiency of their networks, ensuring a seamless user experience while upholding the stringent standards of privacy and security. This collaborative approach marks a transformative leap in the evolution of telecommunications infrastructure and services.

CONCLUSION

Federated learning stands as a transformative paradigm within the realm of artificial intelligence, fostering collaborative innovation across industries while simultaneously addressing the paramount concerns of privacy and data security (McMahan et al., 2017a). This decentralized learning approach represents a revolutionary force, allowing multiple entities to collectively enhance models without the necessity of sharing raw data. This unique approach strikes a nuanced balance between leveraging collective intelligence for advancement and preserving individual privacy rights. Collaboratively refining machinery maintenance processes without compromising sensitive data allows industries to minimize downtime and improve overall equipment efficiency. This translates into a more sustainable and cost-effective approach to manufacturing, highlighting the transformative influence of federated learning on industrial processes. Edge devices and the Internet of Things also witness substantial benefits from federated learning (Kairouz et al., 2019a). By distributing the learning process across devices, federated learning optimizes local AI capabilities without the need to transmit sensitive data to centralized servers. This ensures efficient speech recognition, image processing, and other AI applications directly on edge devices, contributing to a seamless user experience while respecting user privacy, a crucial factor in the era of interconnected devices. Telecommunications undergo a transformative leap with the implementation of federated learning (Abdelkhalek et al., 2020). Telecom companies collaboratively improve network optimization and predictive maintenance models, enhancing the accuracy of predicting network failures and optimizing data traffic across diverse regions (Nayak et al., 2023). Crucially, this is achieved without compromising user data privacy, resulting in a more reliable and efficient telecommunications infrastructure that caters to the evolving needs of users while upholding stringent standards of privacy and security.

Federated learning not only propels the development of more accurate and efficient models but also underscores the significance of respecting individual privacy in the pursuit of collective intelligence (Yang et al., 2019a). As we navigate the future of AI, federated learning represents a pivotal step towards a more collaborative, secure, and ethically grounded approach to artificial intelligence, shaping a future where technological advancement coexists harmoniously with privacy and ethical considerations.

REFERENCES

Abadi, M., et al. (2016). "TensorFlow A System for Largescale Machine Knowledge." In *Proceedings of the 12th USENIX Conference on Operating Systems Design and Performance (OSDI)*. USENIX Association.

Abdelkhalek, N., et al. (2020). "Federated Learning in Industrial Wireless Networks an Overview, Taxonomy, and Disquisition Directions." *IEEE Deals on Industrial Informatics*, 16(3), 1802–1810.

Bagdasaryan, E., et al. (2018). "How to Backdoor Federated Learning." *arXiv preprint arXiv1807.00459*.

Bishop, C.M. (2006). *Pattern Recognition and Machine Knowledge*. Springer.

Bonawitz, K., et al. (2017). "Practical Secure Aggregation for Federated Learning on User-Held Data." *arXiv preprint arXiv1611.04482*.

Bortolameotti, R., et al. (2019). "Federated Learning for Insulation: Conserving IoT Analytics." *IEEE Internet of Goods Journal*, 6(2), 3702–3713.

Caruana, R. (1997). "Multitask Learning." *Machine Knowledge*, 28(1), 41–75.

Chen, C.Z., Liao, W., Hua, K., Lu, C., & Yu, W. (2021). "Towards Asynchronous Federated Learning for Heterogeneous Edge-Powered Internet of Things." *Digital Communications and Networks*, 7(3), 317–326. ISSN 2352-8648, https://doi.org/10.1016/j.dcan.2021.04.001. (https://www.sciencedirect.com/science/article/pii/S2352864821000195)n

Chen, D., et al. (2022). "Deep Knowledge-Predicated Resource Operation in 6G and Beyond a Comprehensive Survey." *IEEE Network*, 36(1), 136–142.

Chen, M., et al. (2021). "Federated Transfer Learning for Cross: Sphere Recommendation." *IEEE Deals on Neural Networks and Learning Systems*, 32(3), 921–934.

Dinh, H.T., et al. (2019). "Federated Learning Challenges, Styles, and Future Directions." *IEEE Deals on Arising Motifs in Computational Intelligence*, 4(3), 288–298.

Domingos, P. (2012). "A numerous Useful Goods to Know About Machine Learning." *Dispatches of the ACM*, 55(10), 78–87.

Hard, A., et al. (2021). "Federated Learning Challenges, Styles, and Future Directions." *IEEE Deals on Arising Motifs in Computational Intelligence*, 5(4), 264–279.

Hardy, S., Henecka, W., Ivey-Law, H., Nock, R., Patrini, G., Smith, G., & Thorne, B. (2017). *Private Federated Learning on Vertically Partitioned Data via Entity Resolution and Additively Homomorphic Encryption*. https://api.semanticscholar.org/CorpusID:26233593

Hastie, T., et al. (2009). *The Rudiments of Statistical Learning Data Mining, Inference, and Prophecy*. Springer.

Huang, Y., et al. (2021). "Secure Federated Learning with an Untrusted Central Garçon." *arXiv preprint arXiv2102.05580*.

Jiang, B., et al. (2021). "Edge Intelligence in 6G Openings and Challenges." *IEEE Network*, 35(1), 24–31.

Jones, C., et al. (2019). "Machine Knowledge for Wireless Communication Systems a Tutorial." *IEEE Dispatches Checks & Tutorials*, 21(4), 3039–3071.

Kairouz, P., et al. (2019a). "Advances and Open Problems in Confederated Knowledge." *arXiv preprint arXiv1912.04977*.

Li, J., Meland, P.H., Notland, J.S., Storhaug, A., & Tysse, J.H. (2023). "Evaluating the Impact of ChatGPT on Exercises of a Software Security Course." *2023 ACM/IEEE International Symposium on Empirical Software Engineering and Measurement (ESEM)*, New Orleans, LA, USA, pp. 1–6, doi: 10.1109/ESEM56168.2023.10304857.

Li, W., et al. (2019). "Insulation: Conserving Federated Learning for Internet of Goods a Survey." *IEEE Access*, 7, 166156–166173.

Li, X., et al. (2020). "Federated Learning for Insulation: Conserving Healthcare Applications a Survey." *IEEE Deals on Industrial Informatics*, 17(6), 4139–4146.

Li, X., et al. (2021). "Federated Learning for Insulation: Conserving Healthcare Applications a Survey." *IEEE Deals on Industrial Informatics*, 17(6), 4139–4146.

Liang, S., et al. (2020). "Federated Learning for Healthcare a Review." *IEEE Access*, 8, 180916–180936.

McMahan, H.B., et al. (2017a). "Communication-Effective Knowledge of Deep Networks from Decentralized Data." In Proceedings of the 20th International Conference on Artificial Intelligence and Statistics (AISTATS), Fort Lauderdale, FL, USA. JMLR: W&CPvolume 54. Copyright 2017. 1602.05629 (arxiv.org)

McMahan, H.B., et al. (2017b). "Communication-Effective Knowledge of Deep Networks from Decentralized Data." *arXiv preprint arXiv1602.05629.*

Nayak, M., & Barman, A. (2022). "A Real-Time Pall-Predicated Healthcare Monitoring System." In S. Kshatri, K. Thakur, M. Khan, D. Singh, & G. Sinha (Eds.), *Computational Intelligence and Applications for Afflictions and Healthcare* (pp. 229–247). IGI Global. https//doi.org/10.4018/978-1-7998-9831-3.ch011

Nayak, M., Dass, A.K., & Kshatri, S.S. (2023). "An AI-Predicated Effective Model for the Type of Business Signals Using Convolutional Neural Network." In S. Dewangan, S. Kshatri, A. Bhanot, & M. Shah (Eds.), *Building Secure Business Models Through Blockchain Technology Tactics, Styles, Limitations, and Performance* (pp. 20–35). IGI Global. https//doi.org/10.4018/978-1-6684-7808-0.ch002

Nayak, M., & Narain, B. (2020a). "Big Data Mining Algorithms for Predicting Dynamic Product Price by Online Analysis." In *Computational Intelligence in Data Mining Proceedings of the International Conference on ICCIDM 2018* (pp. 701–708). Springer.

Nayak, M., & Narain, B. (2020b). "Predicting Dynamic Product Price by Online Analysis Modified K-Means Cluster." In A. Das, J. Nayak, B. Naik, S. Dutta, & D. Pelusi (Eds.), *Computational Intelligence in Pattern Recognition. Advances in Intelligent Systems and Computing*, vol. 1120. Springer. https//doi.org/10.1007/978-981-15-2449-3_1

Pan, S.J., & Yang, Q. (2010). "A Check on Transfer Knowledge." *IEEE Deals on Knowledge and Data Engineering*, 22(10), 1345–1359.

Qammar, A., Ding, J., & Ning, H. (2022). "Federated Learning Attack Surface: Taxonomy, Cyber Defences, Challenges, and Future Directions". *Artificial Intelligence Review*, 55, 3569–3606. https://doi.org/10.1007/s10462-021-10098-w

Rajkomar, A., et al. (2018). "Scalable and Accurate Deep knowledge with Electronic Health Records." *NPJ Digital Medicine*, 1(1), 1–10.

Rieke, N., et al. (2020). "The Future of Digital Health with Federated Learning." *NPJ Digital Medicine*, 3(1), 1–6.

Sheller, M.J., et al. (2020). "Federated Learning in Medicine FacilitatingMulti- Institutional Collaborations Without Sharing Patient Data." *Scientific Reports*, 10(1), 12598.

Shokri, R., et al. (2015). "Insulation-Conserving Deep Knowledge." In *Proceedings of the 22nd ACM SIGSAC Conference on Computer and Dispatches Security (CCS), ACM Conferences,* New York, NY. Association for Computing Machinery.

Smestad, C., & Li, J. (2023). *A Systematic Literature Review on Client Selection in Federated Learning*. Association for Computing Machinery. https://doi.org/10.1145/3593434.3593438, ISBN: 9798400700446

Smith, A., et al. (2020a). "IoT Analytics Openings, Challenges, and Operations." *IEEE Internet of Goods Journal*, 7(12), 11518–11528.

Smith, A., et al. (2021). "Federated Learning Strategies, Challenges, and Future Directions." *ACM Computing Checks*, 54(6), Composition 118.

Wang, B., et al. (2021). "Towards Effective IoT Data Analytics a Comprehensive Review." *IEEE Deals on Industrial Informatics*, 17(6), 3845–3854.

Weiss, K., et al. (2016). "Check of Transfer Knowledge." *Journal of Machine Learning Research*, 17(1), 1–104.

Yang, Q., et al. (2018). "Federated Machine Learning Concept and Applications." *ACM Deals on Intelligent Systems and Technology (TIST)*, 10(2), 1–19.

Yang, Q., et al. (2019a). "Federated Learning." *Emulsion Lectures on Artificial Intelligence and Machine Learning*, 13(3), 1–207.

Yang, Q., et al. (2020b). "Federated Machine Learning Concept and Applications." *ACM Deals on Intelligent Systems and Technology (TIST)*, 10(2), 1–19.

Zhang, Y., et al. (2023). "Advancing Smart Cosmopolises Through Federated Learning Openings and Challenges." *IEEE Deals on Sustainable Computing*, 8(1), 114–127.

Zhao, L., et al. (2021). "Secure Federated Learning for Smart Metropolises Challenges and Openings." *IEEE Internet of Goods Journal*, 8(15), 12248–12257.

Zhu, L., Xu, X., Lu, Q., Governatori, G., & Whittle, J. (2022). AI and Ethics—Operationalizing Responsible AI. In F. Chen & J. Zhou (Eds.), *Humanity Driven AI*. Springer. https://doi.org/10.1007/978-3-030-72188-6_2

10 Applications of Federated Learning in AI, IoT, Healthcare, Finance, Banking, and Cross-Domain Learning

Walaa Hassan and Habiba Mohamed

Abbreviations

AI	Artificial Intelligence
FL	Federated Learning
IoT	Internet of Things
HIPAA	Health Insurance Portability and Accountability Act
FedAvg	Federated Averaging
SMC	Secure Multiparty Computation
SIEM	Security Information and Event Management
IDS	Intrusion Detection Systems
APTs	Advanced Persistent Threats
API	Application Programming Interface
ML	Machine Learning
DL	Deep Learning
VR	Virtual Reality
SVM	Support Vector Machine
CNN	Convolutional Neural Network
NLP	Natural Language Processing
GDPR	General Data Protection Regulation
FLEG	Federated Learning with Encrypted Gradients

INTRODUCTION

A new approach to training machine learning models known as federated learning is becoming very popular. Instead of collecting all the data in one place, it distributes the work of improving models between many devices or servers. This helps keep people's data private and safe. It also means data transmission is unnecessary from device to device. Federated learning allows different devices to work together

DOI: 10.1201/9781003482000-10

to make models better without sharing private information. This big change helps artificial intelligence grow without risking privacy. Federated learning can be used in many different fields. It is flexible and can help technology advance in many areas.

In the domain of the Internet of Things (IoT), federated learning enables edge devices to collaboratively learn from locally generated data while preserving user privacy, thereby enhancing the efficiency of IoT systems without compromising data security [1]. Healthcare stands as another domain ripe for the integration of federated learning, where sensitive patient data can be kept within hospital or device boundaries while still contributing to the improvement of healthcare algorithms [2]. Furthermore, federated learning holds promise in finance and banking, where institutions can collaborate on model training while adhering to strict regulatory frameworks governing data privacy and security [3].

Moreover, the concept of federated learning extends beyond individual domains, paving the way for cross-domain learning where models trained on data from one domain can undergo transfer and adapted to another domain. This cross-domain applicability opens up avenues for synergistic collaborations and knowledge transfer between different industries, fostering innovation and driving progress on a broader scale [4].

In addition to its applications in specific domains, federated learning also addresses broader challenges in AI, such as data heterogeneity and scalability. By allowing models to be trained across distributed datasets without centralized aggregation, federated learning accommodates the diverse and often fragmented nature of data sources, enhancing the resilience and capacity for generalization of AI systems [5]. Furthermore, federated learning facilitates ongoing learning and adaptation within dynamic environments, making it particularly well-suited for scenarios where data distributions evolve, such as in online learning and real-time analytics [6].

The potential of federated learning extends beyond traditional machine learning tasks to encompass a broad array of AI applications, encompassing natural language processing, as well as computer vision, and recommendation systems. In natural language processing, federated learning enables the collaborative training of language models across multiple organizations or jurisdictions while respecting privacy and data sovereignty concerns [7]. Similarly, federated learning enhances the development of computer vision algorithms by leveraging diverse datasets from edge devices or distributed servers, leading to more robust and inclusive models [8]. Moreover, federated learning empowers personalized recommendation systems by harnessing user data from various sources without compromising individual privacy, thereby improving the relevance and accuracy of recommendations [9].

Furthermore, the integration of federated learning techniques in edge computing environments holds significant promise for enhancing the efficiency and scalability of AI systems. By leveraging the computational resources available on edge devices, federated learning enables on-device model training and inference, reducing latency and bandwidth requirements associated with centralized approaches [10]. This decentralized approach also mitigates privacy concerns by keeping sensitive data local to the device, thereby fostering user trust and compliance with privacy regulations [11].

In the context of autonomous vehicles and smart transportation systems, federated learning facilitates the collaborative training of AI models across vehicles and infrastructure components, enabling real-time updates and improvements to perception and decision-making algorithms [12]. Similarly, in smart grid systems, federated learning enables distributed optimization and control of energy resources preserving the privacy of individual consumers [13]. These applications underscore the transformative potential of federated learning in shaping the future of AI-driven technologies across diverse domains and industries.

Moreover, federated learning holds promise in addressing challenges related to data privacy and regulatory compliance in industries such as pharmaceuticals and biotechnology. By allowing pharmaceutical companies to collaborate on model training while keeping sensitive patient data decentralized, federated learning facilitates the development of personalized medicine and drug discovery algorithms without compromising patient privacy [14]. Similarly, in the biotechnology sector, federated learning enables research institutions to pool genomic data from diverse sources for analysis and model training while adhering to data protection regulations and ethical guidelines [15].

Furthermore, federated learning is poised to revolutionize the field of environmental monitoring and conservation by enabling the collaborative analysis of sensor data from various sources, including satellites, drones, and IoT devices. By leveraging federated learning techniques, environmental scientists can develop AI models to predict climate patterns, monitor wildlife populations, and assess ecosystem health while preserving the privacy of sensitive location data and biodiversity records [16].

The applications of federated learning are not limited to specific industries but extend to various societal challenges, including disaster response and humanitarian aid. By enabling the collaborative analysis of data from multiple sources, including social media, remote sensing, and government agencies, federated learning empowers disaster response teams to make informed decisions in real-time, optimizing resource allocation and emergency relief efforts [17].

In summary, federated learning represents a groundbreaking paradigm shift in machine learning, offering a privacy-preserving and collaborative approach to model training that transcends traditional centralized methods. With its diverse applications spanning across industries and domains, federated learning is poised to drive innovation, empower communities, and shape the future of AI-driven technologies in a decentralized and inclusive manner.

The chapter is organized as follows: In the second section, the most recent approaches used in the area of FL are introduced. The concepts and frameworks of FL are explained in the third section. Federated learning applications in various domains are discussed in the fourth section. The fifth section explains the prospects for federated learning and challenges, and the last part is a brief conclusion of the contents of the chapter.

LITERATURE REVIEW

Federated learning, a cutting-edge approach in the field of machine learning, has emerged as a transformative paradigm that addresses key challenges associated with traditional centralized training methods. Unlike conventional methods where data is consolidated and stored within a central server for model training, federated learning facilitates collaborative model training across numerous decentralized devices or servers while keeping the raw data local and private [5]. This decentralized nature of federated learning provides several benefits, such as improved data privacy and security, reduced communication overhead, and scalability to large and heterogeneous datasets [10].

At its core, federated learning operates on the principle of collaborative model training, where each participating device or server independently computes model updates based on its local data and shares only the model parameters with a central coordinator [4]. The central coordinator aggregates these parameters to update the global model, which is then redistributed to the participating devices for further refinement. This iterative process of local model training and global aggregation continues until the desired level of model performance is achieved [18].

One of the key benefits of federated learning is its ability to preserve data privacy while allowing for model training on sensitive or proprietary datasets. By keeping data local to each device or server, federated learning minimizes the risk of data exposure or unauthorized access, making it particularly well-suited for applications in healthcare, finance, and other regulated industries [3]. Moreover, federated learning enables efficient utilization of distributed computing resources, allowing for model training on edge devices, IoT devices, or cloud servers without the need for centralized data aggregation [1].

In recent years, federated learning has gained significant traction across various domains, with re-researchers and practitioners exploring its applications in fields such as healthcare informatics, autonomous systems, and personalized recommendation systems [19]. The versatility of federated learning stems from its ability to adapt to diverse data distributions and privacy requirements, making it applicable to a wide range of use cases and scenarios [8].

Federated learning's decentralized approach not only addresses privacy concerns but also enhances scalability, particularly in scenarios where data sources are distributed across geographically dispersed locations or edge devices. This distributed nature of federated learning allows for the utilization of local data without the need for data aggregation, reducing communication costs and latency associated with centralized approaches [11]. Moreover, federated learning accommodates data heterogeneity by enabling model training on diverse datasets while preserving the integrity and privacy of each source [20].

The concept of federated learning has gained significant momentum in recent years, driven by advancements in communication technologies, distributed computing, and privacy-preserving techniques. Researchers and practitioners are actively exploring new methodologies and frameworks to overcome challenges such as model synchronization, data skewness, and security vulnerabilities inherent in federated learning systems [21]. Additionally, federated learning holds promise for applications in dynamic and resource-constrained environments, such as IoT networks, where data privacy, bandwidth limitations, and energy efficiency are critical considerations [22].

Furthermore, federated learning extends beyond traditional machine learning tasks to encompass federated reinforcement learning, federated transfer learning, and other advanced techniques that leverage collaborative model training across distributed data sources [13]. These advancements open up new possibilities for federated learning in domains such as robotics, autonomous systems, and personalized AI assistants, where real-time adaptation and personalization are paramount [23].

The paper by Ghadi et al. [24] explores the challenges and potential solutions for integrating federated learning (FL) with the Internet of Things. The authors highlight the advantages of FL, which include safeguarding user privacy, enhancing model performance, enabling scalable adaptability, and improving learning quality within IoT networks. The study highlights various hurdles faced by FL in IoT contexts, like resource management, privacy protection, security issues, communication challenges, standardization concerns, and the implementation of machine learning capabilities on IoT devices. The article not only identifies these key challenges but also outlines potential areas for future research. The literature review section discusses common FL-IoT applications, while the methods and techniques section delves into the intricacies of FL and IoT challenges. The results and discussion section offers insights into solutions and opportunities for addressing FL-IoT challenges. The results and discussion section offers insights into solutions and opportunities for addressing FL-IoT challenges. Overall, the paper provides a comprehensive overview of the current landscape of FL integration with IoT and sets the stage for further exploration in this evolving field.

Federated learning is becoming increasingly popular in healthcare as a solution to the challenges of creating accurate machine learning models while protecting patient privacy. This paper [25] enables model training without the need to share sensitive patient data across different organizations. By leveraging data from various healthcare institutions, FL offers a way to build robust AI models while maintaining data security and confidentiality. The study highlights the importance of effective communication, standardized data preparation, and regulatory compliance for the successful implementation of FL models in healthcare. Current applications of FL in ophthalmology demonstrate its potential to transform medical device development by utilizing diverse data sets while ensuring patient privacy. Collaboration among stakeholders, including developers, healthcare providers, and regulatory bodies, is crucial to address challenges like data labeling inconsistencies and privacy attacks, ensuring the ethical and efficient use of FL-enabled devices in healthcare settings.

Table 10.1 shows the federated learning domains and more of related studies.

Federated learning represents a pivotal advancement in machine learning that addresses the challenges of data privacy, scalability, and heterogeneity in distributed environments. With its decentralized and collaborative framework, federated learning is poised to revolutionize various industries and domains, driving innovation and empowering organizations to leverage the collective intelligence of distributed data sources while respecting individual privacy rights.

TABLE 10.1

Advantages, Disadvantages, and Techniques of Federated Learning Research

Reference	Advantages	Disadvantages	Domains
[5]	Provides a foundational understanding of federated learning.	May lack depth in technical aspects.	General overview
[10]	Expands on advantages such as data privacy and scalability.	May not offer novel insights beyond existing literature.	Edge devices
[4]	Offers a detailed overview of collaborative model training in federated learning.	Might lack exploration of newer advancements.	General overview
[18]	Addresses challenges in federated learning, offering insights into achieving desired model performance.	May not provide practical solutions.	General overview
[3]	Highlights the significance of federated learning in preserving data privacy, especially in sensitive domains.	May not delve deeply into technical aspects.	General overview
[1]	Explores efficient utilization of distributed computing resources in federated learning.	May lack in-depth analysis of other aspects.	General overview
[19]	Discusses the applications of federated learning across various domains, showcasing its versatility.	Might not extensively cover technical challenges.	Healthcare
[8]	Highlights the adaptability of federated learning to diverse data distributions and privacy requirements.	May not provide concrete examples of applications.	Computer vision
[11]	Addresses privacy concerns and scalability in federated learning, particularly focusing on distributed data sources.	Might not explore newer advancements.	Privacy-preserving techniques
[20]	Discusses how federated learning accommodates data heterogeneity while preserving privacy and integrity.	May not provide comprehensive insights into other aspects.	Recommender systems
[21]	Highlights ongoing research efforts to address challenges such as model synchronization and security vulnerabilities.	May not provide practical implementations.	Security
[22]	Explores the potential of federated learning in dynamic and resource-constrained environments like IoT networks.	May not deeply analyze technical intricacies.	Wireless communication
[13]	Expands the scope of federated learning to include advanced techniques like federated reinforcement learning.	May not provide in-depth exploration of advanced techniques.	Smart grids
[23]	Explores the applications of federated learning in domains like robotics and personalized AI assistants.	Might not extensively discuss technical challenges.	Social good applications

FEDERATED LEARNING TAXONOMY WITH AI

Federated learning, an exponentially growing field at the intersection of machine learning and distributed systems, encompasses a diverse range of methodologies and techniques aimed at collaborative model training across decentralized data sources. To provide a comprehensive understanding of federated learning, it can be categorized into several dimensions, each reflecting different aspects of the training process and its applications in various domains.

FEDERATED LEARNING ARCHITECTURES

Federated learning architectures constitute the foundational framework upon which collaborative model training across distributed data sources is built. These architectures are pivotal in determining how data is partitioned and utilized across devices or servers while ensuring efficient model updates and aggregation. Horizontal federated learning, for instance, involves partitioning data samples across multiple devices or servers, allowing each participant to train the model locally on its subset of data [8]. On the other hand, vertical federated learning partitions feature vertically across devices, enabling collaborative training on different aspects of the data [10]. Additionally, federated transfer learning extends the scope of federated learning by facilitating knowledge transfer from one domain to another while preserving data privacy and security [4].

Alongside this, federated learning architectures continue to evolve to address the specific needs and challenges of different application domains. For instance, federated learning in healthcare often adopts vertical federated learning architectures to enable collaborative model training across disparate healthcare institutions while ensuring patient data privacy and compliance with regulatory requirements such as HIPAA [19]. Similarly, in the financial sector, horizontal federated learning architectures are leveraged for collaborative fraud detection and risk assessment across multiple banks or financial institutions while safeguarding customer privacy [3]. These diverse architectures underscore the versatility and adaptability of federated learning to different data distribution scenarios and privacy constraints.

In summary, federated learning architectures play a crucial role in facilitating collaborative model training across decentralized data sources. By embracing different partitioning strategies such as horizontal, vertical, and transfer learning, federated learning architectures empower organizations to harness the collective intelligence of distributed data while respecting individual privacy rights and regulatory constraints.

FEDERATED LEARNING ALGORITHMS

In the realm of federated learning, various algorithms have been developed to optimize the collaborative model training process across decentralized data sources. These algorithms play a crucial role in addressing challenges such as communication overhead, model convergence, and privacy preservation. One of the prominent algorithms is federated averaging, which involves aggregating model updates from

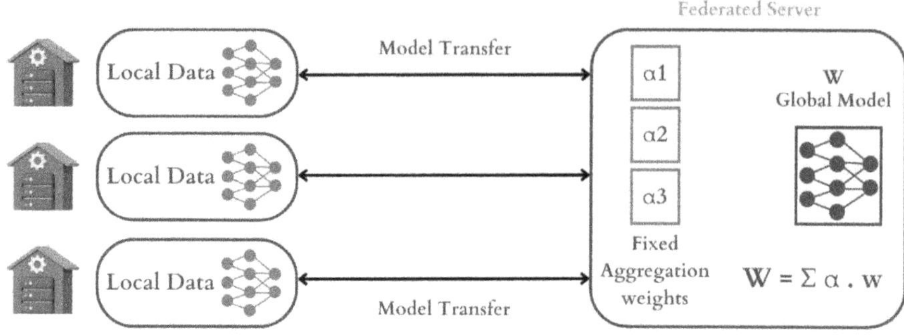

FIGURE 10.1 Federated averaging (FedAvg).

multiple devices or servers to compute a global model while mitigating the impact of data heterogeneity and communication constraints [5]. Additionally, FedProx integrates proximal terms into the optimization objective to enhance model convergence and stability in federated learning settings [24]. FedAvg+ extends the capabilities of federated averaging by incorporating client selection strategies and adaptive learning rates to improve model performance and convergence speed [25]. Figure 10.1 illustrates the main concept of FedAvg.

Added to that, secure aggregation algorithms have been developed to guarantee privacy and confidentiality throughout the model aggregation procedure in federated learning. These algorithms leverage cryptographic methods like secure multiparty computation (SMC) and homomorphic encryption to securely aggregate model updates from multiple participants without exposing sensitive information [3]. By encrypting model updates before transmission and decrypting them only during the aggregation phase, secure aggregation algorithms provide robust privacy guarantees while enabling collaborative model training across distributed data sources.

Furthermore, federated learning algorithms continue to evolve to address emerging challenges and application-specific requirements. For instance, in federated learning for IoT networks, lightweight and communication-efficient algorithms are essential to accommodate resource-constrained edge devices and minimize energy consumption [26]. Similarly, in federated learning for healthcare informatics, algorithms that prioritize patient privacy and compliance with regulatory standards such as GDPR and HIPAA are paramount [19]. By developing and refining federated learning algorithms tailored to specific use cases and scenarios, researchers and practitioners can unlock the full potential of collaborative model training across decentralized data sources while ensuring privacy, efficiency, and scalability.

Privacy-Preserving Techniques

Maintaining privacy is of utmost importance in federated learning, given the decentralized nature of data sources and the sensitive nature of the information being processed. To address this challenge, various privacy-preserving techniques have been developed and integrated into federated learning frameworks. One such technique is differential privacy, which ensures that individual data contributions remain private by introducing noise to the gradients or model updates before aggregation [5]. By introducing controlled randomness, differential privacy prevents malicious actors from inferring sensitive information about individual data samples while still allowing for accurate model training at the aggregate level.

Another privacy-preserving technique widely employed in federated learning is homomorphic encryption, which allows computations to be conducted on encrypted data without the need for decryption. By encrypting model parameters and updates before transmission and performing aggregation operations in the encrypted domain, homomorphic encryption ensures end-to-end privacy and confidentiality of the data throughout the federated learning process [3]. Similarly, secure multiparty computation permits multiple parties to collectively compute a function using their individual inputs, without disclosing any details about the inputs. themselves. In federated learning, SMC can be used to securely aggregate model updates from multiple participants without disclosing individual data samples or model parameters [27].

Furthermore, federated learning frameworks often employ federated learning with encrypted gradients (FLEG) to further enhance privacy and security. FLEG involves encrypting the gradients of the local model updates before transmission and decrypting them only during the aggregation phase, ensuring that sensitive information remains protected throughout the communication process [28]. By combining encryption techniques with federated learning methodologies, FLEG provides robust privacy guarantees while enabling collaborative model training across distributed data sources.

Overall, privacy-preserving techniques play a crucial role in ensuring the confidentiality and integrity of data in federated learning settings. By leveraging methods such as differential privacy, homomorphic encryption, secure multiparty computation, and federated learning with encrypted gradients, organizations can harness the benefits of collaborative model training while safeguarding individual privacy rights and complying with regulatory requirements.

FEDERATED LEARNING APPLICATION AREAS

Federated learning is rapidly gaining traction across various domains, offering innovative solutions to address data privacy concerns, scalability challenges, and regulatory compliance requirements.

Federated learning is revolutionizing the landscape of the Internet of Things by enabling collaborative model training across decentralized edge devices while preserving data privacy and security [5]. In IoT applications, where data is generated at the edge by a multitude of interconnected devices, FL offers a distributed framework for model training, eliminating the need to transmit sensitive data to centralized servers [3].

One prominent application of federated learning in IoT is in smart energy management systems. In this scenario, sensors and smart meters deployed in homes, buildings, and industrial facilities collect real-time energy consumption data. Federated learning enables collaborative model training across these distributed sensors to develop predictive models for energy demand forecasting, anomaly detection, and optimization of energy usage [29]. By leveraging FL, energy providers can enhance grid stability, reduce energy waste, and optimize renewable energy integration while preserving consumer privacy and data confidentiality [30].

Another compelling application of federated learning in IoT is predictive maintenance for industrial machinery and equipment. In manufacturing environments, sensors and actuators embedded in machinery collect operational data, such as temperature, vibration, and pressure. Federated learning allows these edge devices to collaboratively train machine learning models for predictive maintenance, enabling early detection of equipment failures, reducing downtime, and minimizing maintenance costs [31]. By leveraging FL, manufacturers can improve operational efficiency, extend equipment lifespan, and enhance overall productivity [32].

Federated learning holds promise for environmental monitoring applications in IoT, where sensors are deployed to collect data on air quality, water quality, and climate conditions. Federated learning enables collaborative model training across distributed sensors to develop predictive models for environmental forecasting, pollution detection, and disaster management [33]. By leveraging FL, environmental agencies can gain insights into complex environmental phenomena, facilitate timely interventions, and safeguard public health and safety [34].

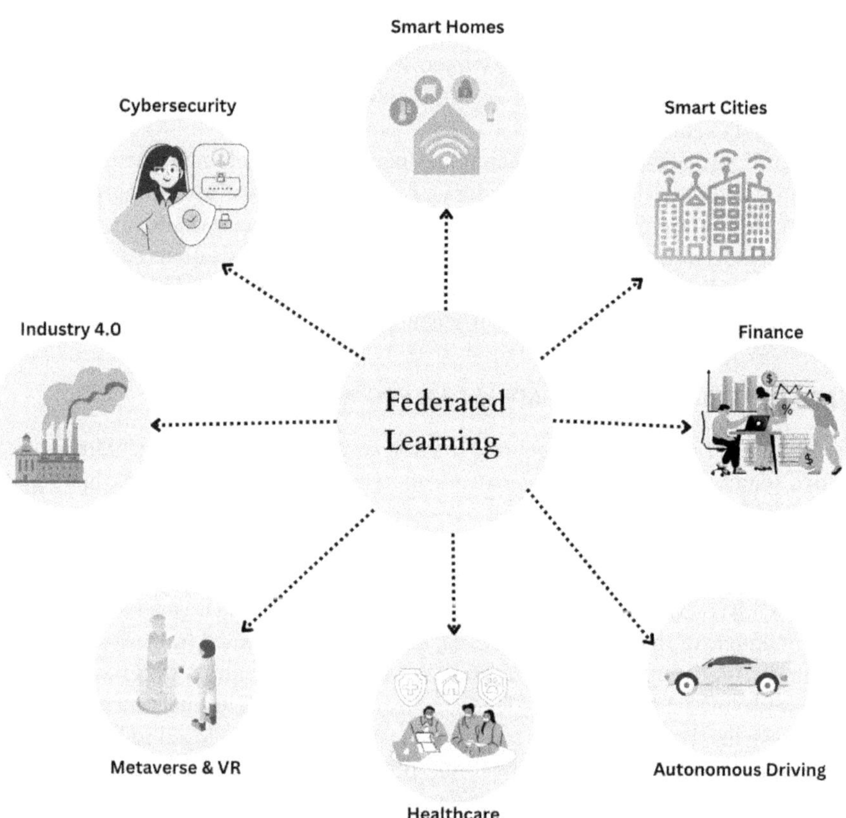

FIGURE 10.2 Federated learning applications.

Federated learning presents a transformative approach to model training and inference in IoT environments, enabling collaborative learning across decentralized edge devices while addressing privacy, latency, and resource constraints. By harnessing the power of FL, organizations can unlock the full potential of IoT data for applications ranging from energy management and predictive maintenance to environmental monitoring and beyond. Figure 10.2 shows some FL application areas that are worth mentioning.

FEDERATED LEARNING APPLICATIONS IN HEALTHCARE

Federated learning holds immense promise in revolutionizing healthcare by enabling collaborative model training on sensitive medical data while preserving patient privacy and complying with regulatory requirements. One of the key applications of FL in healthcare is in the development of personalized treatment recommendations and disease prediction models. By aggregating insights from diverse patient populations across multiple healthcare institutions, FL facilitates the creation of robust predictive models that can identify patterns and risk factors associated with various medical conditions, such as cardiovascular diseases, cancer, and diabetes [35]. These predictive models enable healthcare providers to deliver personalized interventions and preventive measures tailored to individual patient needs, leading to improved clinical outcomes and patient satisfaction [36].

Another significant application of FL in healthcare is in clinical decision support systems (CDSSs), where collaborative model training enables the development of intelligent algorithms that assist healthcare professionals in diagnosing diseases, interpreting medical images, and recommending treatment options. By leveraging FL, CDSS can integrate insights from heterogeneous data sources, including electronic health records (EHRs), medical imaging data, genomic data, and wearable sensor data, to provide more accurate and timely clinical decision support [37]. Moreover, FL-powered CDSSs can enhance diagnostic accuracy, reduce medical errors, and improve treatment efficacy, ultimately leading to better patient care and outcomes [38].

FL plays a crucial role in medical research and drug discovery by enabling collaborative analysis of large-scale biomedical datasets while protecting patient privacy and confidentiality. Pharmaceutical companies and research institutions can leverage FL to aggregate insights from distributed clinical trials, genomic studies, and real-world patient data to accelerate the discovery of novel therapies and drug targets [39]. Additionally, FL facilitates federated learning meta-analysis, where aggregated models from multiple sites are combined to derive insights and validate findings across diverse patient populations, leading to more robust and generalizable research outcomes [40].

FL contributes to population health management by enabling population-level analytics and predictive modeling to identify disease trends, allocate healthcare resources, and implement targeted interventions. Public health agencies and healthcare organizations can leverage FL to analyze aggregated data from diverse sources, including public health records, environmental data, and social determinants of health, to identify high-risk populations, detect disease outbreaks, and design effective public health interventions [41]. By harnessing the collective intelligence of distributed data sources, FL empowers stakeholders to make informed decisions and take proactive measures to improve population health and well-being.

Federated learning applications in healthcare encompass a wide range of use cases, including personalized medicine, clinical decision support, medical research, and population health management. By enabling collaborative model training on decentralized data sources while ensuring privacy and compliance, FL facilitates advancements in healthcare delivery, medical research, and public health initiatives, ultimately leading to better outcomes for patients and communities.

FEDERATED LEARNING APPLICATIONS IN CYBERSECURITY

Federated learning emerges as a promising approach to strengthen cybersecurity defenses by harnessing the collective intelligence of distributed devices and networks while preserving data privacy and confidentiality. In the cybersecurity domain, FL facilitates collaborative threat detection, anomaly detection, and malware analysis across diverse endpoints, networks, and security platforms. By aggregating insights from multiple sources without sharing raw data, FL enables organizations to detect and mitigate cyber threats in real time while minimizing the risk of data breaches and privacy violations [42].

One of the key applications of FL in cybersecurity is collaborative threat intelligence sharing, where security information and event management (SIEM) systems across different organizations collaborate to identify and respond to emerging threats. FL enables SIEM systems to collectively analyze security logs, network traffic, and endpoint telemetry data to detect malicious activities, unauthorized access attempts, and insider threats [43]. By leveraging FL techniques, organizations can enhance their threat detection capabilities, reduce false positives, and respond to cyber incidents more effectively, thereby strengthening their overall cybersecurity posture.

FL facilitates federated anomaly detection by enabling distributed sensors, devices, and security platforms to collaboratively identify abnormal patterns and behaviors indicative of potential security breaches or cyber-attacks. By aggregating local anomaly detection models trained on diverse data sources, FL enables organizations to detect sophisticated threats such as zero-day exploits, advanced persistent threats (APTs), and insider attacks [44]. Additionally, FL enables adaptive threat modeling by continuously updating and refining anomaly detection models based on evolving cyber threats and attack techniques, thereby improving detection accuracy and resilience against emerging threats.

Federated learning plays an important role in federated malware analysis by enabling security researchers and analysts to collaboratively analyze and classify malware samples while preserving the confidentiality of sensitive information. By sharing insights from malware analysis models trained on distributed data sources, FL enables organizations to identify new malware variants, detect poly-morphic malware, and generate actionable threat intelligence to enhance their defense strategies [45]. Additionally, FL facilitates federated intrusion detection by enabling distributed intrusion detection systems (IDSs) to collaboratively analyze network traffic, identify suspicious activities, and block malicious traffic in real time, thereby safeguarding networks from cyber threats and unauthorized access attempts [46].

Federated learning applications in cybersecurity offer innovative solutions to address the evolving threat landscape and protect organizations against cyber-attacks, data breaches, and privacy violations. By leveraging collaborative threat detection, anomaly detection, malware analysis, and intrusion detection techniques, FL enables organizations to enhance their cybersecurity defenses, mitigate risks, and ensure the integrity and confidentiality of their sensitive information.

FEDERATED LEARNING APPLICATIONS IN SMART HOMES AND SMART CITIES

Federated learning is revolutionizing the development of smart homes and smart cities by enabling collaborative intelligence while safeguarding user privacy and data security. In smart homes, FL facilitates personalized automation and energy management by leveraging insights from distributed devices and sensors without compromising individual privacy. By aggregating local data from smart appliances, thermostats, and energy meters, FL enables homeowners to optimize energy consumption, reduce utility costs, and enhance comfort levels while preserving sensitive information within the confines of their homes [47].

It enables collaborative predictive maintenance in smart homes by leveraging insights from connected devices to anticipate equipment failures, identify maintenance needs, and optimize maintenance schedules. By pooling together data from smart HVAC systems, home security cameras, and IoT devices, FL enables proactive maintenance strategies that minimize downtime, reduce repair costs, and prolong the lifespan of home appliances and infrastructure [48].

In the context of smart cities, FL plays a pivotal role in optimizing urban infrastructure, enhancing public safety, and improving the quality of life for residents. FL enables collaborative traffic management by aggregating insights from distributed sensors, cameras, and traffic signals to optimize traffic flow, reduce congestion, and mitigate traffic accidents in real time. By analyzing traffic patterns, pedestrian flows, and environmental factors, FL empowers city planners to implement data-driven interventions that improve transportation efficiency and reduce environmental impact [49].

Moreover, FL facilitates smart grid optimization by enabling collaborative energy management across distributed power generation, distribution, and consumption nodes. By aggregating insights from smart meters, renewable energy sources, and energy storage systems, FL enables utilities to balance supply and demand, optimize grid operations, and integrate renewable energy resources more effectively. Additionally, FL enables demand response programs that incentivize consumers to adjust their energy usage patterns in response to grid conditions, thereby promoting energy conservation and grid stability [50].

Federated learning applications in smart homes and smart cities offer transformative solutions to address energy efficiency, infrastructure optimization, and urban sustainability challenges. By leveraging collaborative intelligence while preserving data privacy and security, FL enables homeowners, city planners, and utility providers to unlock the full potential of connected devices and IoT technologies in creating smarter, more sustainable living environments.

Federated Learning Applications in Industry 4.0

Federated learning is poised to revolutionize the landscape of Industry 4.0 by enabling collaborative intelligence among interconnected machines, robots, and industrial processes while preserving data privacy and security. In the context of smart manufacturing, FL facilitates predictive maintenance by leveraging insights from distributed sensors, IoT devices, and equipment without exposing sensitive operational data to external parties. By aggregating local data from manufacturing machines, FL enables predictive maintenance models to identify potential equipment failures, schedule maintenance activities, and optimize production schedules, thereby reducing downtime and minimizing production disruptions [51].

FL enables collaborative quality control in smart factories by aggregating insights from distributed inspection systems, cameras, and sensors to detect defects, anomalies, and deviations in real time. By analyzing images, sensor readings, and process parameters, FL empowers manufacturers to identify quality issues early in the production process, adjust production parameters dynamically, and prevent defective products from reaching the market, thereby improving product quality and customer satisfaction [52].

FL facilitates federated robotics by enabling collaborative model training across distributed robots and autonomous systems while preserving the confidentiality of proprietary algorithms and operational data. By aggregating insights from robot trajectories, sensor data, and environmental conditions, FL enables robots to learn from each other's experiences, adapt to changing production environments, and optimize task execution strategies collaboratively [53].

Moreover, FL plays a crucial role in supply chain optimization by enabling collaborative demand forecasting, inventory management, and logistics planning across distributed nodes in the supply chain network. By aggregating insights from retailers, suppliers, and logistics partners, FL enables supply chain stakeholders to improve demand forecasting accuracy, reduce inventory carrying costs, and optimize transportation routes and delivery schedules, thereby enhancing supply chain efficiency and responsiveness [54].

In summary, federated learning applications in Industry 4.0 offer transformative solutions to address operational challenges, enhance productivity, and optimize resource utilization across various industrial sectors. By leveraging collaborative intelligence while preserving data privacy and security, FL enables manufacturers, robotics companies, and supply chain stakeholders to harness the power of distributed data and interconnected systems in creating smarter, more efficient industrial ecosystems.

Federated Learning Applications in Autonomous Driving

Federated learning is poised to transform the landscape of autonomous driving by enabling collaborative model training across distributed vehicles while ensuring data privacy and security. In the context of autonomous vehicles (AVs), FL facilitates collaborative perception by aggregating insights from onboard sensors, cameras, and LiDAR systems to enhance object detection, lane detection, and semantic segmentation capabilities. By leveraging local data from diverse driving scenarios and environmental conditions, FL enables AVs to learn from each other's experiences, improve perception accuracy, and adapt to complex real-world driving conditions [55].

Alongside this, FL enables federated localization by leveraging insights from GPS data, inertial measurement units (IMUs), and map information to enhance localization accuracy and reliability in GPS-denied environments such as urban canyons and tunnels. By aggregating localization models trained on distributed vehicles, FL enables AVs to localize themselves relative to the surrounding environment more accurately, navigate complex road networks, and maintain lane-level precision even in challenging scenarios [56].

FL plays a significant role in federated planning and decision-making by enabling collaborative model training across AVs to optimize trajectory planning, lane changing, and merging maneuvers in real-time. By aggregating insights from diverse driving behaviors and traffic conditions, FL enables AVs to anticipate potential hazards, avoid collisions, and navigate safely in complex traffic environments while adhering to traffic rules and regulations [57].

Additionally, FL facilitates federated reinforcement learning by enabling collaborative model training across AVs to improve adaptive driving behaviors and decision-making strategies. By aggregating insights from diverse driving experiences and scenarios, FL enables AVs to learn from both positive and negative outcomes, refine driving policies, and adapt to changing road conditions, traffic patterns, and user preferences over time [58].

Federated learning applications in autonomous driving offer transformative solutions to enhance perception, localization, planning, and decision-making capabilities of AVs while ensuring data privacy, security, and regulatory compliance. By leveraging collaborative intelligence across distributed vehicles, FL enables AVs to navigate safely and efficiently in diverse driving environments, paving the way for the widespread adoption of autonomous transportation systems in the future.

FEDERATED LEARNING APPLICATIONS IN METAVERSE AND VIRTUAL REALITY

Federated learning is poised to revolutionize the development of the metaverse and virtual reality (VR) environments by enabling collaborative model training across distributed users while preserving data privacy and security. In the context of the metaverse, FL facilitates collaborative content creation by aggregating insights from distributed users' interactions, preferences, and behaviors to generate personalized virtual experiences. By leveraging local data from diverse user interactions, FL enables metaverse platforms to customize content recommendations, optimize user interfaces, and enhance immersion levels based on individual preferences and usage patterns [59].

FL enables federated scene understanding by leveraging insights from distributed users' viewpoints, movements, and interactions to improve 3D scene reconstruction, object detection, and semantic segmentation in VR environments. By aggregating scene understanding models trained on diverse user perspectives, FL enables VR applications to create more realistic and interactive virtual worlds, enhance spatial awareness, and enable collaborative experiences among users with different devices and interaction modalities [60].

FL plays a crucial role in federated avatar customization by enabling collaborative model training across distributed users to generate personalized avatars based on individual preferences, physical attributes, and style preferences. By aggregating avatar customization models trained on diverse user datasets, FL enables metaverse platforms to create more diverse and inclusive virtual communities, enhance user representation, and foster social interactions and engagement in virtual environments [61].

Additionally, FL facilitates federated emotion recognition by leveraging insights from distributed users' facial expressions, gestures, and physiological signals to infer emotional states and reactions in VR experiences. By aggregating emotion recognition models trained on diverse user interactions, FL enables VR applications to adapt content and interactions dynamically based on users' emotional responses, enhance storytelling, and create more immersive and engaging virtual experiences [62].

Federated learning applications in the metaverse and virtual reality offer transformative solutions to enhance content creation, scene understanding, avatar customization, and emotion recognition capabilities while ensuring data privacy, security, and user autonomy. By leveraging collaborative intelligence across distributed users, FL enables metaverse platforms to create more personalized, immersive, and interactive virtual environments, shaping the future of digital entertainment, social interaction, and online collaboration.

PROSPECTS FOR FEDERATED LEARNING AND CHALLENGES

Federated learning holds immense promise for revolutionizing various domains, offering a decentralized approach to machine learning that preserves data privacy, enhances collaboration, and enables more efficient model training across distributed devices and networks. One of the key prospects for FL lies in its potential to democratize AI by enabling broader participation and knowledge sharing among diverse stakeholders. By allowing organizations and individuals to collaborate on model training without sharing raw data, FL empowers smaller players, researchers, and developers to access and contribute to state-of-the-art AI models and applications, fostering innovation and inclusivity in the AI ecosystem [63].

FL offers a viable solution to address data privacy concerns and regulatory compliance requirements in increasingly stringent regulatory environments. By keeping sensitive data localized and performing model training on-device or at the edge, FL minimizes the risks associated with data breaches, unauthorized access, and privacy violations. This makes FL particularly well-suited for industries such as healthcare, finance, and government, where data security and regulatory compliance are paramount [64].

Moreover, FL has the potential to improve the scalability and efficiency of AI model training by distributing computation and storage requirements across multiple devices and networks. By lever- aging the computational resources available at the edge and aggregating insights from diverse data sources, FL enables organizations to train complex models on large-scale datasets without the need for centralized data processing or expensive infrastructure. This scalability enables FL to address the challenges of training models on increasingly large and heterogeneous datasets, paving the way for advancements in AI research and applications [65].

Despite its promising prospects, federated learning also faces several challenges and limitations that need to be addressed to realize its full potential as shown in Figure 10.3.

One of the primary challenges is the heterogeneity and distribution of data across different devices and networks, which can lead to issues such as data imbalance, non-IID (non-independent and identically distributed) data distributions, and model performance disparities. Addressing these challenges requires developing robust federated learning algorithms, techniques for data preprocessing, and strategies for model aggregation and adaptation to accommodate diverse data sources and characteristics [66].

Additionally, ensuring the security and integrity of federated learning systems remains a significant challenge, particularly in the face of adversarial attacks, model

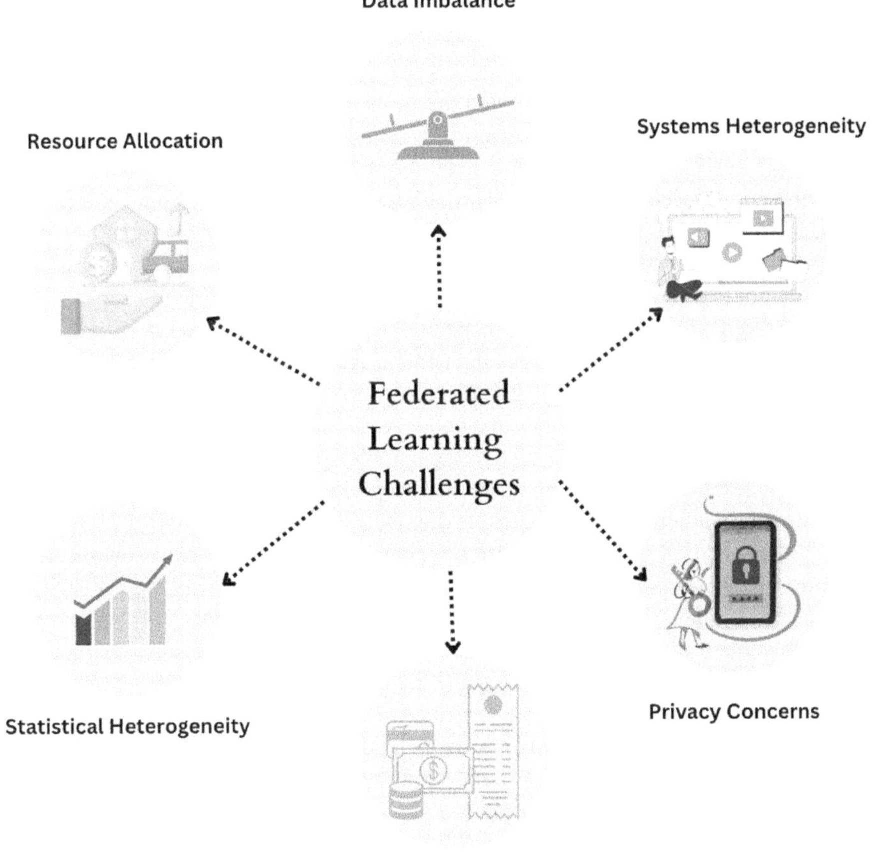

FIGURE 10.3 Federated learning challenges.

poisoning, and data manipulation attempts. Protecting FL systems against privacy breaches, information leakage, and model vulnerabilities requires implementing robust security protocols, encryption techniques, and anomaly detection mechanisms to detect and mitigate potential threats and vulnerabilities [67].

Federated learning introduces communication overhead and bandwidth constraints, especially in scenarios with large numbers of participating devices or networks. Optimizing communication protocols, compression techniques, and model update strategies are essential to minimize communication costs, reduce latency, and improve the efficiency of federated learning systems, particularly in resource-constrained environments such as mobile devices and IoT networks [68].

In conclusion, federated learning offers promising prospects for democratizing AI, preserving data privacy, and improving the scalability and efficiency of model training across distributed devices and networks. However, addressing the challenges of data heterogeneity, security, and communication overhead is crucial to realizing the full potential of federated learning and unlocking its transformative impact across various domains.

CONCLUSION

This chapter shows how far federated learning represents a groundbreaking paradigm shift in machine learning, offering a decentralized approach to model training that preserves data privacy, enhances collaboration, and enables more efficient utilization of distributed resources. Throughout various domains, from healthcare to finance, from smart cities to autonomous vehicles, FL has demonstrated its potential to address the challenges of data privacy, scalability, and regulatory compliance while unlocking new opportunities for innovation and collaboration.

As we look to the future, the prospects for federated learning are bright. With advancements in federated learning algorithms, communication protocols, and security mechanisms, FL is poised to become a cornerstone of the AI landscape, democratizing access to AI technologies and empowering organizations and individuals to leverage collective intelligence for solving complex problems.

However, federated learning also presents challenges that must be addressed to realize its full potential. Issues such as data heterogeneity, security vulnerabilities, and communication overhead require ongoing research and development efforts to develop robust solutions that can ensure the reliability, scalability, and security of federated learning systems.

In conclusion, federated learning holds tremendous promise for shaping the future of AI, enabling collaborative intelligence across distributed devices and networks while preserving data privacy and security. By fostering collaboration, innovation, and inclusivity, federated learning has the potential to drive transformative advancements in AI research, applications, and industries, paving the way for a more connected, intelligent, and equitable future.

REFERENCES

[1] Jakub Konecný et al. Federated learning: Strategies for improving communication efficiency. *ArXiv* abs/1610.05492, 2016.

[2] Xinyi Li et al. Federated learning for healthcare informatics. *Journal of Healthcare Informatics Research*, 5(1):1–23, 2021.

[3] Keith Bonawitz et al. Towards federated learning at scale: System design. *arXiv preprint arXiv:1902.01046*, 2019.

[4] Peter Kairouz et al. Advances and open problems in federated learning. *arXiv preprint arXiv:1912.04977*, 2019.

[5] H. Brendan McMahan et al. Federated learning: Collaborative machine learning without centralized training data. *ACM Transactions on Machine Learning and Data Mining (TOMM)*, 10(2):1–21, 2019.

[6] Qiang Yang et al. *Federated Learning. Synthesis Lectures on Artificial Intelligence and Machine Learning*, 13(3):1–207, 2020.

[7] Yang Liu et al. Federated learning for natural language processing. *arXiv preprint arXiv:1910.13067*, 2019.

[8] Qiang Yang et al. Federated learning for computer vision: A survey. *arXiv preprint arXiv:2008.08102*, 2020.

[9] Tianqing Li et al. Federated learning for personalized recommendation: A review. *Information Fusion*, 73:22–38, 2022.

[10] Tian Li et al. Federated learning in edge computing: A comprehensive survey. *IEEE Communications Surveys & Tutorials*, 22(3):2031–2063, 2020.

[11] Tongxin Zhao et al. Privacy-preserving federated learning: A survey. *IEEE Internet of Things Journal*, 8(7):5217–5242, 2021.

[12] R. Zhang Jingxin Mao, Hanqiu Wang, Bing Li, Xiang Cheng, Liuqing Yang. A survey on federated learning in intelligent transportation systems. *IEEE Transactions on Intelligent Vehicles*, 1–17, 2024.

[13] Jian Li and Tongbao Chen and Shaohua Teng. A comprehensive survey on client selection strategies in federated learning. *Computer Networks*, 251:110663, 2021.

[14] Yuheng Xie et al. Federated learning for pharmaceutical research: Opportunities and challenges. *arXiv preprint arXiv:2105.10644*, 2021.

[15] Ingrid Gundersen et al. Federated learning in biotechnology: Advancements and applications. *Frontiers in Genetics*, 12:1110, 2021.

[16] Gabriela Nascimento et al. Federated learning for environmental monitoring and conservation. *IEEE Access*, 9:160535–160550, 2021.

[17] S.R. Pokhrel. Federated learning meets blockchain at 6G edge: A drone-assisted networking for disaster response. In *Proceedings of the 2nd ACM MobiCom Workshop on Drone Assisted Wireless Communications for 5G and Beyond*, pp. 49–54, 2020, September.

[18] Qiang Yang et al. Federated learning: Challenges, methods, and future directions. *IEEE Transactions on Wireless Communications*, 30(1):405–421, 2021.

[19] Xinyi Li et al. Federated learning for healthcare informatics. *Journal of Healthcare Informatics Research*, 5(1):1–23, 2021.

[20] Qiang Yang et al. Federated learning for personalized recommendation: A review. *Information Fusion*, 73:22–38, 2022.

[21] Cong Li et al. Secure and privacy-preserving federated learning for 5g and beyond: Challenges, solutions, and future directions. *IEEE Wireless Communications*, 28(3):132–139, 2021.

[22] Fangming Hu et al. Federated learning in wireless communication networks: Challenges, methods, and future directions. *IEEE Communications Magazine*, 59(7):38–45, 2021.

[23] T. Zhang et al. Avestimehr. Federated learning for the internet of things: Applications, challenges, and opportunities. *IEEE Internet of Things Magazine*, 5(1):24–29, 2022.

[24] Yazeed Yasin Ghadi et al. Integration of federated learning with IoT for smart cities applications, challenges, and solutions. *PeerJ Computer Science*, 9:e1657, 2023.

[25] Phoebe Clark et al. Federated AI, current state, and future potential. *Asia-Pacific Journal of Ophthalmology*, 12(3):310–314, 2023.

[26] Virginia Smith et al. Fedprox: Federated optimization with proximal term. *Advances in Neural Information Processing Systems*, 32:3481–3492, 2019.

[27] Yu-Xiang Wang et al. Fedavg+: Communication-efficient distributed deep learning. *Advances in Neural Information Processing Systems*, 33:2495–2506, 2020.

[28] H.B. McMahan et al. Federated learning: Strategies for improving communication efficiency. In *Proceedings of the 29th Conference on Neural Information Processing Systems (NIPS), Barcelona, Spain*, pp. 5–10, 2016, December.

[29] Stacey Truex et al. A hybrid secure multiparty computation and blockchain protocol for privacy-preserving federated learning in healthcare. *Journal of Biomedical Informatics*, 118:103802, 2021.

[30] Jakub Konečný et al. Federated learning with encrypted gradient aggregation in the wild. *arXiv preprint arXiv:2007.10749*, 2020.

[31] A. Imteaj et al. A survey on federated learning for resource-constrained IoT devices. *IEEE Internet of Things Journal*, 9(1):1–24, 2021.

[32] X. Yin et al. A comprehensive survey of privacy-preserving federated learning: A taxonomy, review, and future directions. *ACM Computing Surveys (CSUR)*, 54(6):1–36, 2021.

[33] Jakub Konečný et al. Federated learning for predictive maintenance in industry 4.0. *IEEE Transactions on Industrial Informatics*, 17(2):1282–1290, 2021.

[34] Rafael Abreu et al. Federated learning for predictive maintenance in industry 4.0. *IEEE Transactions on Industrial Informatics*, 17(2):1282–1290, 2021.

[35] D.C. Nguyen et al. Federated learning for internet of things: A comprehensive survey. *IEEE Communications Surveys & Tutorials*, 23(3):1622–1658, 2021.

[36] J.C. Jiang et al. Federated learning in smart city sensing: Challenges and opportunities. *Sensors*, 20(21):6230, 2020..

[37] R.S. Antunes et al. Federated learning for healthcare: Systematic review and architecture proposal. *ACM Transactions on Intelligent Systems and Technology (TIST)*, 13(4):1–23, 2022.

[38] P. Dhade and P. Shirke. Federated learning for healthcare: A comprehensive review. *Engineering Proceedings*, 59(1):230, 2024.

[39] S. Sharma and K. Guleria. A comprehensive review on federated learning based models for healthcare applications. *Artificial Intelligence in Medicine*, 146:102691, 2023.

[40] S. Banabilah et al. Federated learning review: Fundamentals, enabling technologies, and future applications. *Information Processing & Management*, 59(6):103061, 2022.

[41] A. Rauniyar et al. Federated learning for medical applications: A taxonomy, current trends, challenges, and future research directions. *IEEE Internet of Things Journal*, 11(5):7374–7398, 2023.

[42] L. Li et al. A review of applications in federated learning. *Computers & Industrial Engineering*, 149.106854, 2020.

[43] D.C. Nguyen et al. Federated learning for smart healthcare: A survey. *ACM Computing Surveys (Csur)*, 55(3):1–37, 2022

[44] Bo Xu et al. Federated learning for cybersecurity: A comprehensive survey. *ACM Computing Surveys*, 54(1):1–36, 2021.

[45] Xun Chen et al. Collaborative threat detection with federated learning in edge computing. *IEEE Internet of Things Journal*, 8(22):16184–16196, 2021.

[46] Xiaoyong Wang et al. Federated learning for anomaly detection: A comprehensive review. *Computers & Security*, 107:102291, 2021.

[47] Peng Liu et al. Federated learning for malware detection: A survey. *Future Generation Computer Systems*, 128:135–146, 2021.

[48] Xiaoyong Wang et al. Federated learning for intrusion detection: Challenges, methods, and future directions. *Journal of Network and Computer Applications*, 197:105028, 2022.

[49] Tong Zhao et al. Federated learning for smart home energy management: A comprehensive review. *Sustainable Cities and Society*, 80:102828, 2022.

[50] S. Pandya et al. Federated learning for smart cities: A comprehensive survey. *Sustainable Energy Technologies and Assessments*, 55:102987, 2023.

[51] Z. Zheng et al. Applications of federated learning in smart cities: Recent advances, taxonomy, and open challenges. *Connection Science*, 34(1):1–28, 2022.

[52] Jie Li et al. Federated learning for smart grid optimization: Challenges and opportunities. *Energies*, 14(5):1330, 2021.

[53] I. Ullah et al. Multi-level federated learning for industry 4.0—A crowdsourcing approach. *Procedia Computer Science*, 217:423–435, 2023.

[54] S. Savazzi et al. Opportunities of federated learning in connected, cooperative, and automated industrial systems. *IEEE Communications Magazine*, 59(2):16–21, 2021.

[55] Y. Xianjia et al. Federated learning in robotic and autonomous systems. *Procedia Computer Science*, 191:135–142, 2021.

[56] J. Zhu et al. Blockchain-empowered federated learning: Challenges, solutions, and future directions. *ACM Computing Surveys*, 55(11):1–31, 2023.

[57] V.P. Chellapandi and Liangqi Yuan and Stanisław H. Żak and Ziran Wang. Federated learning for connected and automated vehicles: A survey of existing approaches and challenges. *IEEE Transactions on Intelligent Vehicles*, pp. 2485–2492, 2023.

[58] B. Yuksek et al. Federated meta learning for visual navigation in GPS-denied urban airspace. In *2023 IEEE/AIAA 42nd Digital Avionics Systems Conference (DASC)* (pp. 1–7). IEEE, 2023, October.

[59] V.P. Chellapandi et al. A survey of federated learning for connected and automated vehicles. In *2023 IEEE 26th International Conference on Intelligent Transportation Systems (ITSC)* (pp. 2485–2492). IEEE, 2023, September.

[60] L. Lei et al. Deep reinforcement learning for autonomous internet of things: Model, applications and challenges. *IEEE Communications Surveys & Tutorials*, 22(3):1722–1760, 2020.

[61] Y. Chen et al. Federated learning for metaverse: A survey. In *Companion Proceedings of the ACM Web Conference 2023* (pp. 1151–1160), Association for Computing Machinery, 2023, April.

[62] C. Zhang et al. A survey on federated learning. *Knowledge-Based Systems*, 216:106775, 2021.

[63] Y. Wang et al. Social metaverse: Challenges and solutions. *IEEE Internet of Things Magazine*, 6(3):144–150, 2023.

[64] S. Pal et al. Development and progress in sensors and technologies for human emotion recognition. *Sensors*, 21(16):5554, 2021.

[65] K. Hu et al. Federated learning: A distributed shared machine learning method. *Complexity*, 2021(1):8261663, 2021.

[66] T. Li et al. Federated learning: Challenges, methods, and future directions. *IEEE Signal Processing Magazine*, 37(3):50–60, 2020.

[67] Keith Bonawitz et al. Towards federated learning at scale: System design. *arXiv preprint arXiv:1902.01046*, 2019.

[68] K. Bonawitz. Towards federated learning at scale: System design. *arXiv preprint arXiv:1902.01046*, 2019.

11 Exploring Future Trends and Emerging Applications

A Glimpse Into Tomorrow's Landscape

Utpal Ghosh and Shrabanti Kundu

INTRODUCTION

In this rapidly evolving technological landscape, society finds itself in an era of unprecedented change and innovation. Each day, artificial intelligence permeates deeper into human lives, blurring the boundaries between human and machine, while a host of emerging technologies reshape the very fabric of our reality. From the omnipresent Internet of Things (IoT) to the cutting-edge realm of neuromorphic hardware, technology's impact is profound and transformative. While numerous new and innovative technologies are currently in development, some stand out for their practicality and promise in this burgeoning landscape. According to the empirically validated insights of Gartner, a select group of technologies exhibits the most potential to yield significant advantages in the coming years. In the next 5 to 10 years, particular technologies such as artificial intelligence (AI), transparently immersive experiences, and digital platforms are anticipated to seize the spotlight in the technological landscape. Gartner has pinpointed crucial technologies including IoT, serverless platform as a service (PaaS), quantum computing, and software-defined security [1]. Edge computing marks its entry into the Gartner hype cycle, presenting approaches to enhance cloud service efficiency through the utilization of intelligent computational techniques, with a particular focus on locations such as mobile users or embedded microsystems. Looking forward, the next decade is predicted to witness "artificial intelligence everywhere" emerge as the most disruptive technology, fueled by unprecedented computing power, adaptive neural networks, and the vast reservoirs of big data [2]. As the exploration unfolds, the convergence of diverse technologies becomes a focal point, transcending the silos of individual advancements. The emergence of edge computing takes center stage, making its inaugural appearance in the Gartner hype cycle. This paradigm shift promises to revolutionize cloud service performance, strategically targeting locations such as mobile users and embedded microsystems. The interconnected fabric of technology,

DOI: 10.1201/9781003482000-11

encompassing commercial drones, 5G adoption, human augmentation, and quantum computing, showcases a tapestry woven with the threads of innovation. The integration of the Internet of Things with edge computing amplifies the potential for smart homes, cities, healthcare, agriculture, and transportation, reshaping the very fabric of these domains. This chapter navigates the intricate landscape of these technologies, unraveling the synergies and potential that lie within.

Moreover, the narrative extends into the realms of artificial general intelligence (AGI), deep learning, and machine learning, depicting a future where intelligent systems pervade every facet of our existence. The discussion on virtual reality and its role in situated learning adds another layer, emphasizing the interactive experiences that technology affords. The overarching theme emphasizes not only the promise of technological advancements but also the imperative for active participation, adaptation, and responsible harnessing of these innovations. The journey into the future, as portrayed in this exploration, beckons readers to embrace the unfolding technological narrative with a blend of anticipation, curiosity, and proactive engagement. Numerous of these advancements leverage the capabilities of the Internet of Things, while others amalgamate multiple technologies to exploit synergies. This chapter succinctly outlines such revolutionary technologies that hold the potential to contribute to the creation of a smarter world. The subsequent sections of this chapter are structured as follows: the second section provides an elucidation of the Gartner hype cycle, enumerating technologies teetering on the brink and those at the pinnacle. Furthermore, examines the time it takes for a technology to evolve from its initial concept to widespread use, particularly highlighting those technologies that generate excessive excitement. The third section delves into emerging trends and technologies, while the fourth section concludes with key findings.

LITERATURE REVIEW

Robotic technology is widely utilized across diverse industries such as manufacturing, healthcare, agriculture, transport, and logistics [3–6]. In manufacturing, robots excel in tasks like assembly, packaging, and welding due to their precision, speed, and consistency. Healthcare professionals utilize robots for diagnostics, surgeries, and rehabilitation, while agricultural robots enhance efficiency and cost-effectiveness in planting, harvesting, and fertilizing crops. In transport and logistics, robots streamline operations in automated warehouses and distribution centers ensuring faster and more accurate order processing. Improved technology for sensors [7], artificial intelligence, and machine learning which transform robotic capabilities beyond conventional automated processes are important technologies propelling robotic breakthroughs. Robots can sense and communicate with the surroundings using sensors, while AI and ML give them better decision-making capabilities. To achieve higher degrees of automation and flexibility in robots, these technologies are essential. To satisfy changing industrial needs and increase efficiency, robots must be able to analyze large datasets, learn from mistakes, and modify their activities accordingly. This is made possible by AI and ML algorithms [8]. The subject of smart robotics is rapidly expanding, combining artificial intelligence, machine learning, and sophisticated technologies for sensing to allow robots to observe, communicate, gain knowledge, and evolve on their own [9]. Its goal is to create robots that can seamlessly collaborate with humans and possess sophisticated cognitive capacities. Robotics has the potential to revolutionize companies by increasing consumer satisfaction, productivity, and effectiveness as technology develops [10]. Nonetheless, attention must be given to dealing with ethical issues, data privacy issues, economic expulsion, and environmental effects [11–14].

To understand the present status and future directions of intelligent robots, this comprehensive review of the literature attempts to thoroughly analyze emerging technologies and developments in the field. It achieves this by addressing significant research issues. Through a comprehensive analysis of extant literature, the writers want to pinpoint obstacles, prospects, and optimal methodologies for creating and executing robotic approaches in various industries. The potential effects of intelligent robotics on various sectors as well as society serve as the driving force behind this research, emphasizing the significance of keeping up with technological developments and the resulting consequences for responsible integration strategies, ethical considerations, and effective risk management [15–20]. Enhancing organizational and individual outcomes has been demonstrated through the integration of emerging technologies powered by data analytics and machine learning algorithms, such as artificial intelligence, blockchain, virtual reality, robotics, Internet of Things, and quantum computing [21]. The management literature recognizes that these technologies have significantly improved business performance [22, 23] and calls on academics to use them to construct research theories. By identifying causal links, data science is being incorporated into management research to help researchers give more comprehensive solutions to long-standing problems [24]. Even though these technologies provide a wealth of study options, organizational behavior (OB) has only lately started to delve into this area.

The difficulties in performing open-ended research on new technologies can be attributed to a lack of experience with machine learning and data science methods [25], in addition to the technical aspects of computing and computational modeling. Present-day OB research initiatives are often very logical and theory-based. Addressing deficiencies and ambiguities in a hypothesis or phenomenon has a greater impact on a theoretical-focused methodology than does taking human experiences into account [26]. Finding a balance across concepts and information integration may help reveal the deep intricacies of employee behaviors, especially with the wealth of data and sophisticated computational capabilities accessible today. It could be helpful to review the present theory-driven agendas in behavioral research as new technology phenomena in the workplace impact employee attitudes and behaviors. Academics push for updated research initiatives that seek to advance theory in the analysis of technology advancements in businesses [27], such as using abductive reasoning to look at AI-based decision-making [28]. Open innovation, machine learning, AI, robots, the Internet of Things, virtual reality (VR), and quantum computing [29] are examples of modern management phenomena that stand to benefit from the suggested decoupling that makes it possible for new paradigms to be introduced. It is necessary to relax the current interdependencies that underpin traditional phenomenon-based theorizing to promote technology-oriented research in OB. To create new paradigms for the subject, academics have previously promoted doing a nontraditional study of management in a variety of situations concentrating on phenomena outside of management [30]. The article [31] addresses organizational neuroscience as a novel paradigm for investigation and provides instructions for incorporating OB ideas into the "black box" of the brain. The goal of micro-organizational academics is to improve scientific knowledge by continuously advocating for the combined application of micro- and macro-level hypotheses, frameworks, and strategies as well as for closing the gap between behavioral management research and practice [32, 33].

GARTNER HYPE CYCLE

A hype cycle provides a thorough depiction of how specific technologies evolve, illustrating their emergence, adoption, maturation, and impact on various applications. It is a valuable perspective for chief information officers (CIOs) and senior IT leaders as they navigate the transition to digital businesses. Figure 11.1 illustrates the Gartner hype cycle for emerging technologies in the year 2023, showcasing the preeminent technologies propelled by industry enthusiasm and evolving trends. The Gartner hype cycle for emerging technologies in 2023 offers valuable insights into the maturity and potential impact of various cutting-edge technologies. This annual report serves as a guide for technology stakeholders, providing a roadmap to navigate the complexities of emerging trends.

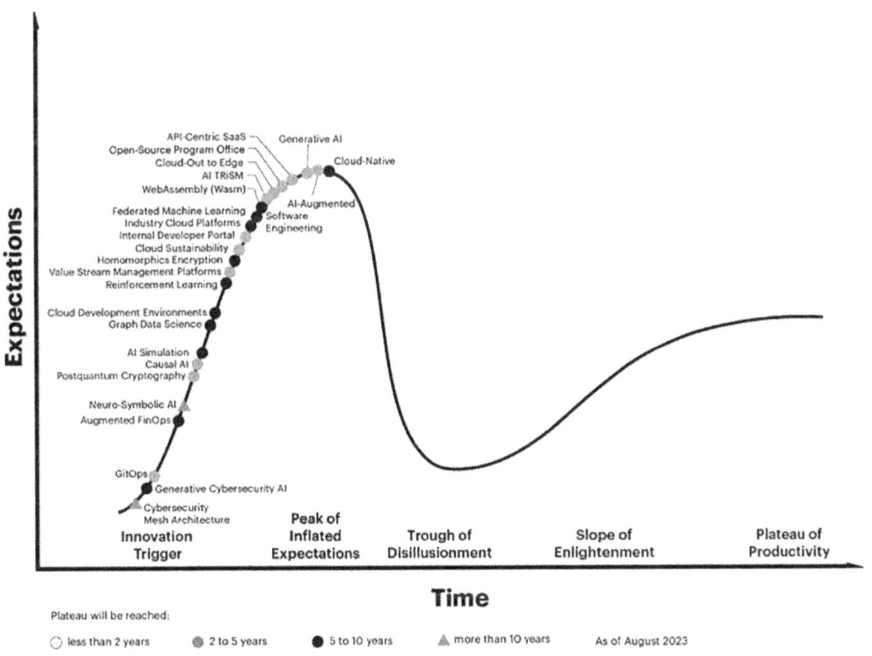

FIGURE 11.1   Gartner hype cycle for emerging technologies, 2023.

Source: www.gartner.com/en/articles/what-s-new-in-the-2023-gartner-hype-cycle-for-emerging-technologies.

A summary of some key highlights from the 2023 edition is as follows:

2.1 Artificial Intelligence: AI remains a central focus, as progress in machine learning, natural language processing, and computer vision propels innovation across various industries. While certain AI applications have reached maturity, others are still in the early stages of development.

2.2 Blockchain: Blockchain technology remains on the radar, with an ongoing exploration of its potential in areas such as decentralized finance (DeFi), supply chain management, digital identity, and smart contracts. Despite the initial hype, practical implementations are gradually gaining traction.

2.3 Quantum Computing: Quantum computing garners significant attention for its transformative potential in solving complex problems that are beyond the capabilities of classical computers. While still nascent, quantum computing is poised to revolutionize fields such as cryptology, optimization, and pharmaceutical research.

2.4 Cybersecurity: With the proliferation of cyber threats, cybersecurity technologies like zero-trust architecture, software-defined security, and decentralized identity solutions are gaining prominence. Organizations are increasingly prioritizing robust cybersecurity measures to protect their digital assets.

2.5 Internet of Things: IoT technologies continue to evolve, with advancements in edge computing, industrial IoT, and connected devices reshaping industries such as manufacturing, healthcare, and smart cities. The integration of IoT devices with AI and analytics drives innovation and efficiency.

2.6 5G Networks: The rollout of 5G networks accelerates the adoption of low-latency connectivity, and speed, enabling advanced applications such as virtual reality, autonomous vehicles, and augmented reality (AR). As 5G infrastructure matures, several new opportunities for innovation emerge.

2.7 Edge Computing: Edge computing architectures gain traction for their ability to process data closer to the source, reducing latency and bandwidth usage. This technology enables real-time analytics, intelligent automation, and enhanced user experiences in various domains.

2.8 Biotechnology: Advances in biotechnology, including gene editing, synthetic biology, and personalized medicine, hold promise for addressing pressing global challenges in healthcare, agriculture, and environmental sustainability. Ethical considerations and regulatory frameworks remain important factors in biotech innovation.

2.9 Digital Twins: Digital twin technologies, which create virtual replicas of physical assets, have seen adoption across industries for predictive maintenance, simulation, and optimization. Digital twins enhance operational efficiency and decision-making by providing real-time insights into asset performance.

2.10 Augmented Reality and Virtual Reality: AR and VR applications continue to expand beyond entertainment, with use cases in training, education, healthcare, and remote collaboration. Immersive experiences offer new opportunities for engagement and interaction.

The Gartner hype cycle for emerging technologies in 2023 reflects a dynamic landscape of innovation and disruption with technologies at various stages of maturity and adoption. By understanding these trends, organizations can strategize effectively and harness the potential of emerging technologies to drive growth and innovation.

PROPOSED METHODS FOR VARIOUS FUTURE TRENDS AND EMERGING APPLICATIONS

The evolving landscape of technology unveils future trends and emerging applications across various domains, reshaping how we live, work, and interact. Artificial intelligence is a key driver, permeating sectors like healthcare, finance, and education, promising automation, efficiency, and personalized experiences. Healthcare sees advancements in telemedicine and personalized medicine, while finance experiences disruption through blockchain and decentralized finance. Transportation transitions to electric and autonomous vehicles, while education adopts immersive technologies for enhanced learning. These innovations offer both challenges and opportunities, promising a more connected, efficient, and technologically enriched future. Adapting to these trends is crucial for individuals, businesses, and policymakers to harness their benefits effectively. Figure 11.2 demonstrates the overall emerging and future trends of recent technologies in various domains.

Various types of future and emerging trends have been briefly discussed as follows:

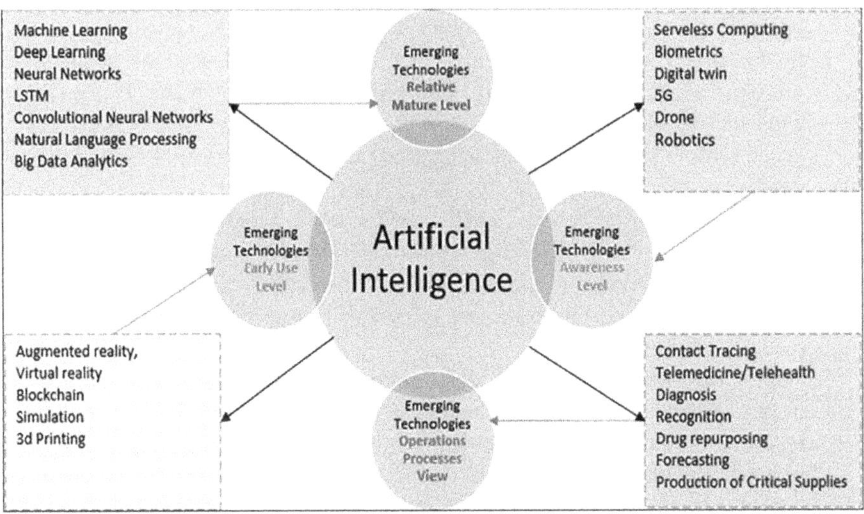

FIGURE 11.2 Schematic framework of emerging and future trends of recent technologies.

EDGE COMPUTING

Edge computing is a paradigm in which data processing and storage are conducted closer to the data source, typically at the edge of the network, rather than relying solely on centralized data centers. This proximity to the data source reduces latency, bandwidth usage, and reliance on cloud services, making it particularly advantageous for applications that require real-time or near-real-time processing. The emergence of the Internet of Things, coupled with advancements in robust cloud services, has paved the way for a groundbreaking computing concept termed edge computing [34]. Operating on both downstream data for cloud services and upstream data for IoT services, edge computing introduces fresh opportunities across diverse application domains [35]. It empowers the handling of vast datasets generated at the IoT edge and optimizes the implementation of ML, deep learning, and augmented discovery of data in harmony with the generation of data. Through a gradual reduction in cost and complexity within the cloud, edge computing facilitates the seamless integration of deep learning functionalities into applications [36].

Key Characteristics and Components of Edge Computing

1. Proximity to Data Source: Edge computing infrastructure is deployed closer to where data is generated or consumed. This could be in locations such as factory floors, retail stores, vehicles, or IoT devices. By processing data at the edge, latency is minimized, and bandwidth requirements are reduced, as data doesn't need to travel long distances to centralized data centers.

2. Distributed Architecture: Edge computing relies on a distributed architecture, where computing resources such as servers, storage, and networking equipment are deployed at various edge locations. This allows for parallel processing and efficient utilization of resources across the network.

3. Scalability: Edge computing architectures are designed to be scalable, allowing organizations to easily deploy and manage edge nodes as needed. This scalability is essential for accommodating growing data volumes and processing requirements, especially in dynamic environments such as IoT deployments.

4. Edge Devices: Edge devices, such as IoT sensors, actuators, and gateways, play a crucial role in collecting and transmitting data to edge computing nodes. These devices often have limited processing capabilities and rely on edge computing infrastructure to perform more complex analysis and decision-making.

5. Edge Computing Platforms: Edge computing platforms provide the necessary software and tools to manage and orchestrate edge computing resources effectively. These platforms may include edge analytics software, security mechanisms, and management tools for monitoring and provisioning edge nodes.

Figure 11.3 depicts an architectural representation of edge computing techniques.

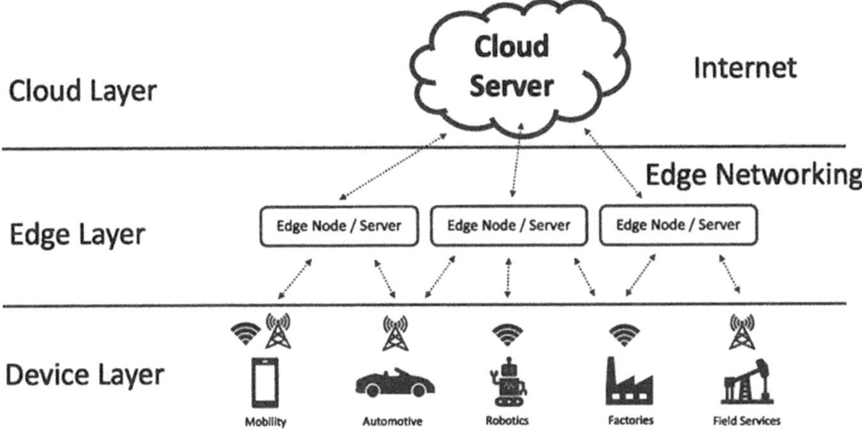

FIGURE 11.3 Architectural layer-wise representation of edge computing.

INTERNET OF THINGS

The Internet of Things encompasses a web of interconnected devices, objects, and systems that interact and exchange data through the internet. Equipped with sensors, actuators, and connectivity capabilities, these devices gather and share data, enabling automation, monitoring, and control of real-world environments. IoT facilitates the networking of physical objects enabling sensors attached to these objects to convey information regarding their status or surroundings. Authors in a recent study [37] discussed key enabling technologies, indicating the driving forces behind future IoT research endeavors. This connectivity opens a vast array of applications across various domains including smart homes, smart cities, smart healthcare, smart agriculture, and smart transportation [38]. The IoT holds vast potential, with expected progress in resource control, energy management, quality of service, interoperability, and interface management as well as security and privacy enhancements [39].

Key Components of the Internet of Things

1. Devices: IoT devices can range from simple sensors and actuators to complex machines and appliances. These devices are embedded with technology to collect data, perform actions, and communicate with other devices or central systems.

2. Connectivity: Connectivity is essential for IoT devices to exchange data with each other or with central systems such as cloud platforms or edge servers. Common connectivity technologies used in the IoT include Zigbee; Bluetooth; Wi-Fi; and cellular networks like 4G LTE, 4G, 5G, and satellite communication.

3. Data Processing: The IoT generates vast amounts of data from connected devices. Data processing techniques such as edge computing and cloud

computing are used to analyze and derive insights from this data. Edge computing involves processing data locally on the device or at the edge of the network, while cloud computing involves processing data in remote data centers.

4. Applications and Services: IoT applications and services leverage data collected from connected devices to offer various functionalities, including smart home automation, industrial automation, healthcare monitoring, environmental monitoring, smart city solutions, asset tracking, and predictive maintenance.

5. Security and Privacy: Security and privacy are critical considerations in IoT deployments to protect sensitive data and prevent unauthorized access to devices and networks. Security measures such as encryption, authentication, access control, and device management are implemented to mitigate cybersecurity risks in IoT systems.

Examples of IoT applications and use cases include smart homes, healthcare, and smart cities. The Internet of Things has the potential to transform industries, improve quality of life, and drive innovation by connecting the physical and digital worlds in ways never before possible. As IoT adoption continues to grow, it will play an increasingly significant role in shaping the future of technology and society. Figure 11.4 showcases the process of an IoT application.

Virtual Reality

Virtual reality is an immersive technology that simulates a realistic, three-dimensional environment using computer-generated imagery and interactive experiences.

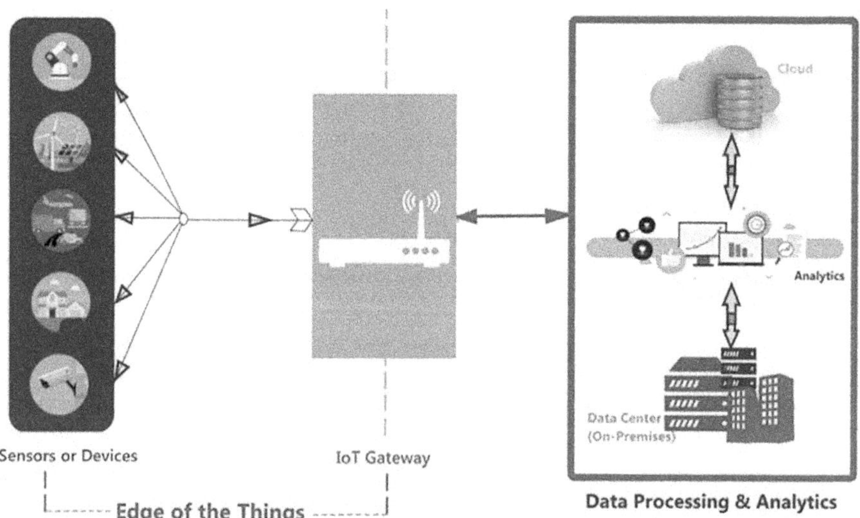

FIGURE 11.4   IoT application processes.

Through VR, users can interact with and explore virtual environments as if they were physically present within them. This technology typically involves the use of a VR headset or goggles, along with motion-tracking sensors and handheld controllers, to enhance the sense of presence and interaction. Virtual reality systems facilitate situated learning by immersing users in interactive environments, objects, and processes. These systems prioritize human interactions and engagement with virtual experiences and environments, fostering learning through hands-on, interactive experiences [40].

Key Components and Features of Virtual Reality

1. Headsets: VR headsets are worn over the eyes and display stereoscopic images to create the illusion of depth and immersion. These headsets may be tethered to a powerful computer or console, or they may be standalone devices with built-in processing capabilities.
2. Motion Tracking: VR systems often utilize motion-tracking sensors to monitor the user's movements and adjust the virtual environment accordingly. This allows users to interact with virtual objects and navigate through virtual spaces naturally and intuitively.
3. Controllers: Handheld controllers are commonly used in VR to provide users with a means of interacting with virtual objects and manipulating the virtual environment. These controllers may feature buttons, triggers, and joysticks for input, and they may also incorporate motion-sensing technology for enhanced interaction.
4. Immersive Environments: Virtual reality environments can range from realistic simulations of real-world locations to fantastical and surreal worlds limited only by the imagination of the creators. These environments may be static or dynamic, allowing users to explore, interact and even create within them.
5. Applications: VR has applications across a wide range of industries and fields including gaming, entertainment, education, training, healthcare, architecture, and design. In gaming and entertainment, VR provides players with immersive experiences that transport them into the virtual worlds of their favorite games and movies. In education and training, VR enables realistic simulations and hands-on learning experiences in fields such as medicine, engineering, and aviation. In healthcare, VR can be used for pain management, rehabilitation, and therapy, while in architecture and design, it allows architects and designers to visualize and explore virtual spaces before they are built.

Virtual reality is a powerful and versatile technology that has the potential to revolutionize how we experience and interact with digital content and the world around us. VR can revolutionize enterprises and improve people's lives, and this ability is only likely to increase as virtual reality gear becomes more widely available and inexpensive and as innovators keep pushing the frontiers of what can be accomplished in virtual worlds. Figure 11.5 depicts an architectural diagram of virtual reality.

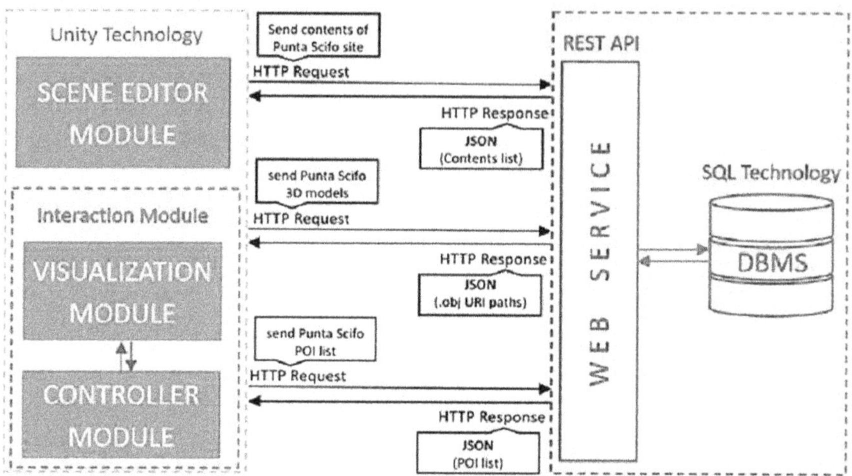

FIGURE 11.5 Architectural block diagram of VR.

COMMERCIAL UAVs

Commercial unmanned aerial vehicles (UAVs), commonly referred to as drones, have witnessed a rapid rise in adoption and application across various industries in recent years. These versatile flying machines, equipped with sensors, cameras, and other payloads, offer numerous advantages in terms of data collection, surveillance, inspection, and delivery. They have become essential in many fields, including science, technology, and society, offering a broad variety of uses, including military tasks, recreational activities, and monitoring. They have effects on safety and confidentiality in both private and public spheres [41]. However, their adaptability is accompanied by constraints, including limited range and power restrictions. To ensure continuous and uninterrupted operation, these devices necessitate periodic recharging at designated intervals [42].

Key Features and Applications of Commercial UAVs

1. Data Collection and Mapping: UAVs are utilized for aerial surveys, mapping, and photogrammetry, providing high-resolution imagery and geospatial data for urban planning, agriculture, forestry, and environmental monitoring.
2. Surveillance and Security: UAVs are employed for surveillance and security purposes in sectors such as law enforcement, border patrol, and infrastructure protection, enabling real-time monitoring of critical areas and rapid response to incidents.
3. Infrastructure Inspection: UAVs facilitate the inspection of infrastructure such as power lines, pipelines, bridges, and buildings, allowing for efficient and cost-effective assessment of structural integrity and maintenance needs.

4. Agriculture and Crop Monitoring: In agriculture, UAVs are utilized for crop monitoring, pest detection, and precision agriculture practices, enabling farmers to optimize crop yields, reduce inputs, and enhance overall farm management.
5. Search and Rescue: UAVs play a vital role in search and rescue operations, providing aerial reconnaissance and thermal imaging capabilities to locate missing persons or disaster survivors in remote or hazardous environments.
6. Delivery and Logistics: Companies are exploring the use of UAVs for the delivery of goods and packages in the last mile, particularly in areas with limited infrastructure or during emergency situations where traditional delivery methods may be impractical.

BENEFITS OF COMMERCIAL UAVS

1. Cost-Effectiveness: UAVs enable faster and more cost-effective data collection and monitoring compared to traditional methods, such as manned aircraft or ground-based surveys.
2. Safety: By removing the need for human operators to be physically present in hazardous or inaccessible environments, UAVs improve safety and reduce the risk of accidents and injuries.
3. Efficiency: UAVs can cover large areas quickly and efficiently, providing timely and accurate data for decision-making and operational planning.
4. Flexibility: UAVs can be deployed in various industries and environments, offering versatility and adaptability to diverse applications and tasks.

In conclusion, commercial UAVs represent a transformative technology with significant potential to revolutionize various industries and sectors. As technology advances and regulations evolve, the adoption and integration of UAVs into commercial operations are expected to continue expanding unlocking new opportunities and driving innovation across different domains. Figure 11.6 demonstrates a workflow diagram of UAVs.

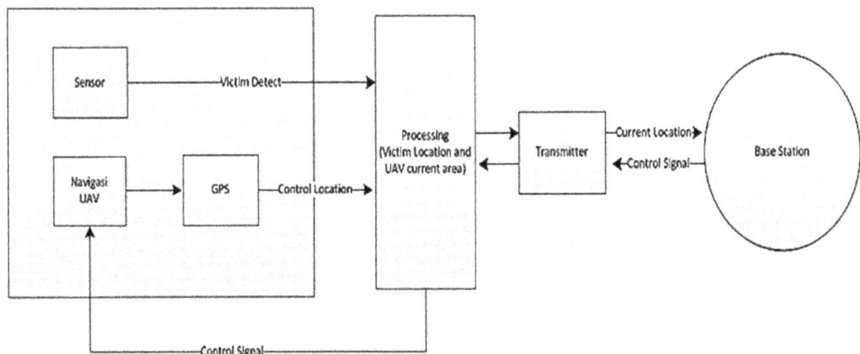

FIGURE 11.6 Workflow architecture of UAVs.

SOFTWARE-DEFINED NETWORKS

Software-defined networking (SDN) is an innovative approach to network management that enhances the agility, flexibility, and programmability of traditional network infrastructures. In SDN, the control plane, responsible for making decisions about where to send traffic, is decoupled from the data plane, which is responsible for forwarding the actual traffic. This decoupling is achieved through the implementation of a centralized controller that manages and directs network traffic. The adoption of software-defined networking in the context of 5G has emerged as a solution to address the limitations of hardware-based cellular architectures. Traditional hardware-centric designs often lack flexibility and are constrained by closed systems. SDN 5G adoption introduces a more efficient approach, enabling the creation of centralized network architectures with programmable capabilities spanning the entire network [43]. In response to this need, the authors in [44] have proposed SoftAir, an SDN architecture tailored specifically for 5G wireless systems. While promising, this approach has raised security concerns, as discussed in [45]. However, there is an anticipation that software-defined networks will play a significant role in the evolution of the mobile industry [46].

Key Components and Characteristics of Software-Defined Networks

1. Controller: In SDN, a central controller serves as the brain of the network making global decisions about how traffic should be directed based on a comprehensive view of the network. The controller communicates with switches and routers in the network through southbound application programming interfaces (APIs) to configure and manage their behavior.
2. Data Plane: The data plane, also known as the forwarding plane, remains responsible for the actual transmission of network packets. Switches and routers in the network forward traffic based on the instructions received from the centralized controller.
3. OpenFlow Protocol: OpenFlow is a standardized communication protocol between the SDN controller and the devices in the network's data plane. It enables the controller to communicate with and manage the flow control to the networking devices.
4. Programmability: SDN allows network administrators and operators to programmatically control and manage the network through software applications. This programmability facilitates automation, making it easier to adapt the network to changing requirements.
5. Flexibility and Scalability: SDN enhances network flexibility, allowing administrators to adjust network behavior via software without altering the physical infrastructure. This architecture is highly scalable, making it suitable for dynamic and large-scale networks.

APPLICATIONS AND BENEFITS OF SOFTWARE-DEFINED NETWORKS

1. Network Virtualization: SDN enables the creation of virtual network overlays, allowing multiple virtual networks to coexist on the same physical network infrastructure.

2. Dynamic Traffic Management: With a centralized controller, SDN allows for dynamic traffic management, optimizing routing decisions based on real-time network conditions.
3. Automation and Orchestration: SDN facilitates automation of network provisioning, configuration, and management tasks, improving operational efficiency.
4. Improved Security: Centralized control provides enhanced visibility and control over network traffic, enabling better security measures and threat detection.
5. Efficient Resource Utilization: SDN enables more efficient use of network resources by dynamically adapting to changing demands, reducing under-utilization and congestion.

In summary, software-defined networking represents a paradigm shift in network management, offering greater flexibility, control, and efficiency, particularly in the face of evolving network requirements and technologies. Figure 11.7 demonstrates the architecture of software-defined networks.

ARTIFICIAL GENERAL INTELLIGENCE

AGI signifies the hypothetical capability of an artificial intelligence system to perceive, learn, and implement knowledge like human intelligence throughout an extensive array of tasks and domains. Unlike narrow AI, which is designed for specific tasks, AGI aims to possess human-like cognitive capabilities, including reasoning, problem-solving, perception, learning, and understanding natural language. Artificial general intelligence finds extensive applications across various domains,

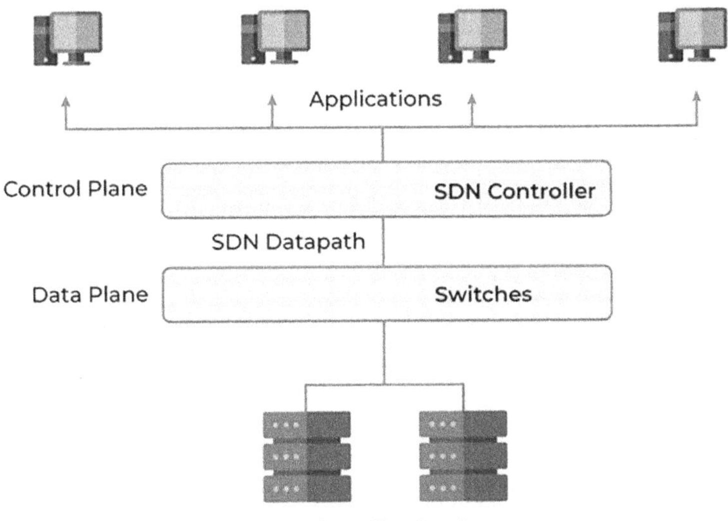

FIGURE 11.7 Block diagram of SDN architecture.

encompassing areas such as cognitive computing, commercial drones, deep learning, deep reinforcement learning, smart robots, machine learning, autonomous vehicles, user interfaces for conversation, smart dust, and smart workspaces. This pervasive presence of AGI is often referred to as "artificial general intelligence everywhere", indicating its widespread integration into numerous high-tech services within cyberspace [2].

Key Features and Characteristics of Artificial General Intelligence

1. Generalization: AGI systems can generalize knowledge and skills learned in one domain to solve problems and tasks in different domains. This capability enables adaptability and versatility across a wide range of contexts.
2. Learning: AGI systems can learn from experience, acquiring new knowledge and skills through observation, interaction, and feedback. This learning process may involve supervised learning, unsupervised learning, reinforcement learning, or a combination of these approaches.
3. Reasoning and Problem-Solving: AGI systems can perform complex reasoning and problem-solving tasks, including logical deduction, inference, planning, and decision-making. These abilities enable AGI to tackle novel and challenging problems autonomously.
4. Natural Language Understanding: AGI systems can understand and generate human language in various forms, including speech, text, and dialogue. This enables seamless communication and interaction with humans in natural language.
5. Perception: AGI systems can perceive and interpret sensory information from the environment including visual, auditory, and tactile inputs. This perception allows AGI to understand and interact with the physical world.
6. Autonomy: AGI systems possess a degree of autonomy, allowing them to operate and make decisions independently without human intervention. This autonomy enables AGI to adapt to changing circumstances and solve problems in real time.

Challenges and Considerations in Artificial General Intelligence

1. Complexity: Achieving AGI entails addressing the complexity of human intelligence which involves various cognitive processes, emotions, and social interactions.
2. Ethical and Societal Implications: The development and deployment of AGI raise ethical concerns related to privacy, autonomy, bias, fairness, accountability, and the potential impact on employment and society at large.
3. Safety and Control: Ensuring the safety and control of AGI systems is critical to prevent unintended consequences or harmful behaviors. Research in AI safety focuses on developing safeguards and control mechanisms to mitigate risks associated with AGI.
4. Interdisciplinary Nature: AGI research requires collaboration across multiple disciplines, including computer science, cognitive science, neuroscience, philosophy, psychology, and ethics, to address the complexity of human intelligence comprehensively.

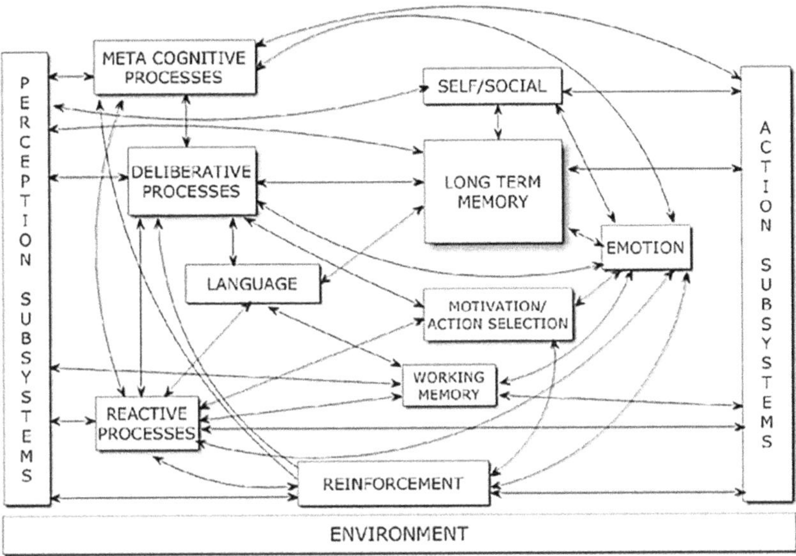

FIGURE 11.8   Workflow diagram of AGI.

In the science of computational intelligence, general intelligence based on artificial intelligence is a revolutionary and inspirational aim that can completely change many facets of business, society, and human existence. While achieving AGI remains a long-term objective, ongoing research and advancements in AI continue to push the boundaries of what is possible, bringing us closer to realizing the vision of truly intelligent machines. Figure 11.8 shows a workflow diagram of artificial general intelligence.

Deep Learning

Deep learning is a subfield of machine learning that involves using a data-driven methods of learning to train neural networks made from data to do particular jobs. Its notable rise stems from its capacity to autonomously discern and represent complex patterns and features embedded within extensive datasets. Deep learning utilizes back-propagation algorithms to uncover complex and intricate data structures within big data. With the development of sophisticated convolutional neural networks that have greatly improved the processing of voice, pictures, video and audio, this strategy has been very successful. Furthermore, neural networks made up of recurrent neurons have proven to be highly effective in processing data that is sequential, including voice and text [47]. Deep learning enables computers to learn from and comprehend the world as a hierarchy of concepts. Unlike traditional methods, deep learning minimizes the need for human intervention in knowledge acquisition, as computers autonomously gather knowledge from their experiences. Through a hierarchical structure of concepts, computers can effectively learn complex ideas from simpler ones leveraging multiple deep layers to enhance their understanding [48, 49].

Key Aspects and Applications of Deep Learning

1. Neural Networks: Deep learning primarily relies on neural networks, which are structured to mimic the human brain's interconnected neurons. It consists of multiple deep layers those enable the model to learn hierarchical representations of data.

2. Image and Speech Recognition: Deep learning has revolutionized image recognition, enabling accurate classification and detection such as recognition of object and facial recognition. In speech recognition, it has enhanced the capabilities of virtual assistants and voice-controlled systems.

3. Natural Language Processing (NLP): Various tasks of NLP such as language translation, sentiment analysis, and chatbots benefit from deep learning models that can understand and generate human-like language patterns. Transformer models like BERT and GPT have significantly advanced NLP applications.

4. Finance and Business: In the financial sector, deep learning is applied to fraud detection, algorithmic trading, credit scoring, and customer service. Businesses leverage deep learning for predictive analytics and recommendation systems.

5. Generative Models: Generative models such as generative adversarial networks (GANs) can create new data samples that resemble existing datasets. This is used in image generation, style transfer, and creating realistic synthetic data.

6. Transfer Learning, Explainability and Interpretability: Transfer learning enables models pre-trained on one task to be adapted for a different but related task. This approach is efficient in scenarios with limited labeled data. Addressing the "black-box" nature of deep neural networks, research is ongoing to enhance model interpretability and explainability, making it easier to understand the decisions made by these complex models.

Deep learning's success can be attributed to the availability of large datasets, powerful computing resources, and advancements in model architectures. Deep Learning is poised to drive further innovations and improvements in various applications, contributing to the advancement of artificial intelligence.

Machine Learning

Machine learning is a dynamic field within the broader realm of artificial intelligence that focuses on developing algorithms and models capable of enabling computers to learn from data. Instead of relying on explicit programming, ML systems utilize statistical techniques to improve their performance on a specific task over time. The essence of machine learning lies in the ability of algorithms to identify patterns, make predictions, or optimize outcomes based on the data they are exposed to. There are various types of machine learning including supervised learning, unsupervised learning, and reinforcement learning, each serving distinct purposes.

Key Concepts and Components of Machine Learning

1. Data: Data is the foundation of machine learning. Algorithms learn from patterns and information within datasets, which can include a wide range of variables and features.
2. Algorithms: Machine learning algorithms are the mathematical models that process data and learn patterns. Common types of algorithms include supervised learning, unsupervised learning, and reinforcement learning.
3. Supervised Learning: In this method, the algorithm is trained using a labelled dataset in which each input pair has an associated collection of output labels. The examples given teach the algorithm how to translate inputs data to the right output.
4. Unsupervised Learning: Using an unlabeled dataset, the algorithm is trained by unsupervised learning. Among the data, the algorithm finds structures, connections, or patterns without specific instructions on the output.
5. Reinforcement Learning: Reinforcement learning is a process where an algorithm learns to make consecutive decisions within an environment to attain a predetermined objective. Through this process, the algorithm is provided with feedback either in the form of rewards or penalties based on the actions it takes.
6. Feature Engineering: Feature engineering involves selecting and transforming relevant features (variables) in the dataset to enhance the performance of machine learning models.
7. Model Evaluation: Machine learning models need to be evaluated to ensure their effectiveness. Common evaluation metrics include accuracy, precision, recall, and F1 score.
8. Training and Testing: Datasets are typically divided into training and testing sets. The model is trained on the training set and then evaluated on the testing set to assess its generalization performance.

APPLICATIONS OF MACHINE LEARNING

1. Image and Speech Recognition: Machine learning is widely used in image and speech recognition systems, powering applications like facial recognition, virtual assistants, and language translation.
2. Finance: In the financial sector, machine learning is utilized for fraud detection, credit scoring, algorithmic trading, and risk assessment.
3. Recommendation Systems: Online platforms use machine learning to build recommendation systems that provide personalized suggestions for users based on their preferences and behavior.
4. Autonomous Vehicles: Machine learning plays a crucial role in the development of autonomous vehicles, enabling them to perceive and respond to the surrounding environment.
5. Natural Language Processing: Machine learning powers natural language processing applications, such as chatbots, sentiment analysis, and language translation.

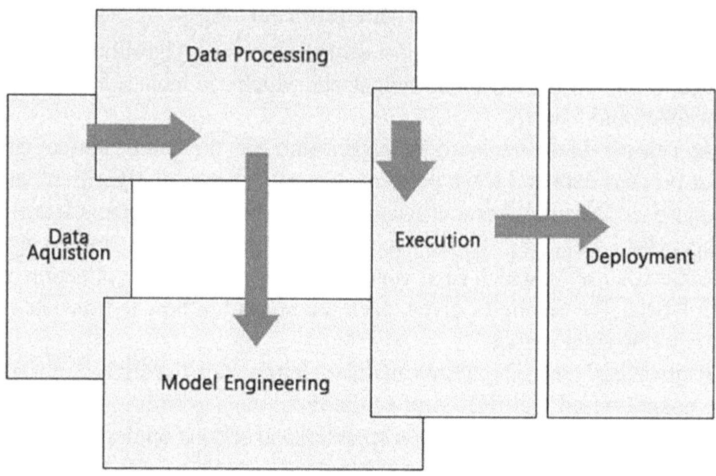

FIGURE 11.9   Representation of operational steps of machine learning.

As technology advances, machine learning continues to evolve, offering new possibilities for automation, prediction, and decision-making across various industries. Figure 11.9 depicts the operational architecture of machine learning.

DIGITAL TWINS

A digital twin is a virtual representation of a physical object, system, or process. It is created using real-time data and simulation models to mimic the behavior, characteristics, and performance of its physical counterpart. Digital twins enable organizations to monitor, analyze, and optimize assets and operations throughout their life cycle, providing valuable insights for decision-making and improving efficiency.

Key Components and Characteristics of Digital Twins

1. Virtual Representation: A digital twin replicates the physical attributes, behavior, and interactions of its real-world counterpart in a digital environment. This virtual representation is created using sensor data, CAD models, and other relevant information.
2. Real-Time Data Integration: Digital twins leverage real-time data from sensors, IoT devices, and other sources to continuously update and synchronize the virtual model with the physical object or system. This allows for accurate monitoring and analysis of current conditions and performance.
3. Simulation and Analytics: Digital twins incorporate simulation and analytics capabilities to simulate scenarios, predict behavior, and analyze performance. By running simulations and what-if scenarios, organizations can assess the impact of changes and optimize operations.
4. Lifecycle Management: Digital twins support the entire lifecycle of assets and systems from design and development to operation and maintenance.

They provide insights into asset performance, health, and maintenance needs, enabling proactive decision-making and optimizing lifecycle costs.

5. Interconnectivity: Digital twins can be interconnected with other digital twins, systems, and applications, forming a network of interconnected assets and processes. This interconnectivity enables holistic analysis, collaboration, and optimization across multiple domains.

6. Applications Across Industries: Digital twins find applications across various industries, including manufacturing, healthcare, transportation, energy, and smart cities. They are used for predictive maintenance, product design and optimization, supply chain management, healthcare simulation, and urban planning, among others.

7. Predictive Maintenance: In manufacturing and asset-intensive industries, digital twins enable predictive maintenance by monitoring equipment health and performance in real time. By detecting anomalies and predicting failures before they occur, organizations can minimize downtime and maintenance costs.

8. Product Design and Optimization: In product development, digital twins allow engineers to simulate and optimize designs, test performance under different conditions, and iterate rapidly. This accelerates the product development process and improves product quality.

9. Smart Cities and Infrastructure: In smart cities and infrastructure projects, digital twins are used to model and simulate urban environments, infrastructure systems, and transportation networks. This enables city planners to optimize resources, improve resilience, and enhance citizen services.

Digital twins serve an essential function in digital transformation initiatives, enabling organizations to gain deeper insights, improve decision-making, and drive innovation across various domains, depicted in Figure 11.10. As technology advances and data connectivity increases, the adoption of digital twins is expected to grow, leading to further optimization and transformation of industries and processes.

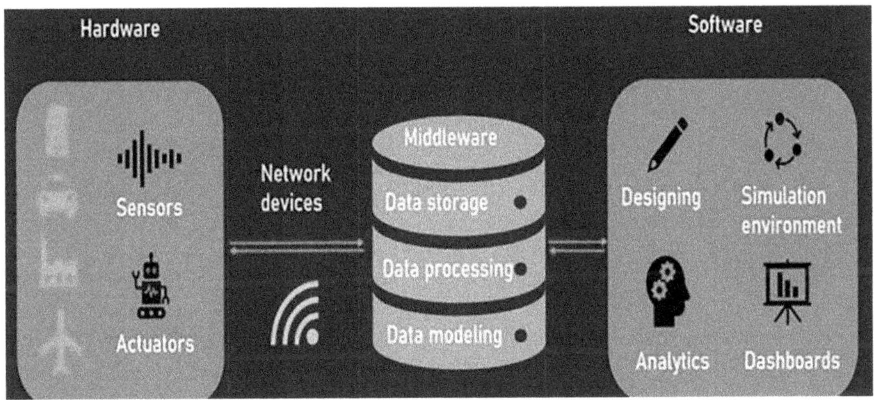

FIGURE 11.10 Operational function of digital twins.

SERVERLESS PLATFORM-AS-A-SERVICE

Serverless Platform-as-a-Service represents a significant evolution in cloud computing architecture, offering a hardware-free environment for orchestrating and managing cloud-based processes. Unlike traditional approaches that rely on infrastructure-based designs, serverless PaaS introduces a paradigm shift by abstracting away the complexities of server management. At its core, serverless PaaS operates on the concept of Function-as-a-Service (FaaS), where computing resources are provided on demand in response to specific events or triggers. This model eliminates the need for developers to provision or manage servers, allowing them to focus solely on writing and deploying code [2]. This approach enhances scalability, agility, and cost-effectiveness, as resources are allocated dynamically based on workload demands. One of the key advantages of serverless PaaS is its ability to streamline the development and deployment of applications without the overhead of infrastructure management. Developers can write code in the form of functions, which are executed in response to events triggered by external sources such as HTTP requests, database changes, or scheduled tasks. Serverless PaaS architectures offer inherent scalability, as resources are provisioned automatically to accommodate fluctuating workloads. This elasticity ensures optimal performance and cost efficiency, as organizations only pay for the resources consumed during code execution. Security is another critical aspect addressed by serverless PaaS. By abstracting away the underlying infrastructure, potential attack surfaces are minimized, reducing the risk of security breaches. Moreover, serverless PaaS providers often offer built-in security features such as encryption, access controls, and monitoring tools to further enhance the security posture of applications. Serverless PaaS represents an approach to cloud computing offering developers a hassle-free and cost-effective way to build and deploy applications at scale. As organizations continue to embrace digital transformation initiatives, serverless PaaS is poised to play a pivotal role in driving innovation and agility in the ever-evolving landscape of cloud computing. Figure 11.11 showcases the PaaS architecture in cloud computing.

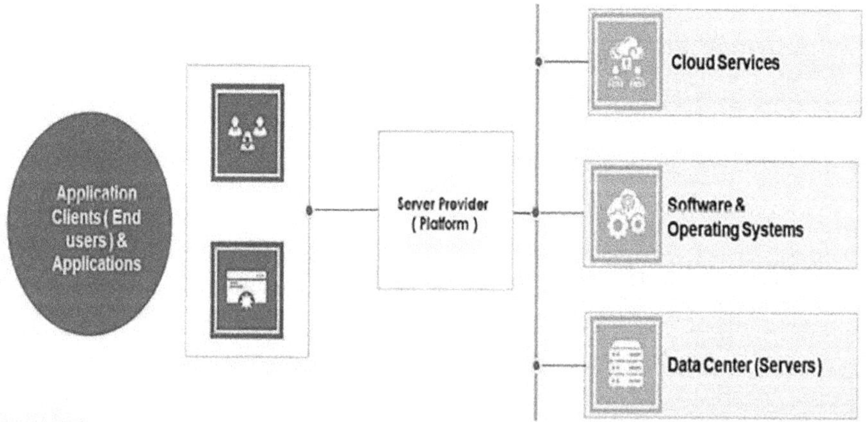

FIGURE 11.11   Serverless PaaS architecture in cloud computing.

COGNITIVE COMPUTING

At its core, cognitive computing is a multidisciplinary field dedicated to crafting computational models and decision-making mechanisms, drawing inspiration from the neurobiological processes of the human brain, cognitive sciences, and psychology [50]. Unlike traditional computing systems, cognitive computing systems aim to mimic human-like intelligence and learning capabilities enabling them to understand, reason, and learn from data in a manner that resembles human cognition. The key objective of cognitive computing is to create systems that can process vast amounts of data, understand natural language, and make context-aware decisions. This involves the utilization of techniques such as machine learning, pattern recognition, natural language processing, and computer vision. These systems continuously learn and adapt, allowing them to improve their performance over time-based on experience and new data. Cognitive computing systems excel in handling unstructured data, which is prevalent in the real world, such as text, images, and audio. By leveraging advanced algorithms, these systems can extract meaningful insights, recognize patterns, and infer context from diverse data sources. One prominent application of cognitive computing is in the development of virtual assistants, chatbots, and conversational interfaces. These systems can understand and respond to user queries in a natural language format, providing a more intuitive and user-friendly interaction. In healthcare, cognitive computing contributes to diagnostic processes by analyzing vast datasets of medical records, images, and research papers. By identifying patterns and correlations, these systems assist healthcare professionals in making more accurate and timely decisions.

The technology also finds applications in finance, customer service, and various industries where data-driven decision-making is crucial. Cognitive computing systems are capable of sifting through enormous datasets, automating routine tasks, and providing valuable insights for strategic planning. The development of cognitive computing represents a significant leap forward in the evolution of artificial intelligence. As technology continues to advance, the integration of cognitive computing into various domains is poised to bring about progressive changes, enhancing efficiency, decision-making, and user experiences across a spectrum of applications, as shown in Figure 11.12.

BLOCKCHAIN

Blockchain stands as a groundbreaking technology that functions as the fundamental mechanism behind cryptocurrencies like Bitcoin, but its applications extend far beyond the realm of digital currencies. Blockchain is a decentralized and distributed ledger that records transactions across a network of computers in a secure and transparent manner. The primary feature of blockchain is its ability to provide a secure and tamper-resistant way of recording transactions. In a blockchain, each transaction is grouped into a block, and these blocks are linked together in a chronological chain. Once a block is added to the chain, it becomes extremely difficult to alter any information within it, ensuring the integrity of the entire transaction history. Key characteristics of blockchain include decentralization, persistency, efficiency,

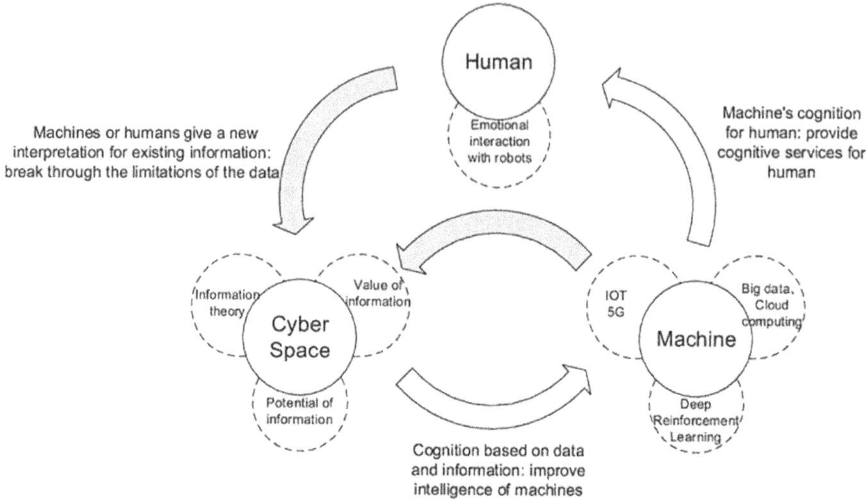

FIGURE 11.12 The application of cognitive computing in various domains.

secrecy, and auditability. Unlike traditional centralized databases, blockchain operates on a peer-to-peer network, where each participant (node) has a copy of the entire ledger. This decentralized nature eliminates the need for a central authority and enhances the security of the system [51].

The concept of decentralization in blockchain introduces trust and transparency. Transactions are verified through a consensus mechanism, often referred to as mining in the context of cryptocurrencies. This ensures that all participants in the network agree on the validity of transactions, reducing the risk of fraud. Blockchain technology finds applications beyond cryptocurrency. It is used for secure and transparent transactions, reducing the reliance on traditional banking systems. Smart contracts, self-executing contracts with the terms of the agreement directly written into code, are another innovative use of blockchain technology to automate and enforce contractual agreements. Industries such as supply chain management, healthcare, and logistics leverage blockchain to enhance transparency, traceability, and security. The immutability of the blockchain ensures that records are tamper-proof, providing a reliable and transparent trail of transactions. Despite its innovative potential, blockchain is not without challenges. Issues related to scalability, energy consumption in certain consensus mechanisms, and regulatory concerns are areas that continue to be explored and addressed. Nevertheless, the underlying principles of blockchain have sparked a paradigm shift in how we envision trust, security, and decentralized systems, making it one of the most groundbreaking technologies of the digital age. Figure 11.13 shows different types of architecture of blockchain.

Human Augmentation

Human augmentation is a technological field focused on enhancing and extending human capabilities, whether naturally or artificially. This emerging area explores

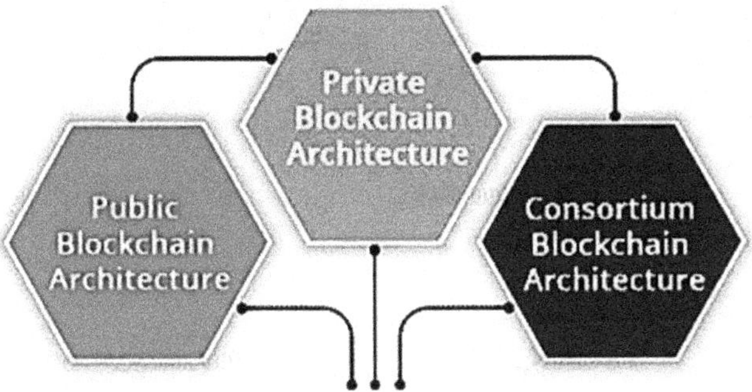

FIGURE 11.13   Various types of blockchain architecture.

various ways to overcome the limitations of the human body, pushing boundaries and augmenting functionalities. The primary objective is to improve and expand human performance, cognition, and physical attributes through the integration of advanced technologies. Technological interventions in human augmentation include the use of wearable devices, implantable technologies, neuroprosthetics, and genetic enhancements. These advancements aim to not only restore impaired functions but also to enhance abilities beyond typical human capacities. Human augmentation technologies often find applications in healthcare, accessibility, sports, and general well-being. From prosthetic limbs with advanced sensory feedback to brain-machine interfaces that enable direct communication between the brain and external devices, human augmentation is at the forefront of merging biology with technology. The ethical considerations surrounding privacy, consent, and the potential for societal disparities are critical aspects that accompany the rapid progress in human augmentation. As the field continues to evolve, researchers and innovators strive to strike a balance between the benefits of improved human capabilities and the ethical implications associated with modifying the essence of what it means to be human. The pursuit of responsible and ethical practices in human augmentation remains pivotal as society navigates this disruptive intersection of biology and technology [51]. Figure 11.14 demonstrates various functions of human augmentation.

Augmented Reality

Augmented reality is a cutting-edge technology that overlays computer-generated information onto the real-world environment, providing users with an enhanced and interactive perception of their surroundings as depicted in Figure 11.15. Unlike virtual reality, which immerses users in a completely simulated environment, augmented reality supplements the physical world with digital elements in real time. Key features of augmented reality include the integration of computer-generated sensory input such as sound, video, graphics, or GPS data into the user's real-time environment. AR applications are diverse and span various sectors, including gaming, education, healthcare, manufacturing, and navigation. AR enhances user experiences by blending the digital and physical realms, offering a seamless interaction

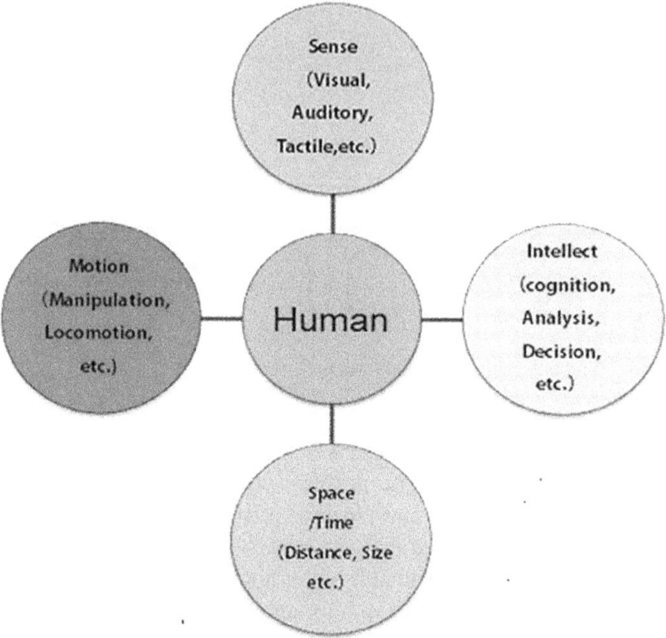

FIGURE 11.14 Functions of human augmentation.

between the virtual and real worlds. This technology can be experienced through specialized AR glasses, smartphones, tablets, or other devices equipped with cameras and sensors [52].

In the educational sector, AR can bring learning materials to life, offering interactive and immersive experiences. In healthcare, it aids in medical training, surgery planning, and patient education. The gaming industry utilizes AR to create engaging and interactive gameplay experiences by integrating virtual elements into the real world. The continuous advancement of AR technology is driven by improvements in hardware, such as better sensors and optics, and sophisticated software development. As AR becomes more widespread, it is expected to revolutionize industries, redefine how users engage with information, and create new opportunities for innovation and collaboration. Despite its evolutionary potential, challenges such as privacy concerns, technological limitations, and the need for seamless integration into everyday life must be addressed. As AR continues to evolve, it holds the promise of reshaping how individuals perceive and interact with their environment, unlocking new possibilities for communication, entertainment, and problem-solving.

QUANTUM COMPUTING

Quantum computing represents a revolutionary approach to computation that harnesses the principles of quantum mechanics to perform operations on data in fundamentally new ways. Unlike classical computers, which process data using bits

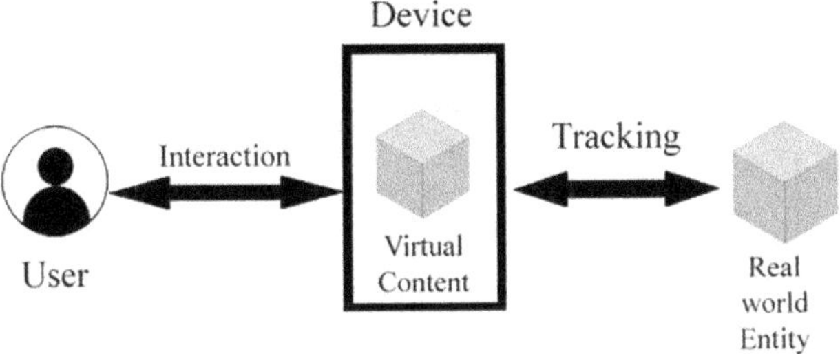

FIGURE 11.15 Functional blueprint for integrating digital content into the real-time environment.

represented as either 0s or 1s, quantum computers use quantum bits or qubits [51], which can exist in multiple states simultaneously due to the principle of superposition. One of the most remarkable features of quantum computing is its ability to perform parallel computations on a vast number of possibilities simultaneously. This parallelism allows quantum computers to solve certain problems much more efficiently than classical computers, especially those involving complex calculations or vast amounts of data. Another key principle of quantum computing is entanglement, which enables qubits to become interconnected in such a way that the state of one qubit instantly influences the state of another, regardless of the distance between them. This phenomenon allows quantum computers to achieve a higher level of computational power and efficiency. With the capacity to reshape multiple sectors such as drug discovery, cryptography, artificial intelligence, materials science, and optimization, quantum computing poses a notable impact. One instance is its potential to dismantle conventional cryptographic algorithms by quickly factoring large numbers, leading to the development of more secure encryption methods. Despite its immense potential, quantum computing is still in its infancy, and many technical challenges remain to be addressed. These challenges include qubit coherence [51] and stability, error correction, scalability, and the development of practical quantum algorithms.

Several companies, research institutions, and governments around the world are investing heavily in quantum computing research and development. Major players in the field include IBM; Google; Microsoft; Intel; and academic institutions such as MIT, Caltech, and the University of Waterloo. As the field of quantum computing continues to advance, it holds the promise of unlocking new frontiers in computation, enabling breakthroughs that were previously thought impossible. While practical quantum computers capable of outperforming classical computers on a wide range of tasks are still years away, the progress being made in this field is both exciting and transformative paving the way for a new era of computing. Figure 11.16 demonstrates the detailed architecture of the quantum computing technique.

SOFTWARE-DEFINED SECURITY

Software-defined security (SDS) is a paradigm in cybersecurity that emphasizes the use of software-based security mechanisms to protect computer systems, networks,

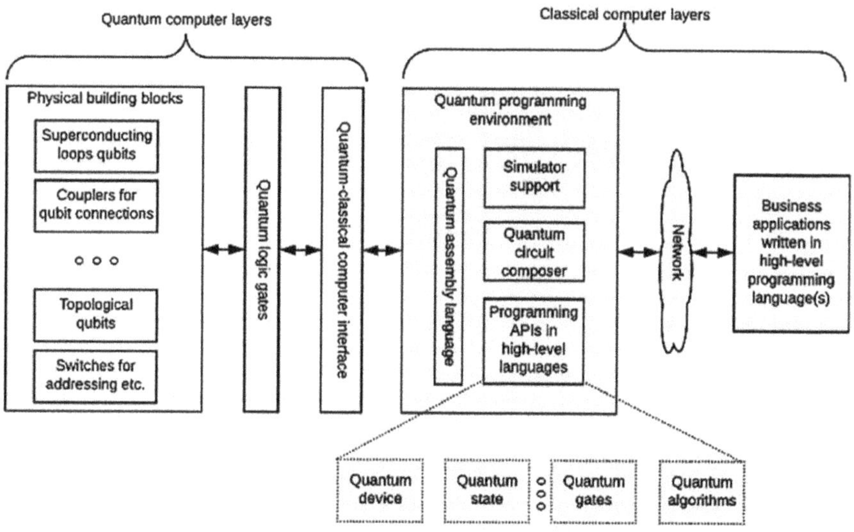

FIGURE 11.16 Schematic diagram of detailed quantum computing process.

and data. Unlike traditional security approaches that rely heavily on hardware appliances and manual configurations, SDS leverages software-defined networking principles to provide a more flexible, scalable, and centralized security solution [53]. At its core, SDS separates the control plane from the data plane, allowing security policies and configurations to be managed centrally through software rather than being tied to specific hardware devices. This decoupling of control enables organizations to dynamically adapt their security posture in response to evolving threats and business requirements. Software-defined security represents a paradigm shift in cybersecurity, offering organizations greater agility, scalability, and resilience in the face of evolving cyber threats. By embracing SDS principles, organizations can enhance their security posture and better protect their digital assets in an increasingly complex and dynamic threat landscape. The principles of SDS are demonstrated in Figure 11.17.

Conclusion and Future Scope

The exploration of future trends and emerging applications offers an intriguing glimpse into the landscape that awaits humanity tomorrow. As society navigates the ever-evolving realm of technology, it stands on the brink of progressive innovations poised to reshape lifestyles, work environments, and social interactions. The multidisciplinary nature of cognitive computing, the decentralized and transparent capabilities of blockchain, the potential for enhancing human capabilities through augmentation, the immersive experiences offered by augmented reality, and the revolutionary possibilities of quantum computing collectively underscore the dynamic trajectory of the technological future. These advancements not only

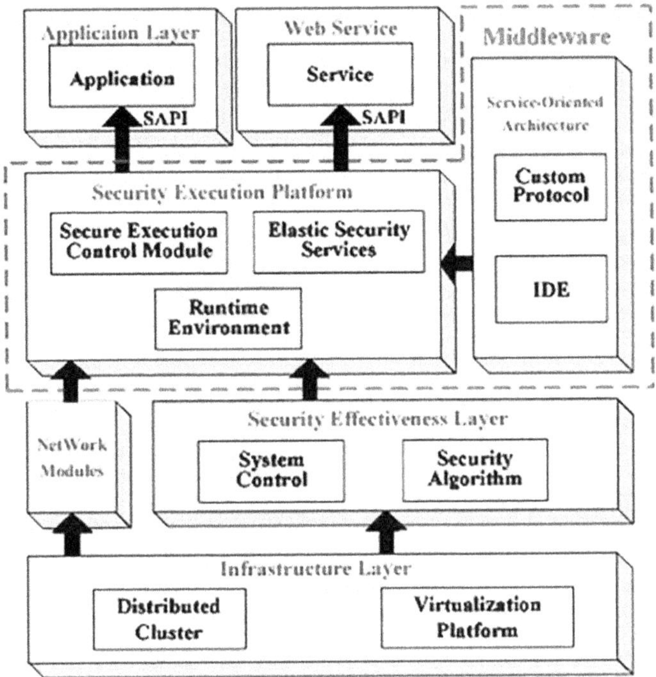

FIGURE 11.17 Software-defined security paradigm.

push the boundaries of what was previously thought possible but also create new opportunities for solving complex problems and addressing challenges across various domains. Furthermore, the paradigm shift towards software-defined security reflects the industry's dedication to crafting flexible, centralized, and efficient solutions to safeguard the increasingly interconnected digital world. The integration of such technologies represents more than a mere compilation of isolated advancements; it signifies the convergence of various disciplines, leading to a comprehensive transformation that transcends individual sectors.

Looking ahead, the rise of autonomous vehicles, smart workspaces, efficient hardware, and virtual assistants further cements the progression towards a more intelligent and interconnected world. These trends hold the promise of enhancing efficiency, convenience, and innovation across diverse industries, paving the way for a future where technology seamlessly integrates into daily life. In this exploration of the future, it becomes clear that staying attuned to emerging trends is not merely an option but a necessity. Adapting to these changes demands active participation, continuous learning, and a collective effort to responsibly harness the benefits they offer. As society stands at the threshold of tomorrow's landscape, the choices made today will shape the trajectory of the digital future. Embracing these advancements with a thoughtful and proactive approach will undoubtedly contribute to a more connected, efficient, and technologically enriched world.

REFERENCES

[1] Lu, Y. (2017). Industry 4.0: A Survey on Technologies, Applications, and Open Research Issues. *Journal of Industrial Information Integration*, 6, 1–10.

[2] Hahanov, V., Gharibi, W., Man, K.L., Iemelianov, I., Liubarskyi, M., Abdullayev, V., & Chumachenko, S. (2018). Cyber-Physical Technologies: Hype Cycle 2017. In *Cyber Physical Computing for IoT-driven Services*, 259–272. Springer, Cham. https://doi.org/10.1007/978-3-319-54825-8_14

[3] Matheson, E., Minto, R., Zampieri, E.G.G., Faccio, M., & Rosati, G. (2019). Human–Robot Collaboration in Manufacturing Applications: A Review. *Robotics*, 8, 100.

[4] Vijayakuymar, G., & Suresh, B. (2022). Significance and Application of Robotics in the Healthcare and Medical Field. *Transaction on Biomedical Engineering Applications and Healthcare*, 3, 13–18.

[5] Oliveira, L.F.P., Moreira, A.P., & Silva, M.F. (2021). Advances in Agriculture Robotics: A State-of-the-Art Review and Challenges Ahead. *Robotics*, 10, 52.

[6] Lee, H.-Y., & Murray, C.C. (2019). Robotics in Order Picking: Evaluating Warehouse Layouts for Pick, Place, and Transport Vehicle Routing Systems. *International Journal of Production Research*, 57, 5821–5841.

[7] Alatise, M.B., & Hancke, G.P. (2020). A Review on Challenges of Autonomous Mobile Robot and Sensor Fusion Methods. *IEEE Access*, 8, 39830–39846.

[8] Soori, M., & Arezoo, B. (2023). Dastres, R. Artificial Intelligence, Machine Learning and Deep Learning in Advanced Robotics, a Review. *Cognitive Robotics*, 3, 54–70.

[9] Raj, M., & Seamans, R. (2019). Primer on Artificial Intelligence and Robotics. *Journal of Organization Design*, 8, 11.

[10] Mosavi, A., & Varkonyi-Koczy, A. (2017). Learning in Robotics. *International Journal of Computer Applications*, 157, 975–8887.

[11] Van Den Hoven Van Genderen, R. (2017). Privacy and Data Protection in the Age of Pervasive Technologies in AI and Robotics. *European Data Protection Law Review*, 3, 338–352.

[12] Elliott, D., & Soifer, E. (2022). AI Technologies, Privacy, and Security. *Frontiers in Artificial Intelligence*, 5, 826737.

[13] Chiacchio, F., & Petropoulos, G. (2018). *Pichler, D. The Impact of Industrial Robots on EU Employment and Wages: A Local Labour Market Approach*. Bruegel Working Paper. Brussels, Belgium: Bruegel.

[14] Wu, C.-J., Raghavendra, R., Gupta, U., Acun, B., Ardalani, N., Maeng, K., Chang, G., Behram, F.A., Huang, J., Bai, C. et al. (2021). Sustainable AI: Environmental Implications, Challenges and Opportunities. *arXiv, arXiv:2111.00364*.

[15] Kamran, S., Farhat, I., Talha Mahboob, A., Gagandeep Kaur, A., Liton, D., Abdul Ghaffar, K., Rimsha, I., Irum, S., & Afifah, R. (2020). The Impact of Artificial Intelligence and Robotics on the Future Employment Opportunities. *Trends in Computer Science and Information Technology*, 5, 50–54.

[16] Salau, A.O., Demilie, W.B., Akindadelo, A., & Eneh, J. (2022). Artificial Intelligence Technologies: Applications, Threats, and Future Opportunities. In *Proceedings of the ACI'22: Workshop on Advances in Computation Intelligence, Its Concepts & Applications at ISIC 2022*, Savannah, GA, 17–19 May, pp. 265–273.

[17] Barfield, W. (2018). Liability for Autonomous and Artificially Intelligent Robots. *Paladyn, Journal of Behavioral Robotics*, 9, 193–203.

[18] Zhu, Q., Rass, S., Dieber, B., & Vilches, V.M. (2021). Cybersecurity in Robotics: Challenges, Quantitative Modeling, and Practice. *Foundations and Trends® in Robotics*, 9, 1–129.

[19] Desai, A., Qadeer, S., & Seshia, S.A. (2018). Programming Safe Robotics Systems: Challenges and Advances. In *Leveraging Applications of Formal Methods, Verification and Validation. Verification, Proceedings of the 8th International Symposium, ISoLA 2018, Limassol, Cyprus, 5–9 November 2018*, Margaria, T., Steffen, B., Eds. Cham, Switzerland: Springer International Publishing, 2018, pp. 103–119.

[20] Robla-Gomez, S., Becerra, V.M., Llata, J.R., Gonzalez-Sarabia, E., Torre-Ferrero, C., & Perez-Oria, J. (2017). Working Together: A Review on Safe Human-Robot Collaboration in Industrial Environments. *IEEE Access*, 5, 26754–26773.

[21] Alpaydin, E. (2020). *Introduction to Machine Learning.* Cambridge, MA: MIT Press.

[22] Eggers, J.P., & Kaul, A. (2018). Motivation and Ability? A Behavioral Perspective on the Pursuit of Radical Invention in Multi-Technology Incumbents. *Academy of Management Journal*, 61(1), 67–93. http://doi.org/10.5465/amj.2015.1123.

[23] George, G., Osinga, E.C., Lavie, D., & Scott, B.A. (2016). Big Data and Data Science Methods for Management Research. *Academy of Management Journal*, 59(5), 1493–1507. http://doi.org/10.5465/amj.2016.4005.

[24] Tonidandel, S., King, E.B., & Cortina, J.M. (Eds.). (2015). *Big Data at Work: The Data Science Revolution and Organizational Psychology* (1st ed., 382). New York: Taylor & Francis. https://doi.org/10.4324/9781315780504

[25] Barnes, C.M., Dang, C.T., Leavitt, K., Guarana, C.L., & Uhlmann, E.L. (2018). Archival Data in Microorganizational Research: A Toolkit for Moving to a Broader Set of Topics. *Journal of Management*, 44(4), 1453–1478. http://doi.org/10.1177/0149206315604188.

[26] Weick, K.E. (1992). Agenda Setting in Organizational Behavior: A Theory-Focused Approach. *Journal of Management Inquiry*, 1(3), 171–182. http://doi.org/10.1177/105649269213001.

[27] Carter, W. (2020). Putting Choice in the Spotlight to Advance Theory on Organizational Adaptation to Technological Discontinuities. *Organization Management Journal*, 17(2). http://doi.org/10.1108/OMJ-04-2019-0720.

[28] Von Krogh, G. (2018). Artificial Intelligence in Organizations: New Opportunities for Phenomenon Based Theorizing. *Academy of Management Discoveries*, 4(4). http://doi.org/10.5465/amd.2018.0084.

[29] Makadok, R., Burton, R., & Barney, J. (2018). A Practical Guide for Making Theory Contributions in Strategic Management. *Strategic Management Journal*, 39(6), 1530–1545. http://doi.org/10.1002/smj.2789.

[30] Bamberger, P.A., & Pratt, M.G. (2010). Moving Forward by Looking Back: Reclaiming Unconventional Research Contexts and Samples in Organizational Scholarship. *Academy of Management Journal*, 53(4), 665–671. http://doi.org/10.5465/amj.2010.52814357.

[31] Becker, W.J., Cropanzano, R., & Sanfey, A.G. (2011). Organizational Neuroscience: Taking Organizational Theory Inside the Neural Black Box. *Journal of Management*, 37(4), 933–961. http://doi.org/10.1177/0149206311398955.

[32] Tenhiälä, A., Giluk, T.L., Kepes, S., Simon, C., Oh, I.S., & Kim, S. (2016). The Research-Practice Gap in Human Resource Management: A Cross-Cultural Study. *Human Resource Management*, 55(2), 179–200. http://doi.org/10.1002/hrm.21656.

[33] Aguinis, H., Boyd, B.K., Pierce, C.A., & Short, J.C. (2011). Walking New Avenues in Management Research Methods and Theories: Bridging Micro and Macro Domains. *Journal of Management*, 37(2), 395–403. http://doi.org/10.1177/0149206310382456.

[34] Shi, W., Cao, J., Zhang, Q., Li, Y., & Xu, L. (2016). Edge Computing: Vision and Challenges. *IEEE Internet of Things Journal*, 3(5), 637–646. http://doi.org/10.1109/JIOT.2016.2579198

[35] Shi, W., & Dustdar, S. (2016). The Promise of Edge Computing. *Computer*, 49(5), 78–81. http://doi.org/10.1109/MC.2016.145

[36] Satyanarayanan, M. (2017). The Emergence of Edge Computing. *Computer*, 50(1), 30–39. http://doi.org/10.1109/MC.2017.9

[37] Gubbi, J., Buyya, R., Marusic, S., & Palaniswami, M. (2013). Internet of Things (IoT): A Vision, Architectural Elements, and Future Directions. *Future Generation Computer Systems*, 29(7), 1645–1660.

[38] Singh, K.J., & Kapoor, D.S. (2017). Create Your Own Internet of Things: A Survey of IoT Platforms. *IEEE Consumer Electronics Magazine*, 6(2), 57–68. http://doi.org/10.1109/MCE.2016.2640718.

[39] Yaqoob, I., Ahmed, E., Hashem, I.A.T., Ahmed, A.I.A., Gani, A., Imran, M., & Guizani, M. (2017). Internet of Things Architecture: Recent Advances, Taxonomy, Requirements, and Open Challenges. *IEEE Wireless Communications*, 24(3), 10–16. http://doi.org/10.1109/MWC.2017.1600421.

[40] Greenwald, S., Kulik, A., Kunert, A., Beck, S., Frohlich, B., Cobb, S.,. . .& Snyder, A. (2017). Technology and Applications for Collaborative Learning in Virtual Reality. https://repository.isls.org/handle/1/210

[41] Rao, B., Gopi, A.G., & Maione, R. (2016). The Societal Impact of Commercial Drones. *Technology in Society*, 45, 83–90. https://doi.org/10.1016/j.techsoc.2016.02.009

[42] Hong, I., Kuby, M., & Murray, A. (2017). A Deviation Flow Refueling Location Model for Continuous Space: A Commercial Drone Delivery System for Urban Areas. In *Advances in Geocomputation*, 125–132. Cham: Springer.

[43] Akyildiz, I.F., Lin, S.C., & Wang, P. (2015). Wireless Software-Defined Networks (W-SDNs) and Network Function Virtualization (NFV) for 5G Cellular Systems: An Overview and Qualitative Evaluation. *Computer Networks*, 93, 66–79. https://doi.org/10.1016/j.comnet.2015.10.013.

[44] Akyildiz, I.F., Wang, P., & Lin, S.C. (2015). SoftAir: A Software Defined Networking Architecture for 5G Wireless Systems. *Computer Networks*, 85, 1–18. https://doi.org/10.1016/j.comnet.2015.05.007.

[45] Chen, M., Qian, Y., Mao, S., Tang, W., & Yang, X. (2016). Software-Defined Mobile Networks Security. *Mobile Networks and Applications*, 21(5), 729–743.

[46] Chen, T., Matinmikko, M., Chen, X., Zhou, X., & Ahokangas, P. (2015). Software Defined Mobile Networks: Concept, Survey, and Research Directions. *IEEE Communications Magazine*, 53(11), 126–133. http://doi.org/10.1109/MCOM.2015.7321981.

[47] LeCun, Y., Bengio, Y., & Hinton, G. (2015). Deep Learning. *Nature*, 521(7553), 436. http://doi.org/10.1038/nature14539

[48] Kim, K.G. (2016). Book Review: Deep Learning. *Healthcare Informatics Research*, 22(4), 351–354.

[49] Schmidhuber, J. (2015). Deep Learning in Neural Networks: An Overview. *Neural Networks*, 61, 85–117. https://doi.org/10.1016/j.neunet.2014.09.003.

[50] Gutierrez-Garcia, J.O., & López-Neri, E. (2015). Cognitive Computing: A Brief Survey and Open Research Challenges. In *Applied Computing and Information Technology/2nd International Conference on Computational Science and Intelligence (ACIT-CSI), 2015 3rd International Conference on*, 328–333. IEEE. http://doi.org/10.1109/ACIT-CSI.2015.64.

[51] Zheng, Z., Xie, S., Dai, H.-N., Chen, X., & Wang, H. (2018). Blockchain Challenges and Opportunities: A Survey. *International Journal of Web and Grid Services*, 8 Inderscience Enterprises Ltd., 14(4), 352–375.

[52] Dunleavy, M., & Dede, C. (2014). Augmented Reality Teaching and Learning. In *Handbook of Research on Educational Communications and Technology* (pp. 735–745). Springer. https://doi.org/10.1007/978-1-4614-3185-5_59

[53] Al-Ayyoub, M., Jararweh, Y., Benkhelifa, E., Vouk, M., & Rindos, A. (2015). Sdsecurity: A software-defined security experimental framework. In *Communication Workshop (ICCW), 2015 IEEE International Conference on*, 1871–1876. IEEE. http://doi.org/10.1109/ICCW.2015.7247453

12 Securing Federated Deep Learning
Privacy Risks and Countermeasures

*Atharva Saraf, Shaurya Sameer Talewar,
Susanta Das, Khushbu Trivedi, and
Ahmed A. Elngar*

INTRODUCTION

Artificial intelligence has rapidly gained global acceptance, driving a surge in demand for powerful ML models. ML's ability to uncover hidden insights and detect complex patterns has revolutionized scientific research and real-world applications [1]. On the other hand, this pursuit frequently comes into conflict with growing worries over the privacy and security of data. The act of centralizing large volumes of sensitive data, such as medical records or financial transactions, leaves data vulnerable to the possibility of breaches and access violations by unauthorized parties. Furthermore, tough data privacy legislation such as the General Data Protection Regulation (GDPR), which is the EU's data protection regulation and went into effect on May 25, 2018, and is aimed at safeguarding the privacy and security of EU citizens' data [2], and the California Consumer Privacy Act (CCPA) further restrict data collecting and sharing, which hinders the creation of novel artificial intelligence solutions [2].

Under these laws, operators are strictly required to clearly explain user agreements and are not allowed to trick or coerce consumers into giving up their privacy rights. Users must give consent before the operator can train models, and they have the right to delete their personal information. Additionally, network operators are prohibited from releasing, modifying, or deleting the personal information of the consumers they collect [3].

Federated learning comes as a groundbreaking approach that reshapes the landscape of collaborative learning. McMahan et al. define federated learning as a "method facilitating collaborative model training between devices and servers without the need to exchange raw data" [4, 5]. This is different from traditional methods, which are dependent on centralized data storage. By using a decentralized learning framework, we get remarkable benefits while ensuring the protection of individual

DOI: 10.1201/9781003482000-12

TABLE 12.1

Notation

Symbol	Description
K	The total number of client participants in the federated learning process.
C_k	The Kth client in the federated learning system, where k = 1, 2,. . . ., K.
X_k	The dataset of features held by the client C_k.
Y_k	The dataset of labels corresponding to X_k for client C_k.
θ_k	The global model parameters at iteration t.
E	The number of local epochs each client C_k runs before updating the global model.
θ_k^t	The local version of the global model parameters at client C_k after local training at iteration t.
$\mathcal{L}$	The cross-entropy loss function.
$F_{\theta_k^t}(x_k)$	The prediction made by the local model with parameters θ_k^t on the mini-batch x_k
x_k, y_k	A mini-batch of data (feature and labels) sampled from (X_k, Y_k) during local training.
$\theta^{(t+1)}$	The updated global model parameters after aggregating updates from all clients at iteration $t+1$.
D	The dataset consisting of N samples, where each sample is represented as (X_i, Y_i).
N	The total number of samples in the dataset D.
X_i	The feature of the ith sample in the dataset D.
Y_i	The label corresponding to the ith sample in the dataset D.
Θ	The joint machine learning (ML) model parameters.
$l(\Theta; D)$	The loss function for vertical federated learning.
$f(\Theta; X_i, Y_i)$	The loss function applied to the joint ML model Θ using sample X_i and label Y_i.
$\gamma(\Theta)$	The regularization term applied to the joint ML model Θ.
λ	The hyperparameter controlling the strength of the regularization term $\gamma(\Theta)$ in the loss function.

privacy. Unlike the traditional method of machine learning, which requires gathering and storing the data in a central location/server for model training, the decentralized method of federated learning does not need storage in a central server or location, as shown in Figure 12.1 [3].

Centralized data storage has many disadvantages, including concerns regarding data breaches and unauthorized access, which could lead to the compromise of sensitive information and privacy. Scalability is also an issue: as the volume of data grows, centralized infrastructure struggles to keep the high demand of training complex models requesting vast datasets [6]. The advantages of federated learning are schematically shown in Figure 12.2.

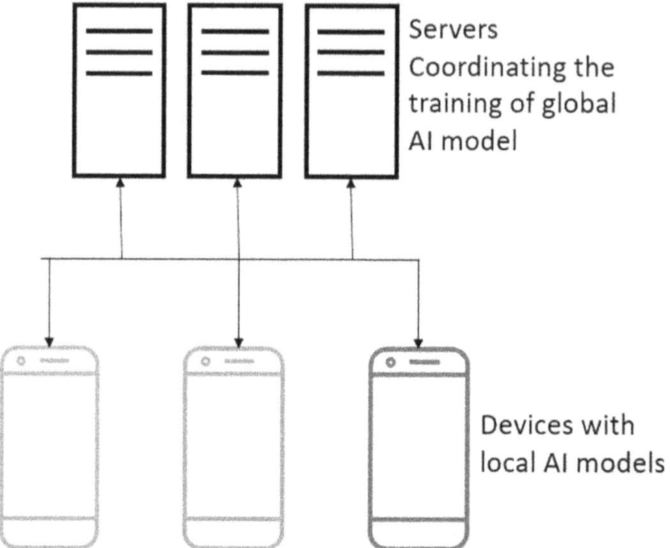

FIGURE 12.1 General federated learning framework.

FIGURE 12.2 Advantages of federated learning.

BENEFITS OF DECENTRALIZED METHOD

- Enhanced Privacy: Individuals get control over their data, which reduces threats to their privacy that relate to traditional data centralization. It is possible to maintain the confidentiality of sensitive information on user devices, which helps in building trust and transparency in the machine learning process.

- Unleashed Scalability: Federated learning solves the challenge of training models in vast and distributed datasets that are located in different parts of the world. This is accomplished by combining the collective power of a large number of devices [7].
- Enhanced Security: With no centralized data storage, the attack surface is reduced, which makes it more difficult for breaches to penetrate huge information sets. Even if a single device is breached, only a small portion of the data is exposed, which improves the overall security and resilience of the system [7].

Organizations can participate in collaborative learning while following the legal requirements because of the decentralized nature of federated learning, which is in line with data privacy standards such as the General Data Protection Regulation and the California Consumer Privacy Act [2].

Federated learning not only leverages the advantages of nature but also opens up new avenues such as on-device and edge learning, collaborative social research, and broader applications. [7].

As artificial intelligence increasingly interferes with our lives, data privacy and security concerns are increasing. Traditional machine learning methods involve centralizing data, which increases privacy risks. Federated learning offers a new approach by decentralizing model training and allowing collaborative work by sharing raw data. In this chapter, we dive into the fundamental workings of federated learning and its potential to protect data while improving AI.

BASIC CONCEPTS OF FEDERATED LEARNING

DEFINITION

Federated learning (FL) is a method of collaborative model training across various decentralized data sources or devices, ensuring data privacy and security. In this method, the raw/original data stays restricted on individual devices or servers, and the computation for model training occurs locally on these data sources [8].

The key components and processes that constitute FL are as follows [8]:

3.1.1. Decentralized Data Sources: Different data sources like smartphones, Internet of Things (IoT) devices, or remote servers have their own private datasets, often containing sensitive information. This prevents direct data sharing.

3.1.2. Global Model Initialization: The process begins with setting up a global machine learning model, typically a pre-trained neural network, as the starting point for collaboration.

3.1.3. Local Model Training: Each data source trains its model locally using its data and the global model as a reference.

3.1.4. Local Model Updates: After training, each data source calculates updates based on the difference between its local model and the global model. These updates reflect the knowledge gained from the local data.

3.1.5. Model Aggregation: Updates from all the data sources are securely combined to create an updated global model. Techniques like advanced cryptographic methods protect individual update privacy.

3.1.6. Model Deployment: The updated global model is deployed back to all participating data sources, refining their models without revealing local data.

3.1.7. Iterative Collaboration: This iterative process involves multiple rounds of training, updates, and aggregation until a satisfactory model is achieved.

CLASSIFICATION OF FEDERATED LEARNING

Federated learning methods are usually subdivided into horizontal federated learning (HFL), vertical federated learning (VFL), and federated transfer learning (FTL) [9].

Horizontal Federated Learning

Horizontal federated learning involves combining samples. It's applicable when there's a significant overlap of participant data features but a small overlap of user data. The data used for joint model training is where both parties share the same data characteristics but have different users. Horizontal federated learning has broad application scenarios, as shown in Figure 12.3 [9].

HFL Problem

The most popular horizontal federated learning method is FedAvg. Consider K clients $\{C_k\}_{k=1}^{K}$, each with dataset (X_k, Y_k), where each X_k has the same feature space. At iteration t in FedAvg, the server sends a global model with parameters θ^t to all clients. Each client C_k updates θ^t for given local epochs E to obtain the local version of the global model θ_k^t by minimizing the cross-entropy loss $\mathcal{L}\left(F_{\theta_k^t}(x_k), y_k\right)$

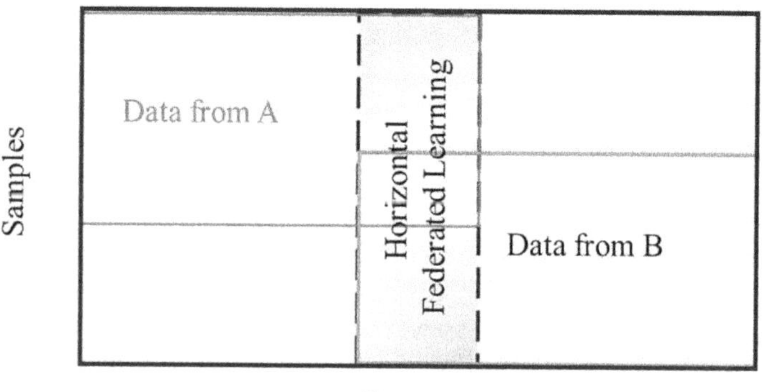

FIGURE 12.3 Horizontal federated learning.

evaluated at mini-batch (x_k, y_k) of (X_k, Y_k). Then it returns θ_k^t to the server that will aggregate each θ_k^t to obtain θ^{t+1} as [10]:

$$\theta^{t+1} = \frac{1}{K}\sum_{k=1}^{K}\theta_k^t.$$
(1)

Vertical Federated Learning

Vertical federated learning involves combining features across users in different formats. It's applicable when there's more overlap of users but fewer overlaps of data features. The data used for joint model training is the part where data characteristics for the same users on both sides are not identical, as shown in Figure 12.4 [9].

3.2.2.1. VFL Problem

A vertical federated learning system aims to collaboratively train a joint machine learning model using a dataset $D = \{(X_i, Y_i)\}_{i=1}^{N}$ with N samples while preserving the privacy and safety of local data and models. We formulate the loss of vertical federated learning as follows.

$$\min_{\Theta} l(\Theta; D) \triangleq \frac{1}{N}\sum_{i=1}^{n} f(\Theta; X_i, Y_i) + \lambda\sum_{k=1}^{K} \gamma(\Theta)$$
(2)

where Θ denotes the joint ML model, $f(.)$ and $y(.)$ denote the loss and regularize functions, and λ is the hyperparameter that controls the strength of γ [11, 12].

Federated Transfer Learning

Federated transfer learning becomes relevant when there's limited overlap in features and samples among participants, like collaboration between banks and supermarkets

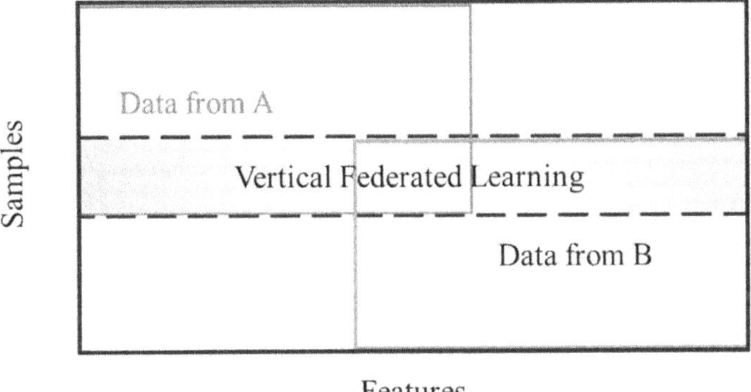

FIGURE 12.4   Vertical federated learning.

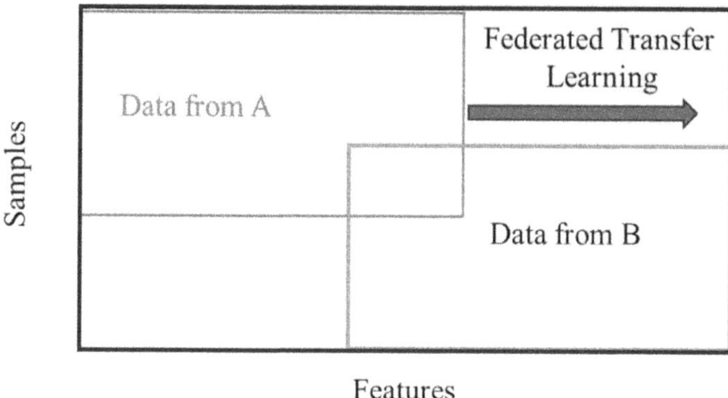

FIGURE 12.5   Federated transfer learning.

TABLE 12.2

Comparison of Three Federated Learning Frameworks

Federated Learning Types	User Overlap	Data Overlap	Application Scenarios (Taking Banks and Supermarkets as Examples)
Horizontal federated learning	Less	Less	Between banks in different regions
Vertical federated learning	Many	Many	Between banks and supermarkets in same area
Federated transfer learning	Less	Less	Between banks and supermarkets in different areas

across different regions. This approach is primarily used in deep neural network scenarios to address challenges related to insufficient data or labels during joint modeling, as shown in Figure 12.5 [9].

Table 12.2 shows a comparison of three federated learning frameworks.

FEDERATED VS TRADITIONAL LEARNING

Federated learning offers several advantages over traditional machine learning; see Table 12.3.

3.3.1. **Enhanced User Privacy:** Federated learning ensures user data privacy by storing data locally, without sharing it among participants [13]. This approach, in conjunction with the data privacy, mandates outlined in the General Data Protection Regulation [2].

3.3.2. **Adaptability to Large-Scale Data:** Leveraging large-scale training data enhances model quality. Federated learning achieves better accuracy compared to traditional methods while also reducing the need for

TABLE 12.3

Comparison Between Federated Learning, Traditional Machine Learning and Deep Learning

	Federated Learning	Traditional Machine Learning	Traditional Deep Learning
Safety	High	Low	Low
Amount of data	Large	Small or medium	Small or medium
Sharing	High	None	None
Model building process	Multi-party	Single-party	Single-party

extensive equipment during training and speeding up model training with large datasets [13].

3.3.3. **Increased Flexibility of Data Sources:** Federated learning enables the inclusion of data sources that were previously unable to participate due to specific constraints. These sources can now store data locally while contributing to overall model training, thereby improving model generalization.

LITERATURE REVIEW

In this literature review, we dived into the advantages, disadvantages, and practical implementations of federated learning in various domains, with a particular focus on its applications in medical image analysis, personalized solutions, and the integration of blockchain technology to enhance privacy and security, as shown in Table 12.4.

DECENTRALIZED LEARNING

Decentralized training is the concept that lies at the core of the new approach that federated learning takes. The usual paradigm of centralizing data for training is overturned by this innovation, which instead gives devices and servers the ability to learn together without revealing any of their sensitive information between themselves. According to Yang et al. [7], this change introduces several distinct issues that are worth investigating, in addition to offering compelling advantages.

PROTECTION OF PERSONAL INFORMATION

Privacy concerns are raised when data is centralized because it provides a honeypot for potential attackers. Federated learning reduces the likelihood of these dangers occurring by storing data either on the device itself or on local servers, ensuring that information is never exposed to the outside world [1, 5, 15–17]. The only updates to the model that are shared for aggregation are those that are produced from local training on the data. Imagine that a language model is being trained on millions of

TABLE 12.4

Advantages, Disadvantages, and Practical Implementations of Federated Learning in Various Domains

S.no	References	Advantages	Disadvantages	Practical Implementations
1	[14]	• Federated learning enables collaborative model training without sharing patient data. • Differential privacy ensures strong privacy guarantees in medical image analysis • Federated learning algorithms like FedAvg are scalable and effective for large datasets. • FedDG offers privacy-preserving FL with frequency space interpolation across clients. • FL model can be trained across institutions without centralizing data.	• Limited availability of large-scale medical datasets hinders machine learning applications. • Challenges due to confidentiality, privacy concerns, and regulations around sharing medical data. • Difficulty in sharing histopathology images due to data size constraints.	• Demonstrates viable federated learning for medical image analysis. • Addresses privacy concerns in sharing medical data for machine learning. • Achieves superior results in histopathology image analysis through collaborative efforts. • Validates federated learning with lung cancer images from TCGA dataset.
2	[15]	• FL divides computational power, reduces communication burden, and preserves privacy. • Personalization solutions improve fairness, robustness, and performance in FL. • Blockchain technology addresses privacy, data computation, incentives, and security challenges.	• Privacy approaches may reduce performance by adding noise to data. • Fine-tuning approaches require additional client computing resources.	• Enhances understanding of client-side challenges in federated learning. • Offers insights into state-of-the-art solutions for client-side challenges. • Highlights the impacts of solutions on related challenges.
3	[16]	• FL divides computational power, reduces communication burden, and preserves privacy. • Personalization solutions improve fairness, robustness, and performance in FL. • Blockchain technology addresses privacy, data computation, incentives, and security challenges.	• Proof of work in blockchain leads to energy waste and centralization. • Non-IID data samples impact machine learning model accuracy.	• Enhancing FL performance and security with blockchain integration. • Summarizing studies on incentive mechanisms and industrial applications of BCFL. • Providing insights for improving FL performance through blockchain integration.

(Continued)

TABLE 12.4 *(Continued)*

Advantages, Disadvantages, and Practical Implementations of Federated Learning in Various Domains

S.no	References	Advantages	Disadvantages	Practical Implementations
4	[17]	• FL generates robust models without sharing data, ensuring privacy and security. • FL enables training across multiple devices without exchanging actual data. • FL provides higher security and access privileges to data.	• Imperfect training data leads to highly inaccurate models. • Data skewing issues due to factors like imbalance and missing classes. • Communication challenges, slower due to vast device quantity. • Privacy concerns over raw data transmission in federated learning.	• Provides insights on federated learning–enabling technologies, protocols, and applications. • Highlights challenges and advantages of federated learning systems. • Offers a framework for personalized models in healthcare using federated learning. • Explores the potential and applicability of federated learning in various industries.
5	[5]	• Scalable production system for federated learning on mobile devices. • Enables model training on a large corpus of decentralized data.	• Potential bias due to device training restrictions and deployment limitations.	• Addresses device availability, connectivity, storage, and computer resource challenges. • Focuses on reducing bandwidth through compression techniques and quantized representation. • Evaluates models in live AB experiments to detect bias issues.
6	[11]	• VF2Boost is 12.8–18.9× faster than existing implementations. • Supports much larger datasets with tens of millions of instances. • Customized operations speed up cryptography operations in the system. • Achieves the same model accuracy as non-federated training. • Employs a concurrent training protocol to reduce idle periods.	• Sequential key procedures lead to frequent idle waiting periods. • Cryptography operations are time consuming and dominate overall training time.	• Addresses device availability, connectivity, storage, and computer resource challenges. • Focuses on reducing bandwidth through compression techniques and quantized representation. • Evaluate models in live AB experiments to detect bias issues.

(Continued)

TABLE 12.4 (Continued)

Advantages, Disadvantages, and Practical Implementations of Federated Learning in Various Domains

S.no	References	Advantages	Disadvantages	Practical Implementations
7	[18]	• Addressing communication overhead in federal training for high-efficiency demand.	• Lack of fully secure transmission may lead to privacy concerns. • Insider and outsider adversaries pose threads to privacy during training.	• VF 2 Boost enhances vertical federated learning speed significantly. • Enables cross-enterprise collaboration with privacy guarantees. • Optimizes training workflow, supports larger datasets, and speeds up operations. • Improves CPU utilization and reduces public network transmission.
8	[6]	• FATE is an industrial-grade platform for collaborative learning with data protection. • Supports secure computation protocols, machine learning algorithms, and visualization tools. • Developed to bridge data silos, build models, and protect user privacy.	• Limited scalability due to computational heaviness and communication requirements. • May require significant coordination costs for large-scale training.	• Addresses data privacy and sharing concerns in federated learning. • Highlights the need for communication-efficient methods in the medical industry. • Discusses privacy risks related to indirect privacy leakage during training.
9	[19]	• Improved privacy, localized data processing, and real-time decision-making capabilities. • Tenser Flow Lite benefits: improved performance, reduced model size, offline inference.	• Tensor Flow Lite imposes model size limitations for mobile device deployment. • Limited support for custom operations in Tensor Flow Lite can be challenging.	• FATE enables secure collaborative machine learning in industrial applications. • Supports secure computation protocols and machine learning algorithms for enterprises.
10	[3]	• User privacy protection is ensured by locally stored federated learning data. • Model training adapted to large-scale data for improved training quality.	• Data poisoning and model poisoning are significant disadvantages of federated learning. • Federated learning faces security hazards from malicious users and criminals.	• Enhances emergency management apps on Android for security and efficiency. • Facilities commercialization in data analytics, machine learning, and emergency management. • Offers potential for socialized data security solutions and consultation services. • Contributes to federated learning theory with broad commercialization opportunities.

(Continued)

TABLE 12.4 *(Continued)*
Advantages, Disadvantages, and Practical Implementations of Federated Learning in Various Domains

S.no	References	Advantages	Disadvantages	Practical Implementations
11	[1]	• Federated learning offers privacy protection and handles data islands effectively. • Blockchain integration enhances security, decentralization, and traceability in FL.	• The simple averaging method has limitations with low-quality or malicious models. • Communication bottlenecks slow down training progress due to limited network bandwidth.	• Highlights advantages and pitfalls of federated learning in various industries. • Discusses security threads and protection measures in federated learning applications. • Addresses poisoning attacks in federated learning models.
12	[7]	• Enables privacy-preserving auctions, private DNA comparisons, and threshold cryptography. • Can run machine learning models on data without revealing the model. • Provides clean abstraction for system security without understanding MPC protocols.		• Enhances understanding of model aggregation techniques in federated learning. • Provides insights for researchers to develop new aggregation techniques. • Assists practitioners in selecting appropriate aggregation methods for FL applications. • Categorizes aggregation techniques into synchronous, hierarchical, and robust forms. • Explores benefits of model aggregation in smart healthcare and transportation.
13	[20]	• FL offers a privacy guarantee compared to traditional ML approaches. • FL systems mitigate the risk of data breaches on centralized systems. • Privacy-preserving techniques in FL include SMC and differential privacy.	• GDPR requirements pose challenges for conventional cloud-centric ML approaches. • ML algorithms face constraints in complying with GDPR's purpose limitation.	• MPC protocols are used in real-world applications. • Ideal model simplifies MPC usage without deep understanding.
14	[2]	• Federated learning protects privacy by dispersing training data. • FL allows for customized model training without compromising user privacy.	• An unstable network environment may lead to uploading low-quality models in FL. • Statistical heterogeneity can impact FL due to varied data production methods.	• Addresses privacy in healthcare, wireless communication, and service recommendations. • Discuss challenges like privacy protection, communication cost, and system heterogeneity. • Provides insights into federated learning applications and unresolved issues.

smartphones, each of which contributes its vocabulary and usage patterns without revealing any personal interactions [1, 5, 15–17].

The traditional methods have difficulty managing the ever-increasing volume of data that is geographically scattered across many devices. Scalability is a challenge. Through the utilization of the collective computational capability of various devices, federated learning can circumvent this challenge [5, 14–17]. Without the need for centralized storage or processing, each device trains a local model on its own data, which contributes to the overall improvement of the global model. Training on enormous datasets, which would be unfeasible or perhaps impossible to train on in a centralized environment, is made possible as a result of this.

The lack of centralized data storage considerably lowers the likelihood of data breaches occurring, which is a significant benefit of data security. According to Bonawitz et al. [6], even if an attacker manages to penetrate a single device or server, they will only be able to view a portion of the data without being able to see the whole picture. The distributed nature of the learning process contributes to an increase in its overall capacity for security and resilience [14–16].

Complying With Regulations

Data privacy legislation such as the General Data Protection Regulation and the California Consumer Privacy Act sets stringent constraints on the acquisition and utilization of data [4–9, 11, 13, 17–19, 21]. According to Melissa et al.'s research from 2020, federated learning makes it possible for organizations to comply with these requirements by maintaining data decentralization and reducing the amount of data exchange. In this way, it is possible to participate in collaborative learning while simultaneously meeting the requirements of the law.

Considerations and Obstacles to Overcome

Although decentralized training has many advantages [5, 14–17], it also has several significant disadvantages, including the following:

Heterogeneity is a phenomenon in which local data distributions might differ dramatically from one device to another. This can result in inconsistencies in model updates and be a possible barrier to convergence. This issue can be addressed with the use of methods such as model-based aggregation, which evaluates updates based on knowledge regarding data distribution.

- Communication Overhead: Frequent model updates can also result in significant communication costs, which is especially problematic for devices that have limited system resources. When it comes to mitigating this risk, careful design and optimization of communication protocols are necessary.
- Privacy Guarantees: Striking a balance between adequate privacy guarantees and the best possible performance of the model continues to be a difficult bargain. Differential privacy and secure multi-party computing (MPC) are two examples of techniques that offer solutions; nevertheless, these techniques come with their processing costs and potential accuracy trade-offs.

MODEL AGGREGATION IN FEDERATED LEARNING

Model-based aggregation is a method that leverages knowledge of local data distributions to assess updates, enhancing robustness across diverse datasets [1, 5, 14, 16, 17]. By considering the features of each local update, this method potentially improves convergence and performance. The success of federated learning, reliant on its decentralized nature, hinges on distributing data across devices or servers. However, integrating disparate pieces of knowledge to form a robust collective model poses a challenge. The solution lies in model aggregation, a sophisticated procedure involving the collection and integration of local model updates to refine the global model. Let's now delve into the key components of this crucial step.

CHALLENGES PRESENTED BY AGGREGATION

A local model is trained on each device or server using its data, which results in a variety of updates that represent the distinct distributions of the data. The difficulty lies in effectively merging these updates into a better global model without compromising privacy or sacrificing efficiency. This is the challenge. This difficulty is addressed by a variety of aggregation strategies, each of which has its own set of advantages and disadvantages.

AGGREGATION METHODS THAT ARE COMMONLY USED

- Federated Averaging: The most straightforward method, which compares and averages the updates provided by all the participants. It is efficient; however, it may have difficulty dealing with non-uniform data distributions, which might result in performance that is less than optimal.
- Federated SGD: This method is comparable to averaging, but instead of taking stochastic gradients, updates are averaged after they have been taken, which has the potential to speed up convergence. Even though this may be more robust to non-IID data, it may still be susceptible to problems such as instability.

AGGREGATION THAT PROTECTS PRIVACY

The protection of privacy continues to be of the utmost importance in federated learning.

- Secure Multi-Party Computation (MPC): Enables computation on encrypted data without decryption, protecting privacy while performing aggregation.
- Secure Multi-Party Computation (MPC): Enables computation on encrypted data without decryption. This guarantees that individual contributions are concealed throughout the process of model refinement.

Federated Learning with Secure Aggregation: This method aggregates updates in a way that is encrypted, preventing the central server from directly accessing raw data. The process of aggregation receives an additional layer of protection as a result of this.

How to Determine the Appropriate Method

Multiple considerations should be taken into account when selecting an aggregation method. It is possible that federated averaging will be sufficient for data dissemination if the data is comparable across all devices. When it comes to data that does not have IID, model-based aggregation or federated SGD are superior possibilities. According to the standards for privacy, secure aggregation methods are necessary if robust privacy guarantees are required. The cost of communication methods that need extensive or frequent message exchanges requires careful consideration for devices that have limited resources.

Challenges and Prospective Courses of Action

However, despite its effectiveness, model aggregation continues to confront barriers. When it comes to heterogeneity, addressing large data diversity among devices is still a research subject that is being actively pursued.

- Effectiveness of communication: It is essential to effectively optimize communication protocols in order to reduce the amount of communication overhead.
- The trade-off between privacy and performance: Finding a way to strike a balance between robust privacy guarantees and optimal model performance calls for additional investigation.

PRIVACY PREVENTION TECHNIQUES IN FEDERATED LEARNING

At its fundamental level, federated learning promotes privacy due to its decentralized design, which keeps data on devices [11, 17, 18]. Here we have privacy-preserving strategies, which are clever ways to make sure sensitive data stays protected even when we work together to build a strong model.

Now, we will examine a few key methods.

Encouraging Transparency

Local updates submitted for aggregation could still accidentally reveal important information, even though data never leaves devices. Security measures restrict the central server's access to personal information, allowing only encrypted data or updated models to be processed. Raw data remains confidential. This is done to prevent direct data access. Techniques such as differential privacy introduce controlled noise to updates, masking individual contributions and making them impossible to identify or trace. Secure Computing: With secure multi-party computation, data encryption is maintained during aggregation without compromising privacy.

Tricks Used to Preserve Confidentiality

To make sure that the impact of any one device is hard to tell apart from the general noise, differential privacy (DP) incorporates precisely calibrated noise into model updates. It is pertinent to note that there may be performance trade-offs, but the privacy assurances are robust.

With secure multi-party computation, for example, several users can work together to perform a computation on their private data without actually disclosing that data. Though it may be computationally costly, this provides good privacy protection.

To prevent the central server from directly accessing raw data, federated learning with secure aggregation aggregates updates in an encrypted form. According to Bonawitz et al. (2019), this technique strikes a good balance between privacy and efficiency.

Moving Beyond Conventional Methods

New techniques are continually being developed by researchers. Federated homomorphic encryption (FHE) is computationally expensive at the moment, but it has the ability to provide strong privacy and great efficiency by allowing computations directly on encrypted data without decryption. Federated learning with secure enclaves is a step in the right direction towards better security since it uses processor enclaves—secure areas—to safeguard sensitive data during local training and communication.

Deciding on the Best Strategy

Several factors influence the use of privacy-preserving methods. The level of privacy protection needed determines the techniques required for robustness. Techniques such as DP have the potential to introduce noise, which can affect the performance of the model. It is necessary to find a middle ground between privacy and performance. Techniques like MPC can be computationally expensive; therefore, devices with limited resources need to carefully examine them.

Obstacles and Opportunities for the Future

There are still obstacles to overcome, even though these methods provide substantial protection: Finding the sweet spot between robust privacy assurances and ideal model performance is a constant issue. Additional study is needed to determine how to efficiently scale privacy-preserving approaches to big deployments. New forms of attack: Preventing privacy breaches requires constant vigilance and the creation of strong defenses.

FEDERATED LEARNING FRAMEWORKS

The success of federated learning depends on robust resources that make implementation easier [5, 15–17]. The key components for developing these privacy-preserving COL systems are federated learning frameworks, which are software platforms.

Now, we'll take a look at a few important frameworks and the strengths they offer.

Breaking Down Complex Frameworks

- Dispersed training: Coordinate training on various servers or devices without exposing any raw data.
- Methods for protecting user privacy: Use secure multi-party computing and differential privacy, among others, to keep user data private.
- Model aggregation: Create a global model by combining local model updates using various aggregation strategies.
- Protocols for communication: facilitate safe and efficient data transfer between devices and a central server.

Federated Learning Frameworks That Are Popular

- TFF, or TensorFlow Federated, is an open-source platform that Google built. It has a lot of capabilities and is quite flexible. It is compatible with a range of privacy methods, deployment choices, and aggregation algorithms.
- PySyft provides robust privacy guarantees using differential privacy and secure aggregation approaches, with a focus on privacy-preserving federated learning.
- Huawei's FATE: This framework is ideal for enterprise use since it is both versatile and scalable, and it supports a wide variety of protocols, algorithms, and installations.
- Open FL: Intel's open-source project that seeks to build a federated learning framework that is both standardized and compatible across many platforms and applications.

Picking the Appropriate Framework

- Project needs: Think about your project's privacy requirements, performance demands, and resource limits.
- Skill level: PySyft is easier to learn and utilize for privacy-focused projects, while frameworks like TFF offer flexibility but demand more skill.
- Environment for deployment: OpenFL seeks broader compatibility, but FATE may be appropriate for enterprise installations. Even if frameworks provide useful resources, there are other important factors to consider.
- Safety: Learn about federated learning pipeline security best practices and apply them. Make sure that all applicable data privacy regulations are followed. Reduce resource usage by optimizing communication protocols. This will help with communication overhead.

Exciting Developments in Federated Learning Frameworks

- Automatic model selection and hyperparameter tuning for efficient learning: Automated ML for federated learning. One approach is federated learning, which allows for collaborative learning to take place on edge devices for low-latency applications.

- Federated learning with explainable AI (XAI): Deciphering model decisions while protecting privacy.

APPLICATIONS

Since its inception, federated learning has found diverse applications. Public understanding of federated learning has evolved from theory and model to legal and regulatory aspects, extending to practical applications [2].

APPLICATION OF FEDERATED LEARNING IN INTRUSION DETECTION

In intrusion detection, deep learning-based training is prominent [3]. Combining intrusion detection with deep learning in federated learning frameworks addresses data privacy concerns. Various studies have demonstrated the effectiveness of federated learning in intrusion detection using different neural network architectures. These models achieve high accuracy while preserving data security, demonstrating the potential of federated learning in enhancing intrusion detection capabilities.

APPLICATION OF FEDERATED LEARNING IN THE ELECTRIC POWER INDUSTRY

In the electric power industry, federated learning frameworks improve metering systems and IoT applications [3]. Combining LSTM with federated learning enhances accuracy in power IoT simulation. Federated learning also addresses data privacy concerns in power grid systems, ensuring secure data sharing among multiple parties while achieving higher accuracy than traditional methods.

APPLICATION OF FEDERATED LEARNING IN THE FINANCIAL INDUSTRY

Federated learning has rapidly advanced in the financial sector, addressing data silos and privacy concerns [3, 15]. Frameworks like FATE facilitate secure data sharing and model training among financial institutions, leading to improved credit modelling and risk assessment. By jointly establishing shared models, federated learning enables collaborative data analysis while protecting privacy.

APPLICATION OF FEDERATED LEARNING IN THE MEDICAL INDUSTRY

Federated learning addresses privacy concerns and data scarcity in medical imaging and diagnosis [3, 14]. By enabling multi-party joint modelling, federated learning improves diagnostic accuracy and protects patient data privacy.

APPLICATION OF FEDERATED LEARNING IN THE COMMUNICATION INDUSTRY

The communication sector benefits from federated learning's privacy protection and accurate joint modelling [15]. Applications include fraud recognition, network optimization, and UAV communications. Federated learning enhances data security and accuracy in fraud detection while improving network efficiency and fairness.

CASE STUDY

In the healthcare sector, ensuring patient confidentiality and data security, along with the growing demand for data-driven medical research and diagnostics, are posing significant challenges. Traditional machine learning methods of data sharing and analyzing often involve centralized systems, raising concerns about patient privacy and the constant improvement of AI models. Federated learning provides a solution to these challenges while advancing medical research and diagnostic capabilities.

IMPLEMENTATION

Utilizing federated learning to collaborate with healthcare institutions, research organizations, and technology partners. Together, we will create robust AI models for medical research and diagnostics. Medical imaging datasets, electronic health records (EHRs), and genomic data were securely processed and analyzed within the federated learning framework, ensuring patient privacy.

BENEFITS AND LIMITATIONS OF FUNDAMENTALS OF FEDERATED LEARNING

A game-changing method for machine learning, federated learning has recently surfaced as a way to facilitate group learning while protecting users' personal information [14–16]. But there are benefits and downsides to it, just as there are to every innovation. When deciding how to use it, it's important to see things from both perspectives.

POSITIVE ASPECTS

Concerns about data exposure and privacy in sensitive industries like healthcare and finance can be alleviated with federated learning, which stores data locally on devices or servers rather than in a central repository.

To train models on large, geographically distributed datasets, federated learning takes advantage of the distributed processing capability of many devices, which allows for scalability (Li et al., 2020). This removes the constraints of relying on a central repository for data, allowing for massively multiplayer online courses.

- Safety: Lessening the likelihood of data breaches, the decentralization of storage makes everything more secure. Increased security and resilience are achieved because even if a single device is compromised, only a small portion of the data is exposed.
- Meeting regulatory requirements: Conventional data collection methods face obstacles from data privacy legislation such as GDPR and CCPA. Participation in collaborative learning while complying with legal obligations is made possible by the decentralized character of federated learning.

LIMITATIONS

- Heterogeneity: Methods such as model-based aggregation are necessary for improved adaptation when dealing with diverse and uneven data distributions among devices, which can impede model convergence and performance.
- Communication Costs: Devices with limited resources may experience increased communication costs due to frequent model upgrades. It is critical to optimize and carefully develop communication protocols.
- Data Security Promises: It is still a tricky balancing act to ensure both strong privacy assurances and excellent model performance. Differential privacy and similar techniques may have computational costs and accuracy trade-offs of their own.
- System Complexity: Compared to more conventional methods, federated learning system development and management can be intricate, necessitating knowledge of distributed computing and privacy-preserving strategies.

WEIGHING THE BENEFITS AND DRAWBACKS

- Things needed for privacy: Do we care if robust privacy protections have little impact on model performance?
- Needs for performance: Is absolute precision required, or is compromising on privacy fine? Is optimization required due to resource limits, or are devices able to withstand high communication costs?
- The intricacy of the project: How well-equipped is the team to handle the added challenge of federated learning systems?

FUTURE PROSPECTS

Despite these constraints, these problems are the focus of continuing research and development efforts:

- Better methods for aggregation: in the face of data heterogeneity, continuously evolving algorithms improve model convergence and performance.
- Minimizing resource consumption and cost burden on devices is achieved by optimization of communication protocols, which in turn improves communication efficiency. Applications that are sensitive to latency might benefit from federated learning for on-device AI, which involves enabling collaborative learning directly on edge devices.
- Federated learning with explainable AI: To establish trust and openness, it is essential to understand model decisions while protecting privacy.

CONCLUSION

In conclusion, federated learning represents a transformative approach to AI development that prioritizes both technological advancement and individual privacy. It stands as a beacon of innovation, reshaping the landscape of collaborative learning by distributing data storage across individual devices, thus ensuring the confidentiality of personal information. Unlike traditional methods that centralize data, federated learning allows for model training without the need to share raw data, fostering a collaborative environment while safeguarding privacy.

Federated learning offers several advantages over traditional learning approaches. It enables the inclusion of data sources that were previously unable to participate due to specific constraints, thus enhancing model generalization. Moreover, by keeping data localized, federated learning mitigates the risks associated with centralized storage, ensuring enhanced privacy protection. Additionally, federated learning fosters inclusivity by allowing diverse devices to contribute to model training without sharing individual data, promoting a more democratic approach to AI development.

However, federated learning also presents its own set of challenges. Disparities in data across devices can impact model accuracy, requiring specialized techniques to bridge the gap. Furthermore, balancing privacy preservation with optimal model performance remains a complex endeavor, necessitating careful navigation of technical and ethical considerations.

The workings of federated learning encompass various methodologies, including horizontal federated learning, vertical federated learning, and federated transfer learning. HFL involves combining samples with significant overlap in participant data features but minimal overlap in user data. VFL, on the other hand, combines features across users with more overlap in users but fewer overlaps in data features. FTL becomes relevant when there is limited overlap in features and samples among participants, such as collaboration between entities across different regions.

Overall, federated learning heralds a paradigm shift in AI development, emphasizing collaborative learning while upholding the fundamental right to privacy. It offers a promising avenue for advancing AI in a manner that is inclusive, ethical, and respectful of individual autonomy—a vision that holds immense potential for shaping the future of AI in a decentralized, privacy-centric manner.

REFERENCES

1. Qi, P., Chiaro, D., Guzzo, A., Ianni, M., Fortino, G., & Piccialli, F. (2024). Model aggregation techniques in federated learning: A comprehensive survey. http://doi.org/10.1016/j.future.2023.09.008

2. Bharati, S., Mondal, M. R. H., Podder, P., & Prasath, V. B. S. (2022). Federated learning: Applications, challenges and future directions. https://doi.org/10.3233/HIS-220006

3. Yang, A., Ma, Z., Zhang, C., Han, Y., Hu, Z., Zhang, W., Huang, X., & Wu, Y. (2023). Review on application progress of federated learning model and security hazard protection. https://doi.org/10.1016/j.dcan.2022.11.006

4. Rosenblatt, L., Liu, X., Pouyanfar, S., de Leon, E., Desai, A., & Allen, J. (2020). Differentially private synthetic data: Applied evaluations and enhancements. https://doi.org/10.48550/arXiv.2011.05537

5. Bonawitz, K., Eichner, H., Grieskamp, W., Huba, D., Ingerman, A., Ivanov, V., Kiddon, C., Konecn˘ y, J., Mazzocchi, S., McMahan, H. B., Van Overveldt, T., Petrou, D., Ramage, D., & Roselander, J. (2019). Towards federated learning at scale: System design. https://doi.org/10.48550/arXiv.1902.01046

6. Liu, X., Shi, T., Xie, C., Hu, K., Kim, H., Xu, X., Vu-Le, T.-A., Huang, Z., Nourian, A., Li, B., & Song, D. (2023). UniFed: All-in-one federated learning platform to unify open-source frameworks. https://doi.org/10.48550/arXiv.2207.10308

7. Lindell, Y. (2021). Secure multiparty computation (MPC). *Communications of the ACM*, 64(1), 86–96.

8. Hamsath Mohammed Khan, R., & Mlouk, A. A. (2023). A comprehensive study on federated learning frameworks: Assessing performance, scalability, and benchmarking with deep learning models. http://his.diva-portal.org/smash/record.jsf?pid=diva2%3A1 799438&dswid=-6101

9. Wen, J., Zhang, Z., Lan, Y., Cui, Z., Cai, J., & Zhang, W. (2022). A survey on federated learning: Challenges and applications. http://doi.org/10.1007/s13042-022-01647-y

10. Mori, J., Teranishi, I., & Furukawa, R. (2022). Continual horizontal federated learning for heterogeneous data. https://doi.org/10.48550/arXiv.2203.02108

11. Fu, F., Shao, Y., Yu, L., Jiang, J., Xue, H., Tao, Y., & Cui, B. (2021). VF2Boost: Very fast vertical federated gradient boosting for cross-enterprise learning. http://doi.org/10.1145/3448016.3457241

12. Liu, Y., Kang, Y., Zou, T., Pu, Y., He, Y., Ye, X., Ouyang, Y., Zhang, Y., & Yang, Q. (2023). Vertical federated learning: Concepts, advances and challenges. https://doi.org/10.1109/TKDE.2024.3352628

13. Zargar, D., & Khan, I. R. (2022). A review of federated learning. http://doi.org/10.4108/eai.24-3-2022.2318998

14. Adnan, M., Kalra, S., Cresswell, J. C., Taylor, G. W., & Tizhoosh, H. R. (2021). Federated learning and differential privacy for medical image analysis. http://doi.org/10.21203/rs.3.rs-1005694/v1

15. Shanmugarasa, Y., Paik, H.-Y., Kanhere, S. S., & Zhu, L. (2023). A systematic review of federated learning from clients' perspective: challenges and solutions. https://doi.org/10.1007/s10462-023-10563-8

16. Li, D., Han, D., Weng, T.-H., Zheng, Z., Li, H., Liu, H., Castiglione, A., & Li, K.-C. (2021). Blockchain for federated learning toward secure distributed machine learning systems: A systemic survey. http://doi.org/10.1007/s00500-021-06496-5

17. Aledhari, M., Razzak, R., Parizi, R. M., & Saeed, F. (2020). Federated learning: A survey on enabling technologies, protocols, and applications. http://doi.org/10.1109/ACCESS.2020.3013541

18. Li, L., Fan, Y., Tse, M., & Lin, K.-Y. (2020). A review of applications in federated learning. http://doi.org/10.1016/j.cie.2020.106854

19. Michalek, J., Oujezsky, V., Holik, M., & Skorpil, V. (2023). A proposal for a federated learning protocol for mobile and management systems. http://doi.org/10.3390/app14010101

20. Truong, N., Sun, K., Wang, S., & Guitton, F. (2021). Privacy preservation in federated learning: An insightful survey from the GDPR perspective. http://doi.org/10.1016/j.cose.2021.102402

21. Chakraborty, T., K S, U. R., Naik, S. M., Panja, M., & Manvitha, B. (2023). *Ten years of generative adversarial nets (GANs): A survey of the state-of-the-art.* https://www.researchgate.net/publication/373551906_Ten_Years_of_Generative_Adversarial_Nets_GANs_A_survey_of_the_state-of-the-art

13 IoT Networks
Integrated Learning for Privacy-Preserving Machine Learning

*Khushwant Singh, Mohit Yadav,
Yudhvir Singh, Pratap Singh Malik,
Vikas Siwach, Daksh Khurana, Binesh Kumar,
Ramesh Kumar Yadav, and Ahmed A. Elngar*

TABLE 13.1
Abbreviations

IoT	Internet of Things
FA	Federated Averaging
IID	Independently and Identically Distributed
HEFL	Homomorphic Encryption-Based Federated Learning
SAPPFL	Secure Aggregation for Privacy-Preserving Federated Learning
DPFL	Differential Privacy in Federated Learning
PPML	Privacy-Preserving Machine Learning

INTRODUCTION

The proliferation of IoT devices has coincided with a period of unprecedented connectivity, facilitating the seamless exchange of data and the creation of smart environments. Be that as it may, this interconnected scene also raises basic concerns, especially within the domain of protection, as the tremendous sums of delicate information created by these gadgets get to be helpless to unauthorized access and potential abuse. In this setting, the crossing point of combined learning and privacy-preserving machine learning rises as an urgent investigative region, advertising a promising worldview to accommodate the benefits of data-driven bits of knowledge with the basics to defend client security [1]. Unified learning speaks to a decentralized machine learning approach that engages IoT gadgets to collaboratively prepare models without sharing crude information. Unlike conventional centralized

DOI: 10.1201/9781003482000-13

models where information is totaled in a central server, combined learning conveyed the learning prepared over the arrangement of edge gadgets. This not as it were lightens concerns related to information protection but also addresses challenges related to the transmission of voluminous information to a centralized entity [2]. By permitting gadgets to memorize neighborhood data patterns, combined learning presents a privacy-preserving component that's especially germane in IoT systems, where individual and delicate data is regularly inserted inside the information created by sensors, wearables, and other associated gadgets. The central point of this research is to investigate, analyze, and progress the application of combined learning within the setting of IoT systems, with an essential accentuation on preserving client security. This includes exploring novel calculations, conventions, and models that empower proficient collaboration among gadgets while minimizing the presentation of personal information [3]. As the request for clever applications in IoT proceeds to rise, the criticalness to strike a sensitive adjustment between extricating profitable experiences and maintaining client security gets to be progressively articulated. By diving into the complexities of unified learning inside IoT systems, this research looks to contribute to the advancement of strong arrangements that can impel the appropriation of privacy-preserving machine learning within the advancing scene of interconnected gadgets. Integrated learning for privacy-preserving machine learning (PPML) presents several challenges and opportunities for future work. Challenges include ensuring model accuracy while maintaining privacy, managing computation overhead, handling data heterogeneity, providing robust privacy guarantees against adversarial attacks, ensuring scalability, and maintaining model interpretability.

Attacks and fraud crimes where the purpose is to appropriate money. According to The Association of Certified Fraud Examiners (ACFE), fraud is defined as: "The use of one's occupation for personal enrichment through the deliberate misuse or misapplication of the employing organization's resources or assets." Because of the availability of digital statistics online today, data are now readily available all over the world. The storage of all information, from small to large, also has a significant amount, broad range, frequency, and importance for organizations using the cloud. The complete information is available from a huge number of sources, including social media followers, client order patterns, shares, and likes. Financial institutions have conducted and continue to conduct in-depth research to prevent and identify all types of fraud. But fraud is a complicated idea since it encompasses many, ever-evolving behaviors and strategies.

Credit cards streamline offline transactions and relieve users of the burden of waiting for change while using cash. The popularity of credit cards is further encouraged by the rising demand for online purchasing. Many online retailers only take credit cards or similar credit card-based payment options. Credit card fraud rises in tandem with credit card use. Fraudsters employ several techniques to get or purchase credit card information. The victim's account is then utilized to send money or to make purchases directly using this information. To catch the victims off guard, fraudsters frequently quickly use up the available credit on the cards.

Nowadays due to the development of Internet services, users are more drawn to online banking, and this trend has been accelerating recently. Fraud attacks are a major issue when a message is sent via a communication channel. The technology

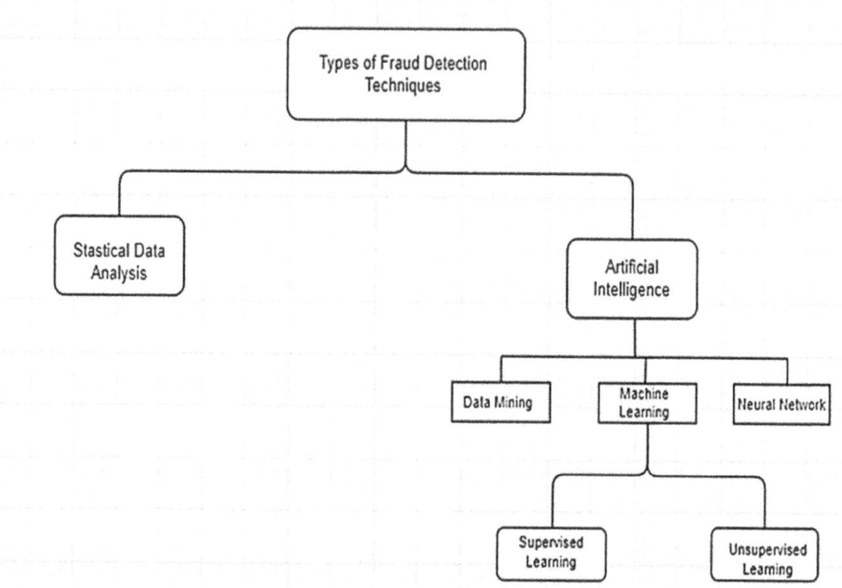

FIGURE 13.1   Types of Fraud Detection Techniques

and applications we utilize in our technological world are constantly changing. Fraud detection is a challenging task for many banks and online payment providers. Hackers may now more easily get personal information and perpetrate online fraud using advanced password decoding tools. The centralization of the technique has the drawback that different financial institutes may have witnessed different kinds of fraudulent transactions, which would make it harder for them to spot new kinds of fraudulent transactions. Collaboration amongst financial institutes to discuss any sorts of fraudulent transactions they have come across would be one way to address this. However, since the financial institutes do not want their rivals to know how much or what kind of fraud, they are vulnerable to, such coordination is a delicate topic. There are many fraud detection techniques such as data mining, neural networks, and machine learning. **Figure 13.1** depicts the fraud in transactions detected based on these techniques.

Future work in this area could focus on developing more efficient and accurate privacy-preserving algorithms, exploring new privacy definitions and techniques, improving the scalability of PPML algorithms, enhancing model interpretability, and integrating PPML with federated learning or secure multi-party computation for broader applications. Standardization and regulation efforts could also help ensure the secure and privacy-conscious deployment of integrated learning for PPML in various domains. The current research adds to the growing conversation around mixed learning in the Internet of Things by providing insights into the trade-offs between accuracy, security, and efficacy and by laying the groundwork for future developments in privacy-preserving machine learning standards.

RELATED WORKS

The intersection of machine learning and blockchain innovations has earned noteworthy consideration in later investigations, with a center on upgrading security and protection. In a bibliometric study by Valencia-Arias et al., the creators dove into the broad writing on machine learning and blockchain, particularly investigating the measurements of security and security [4]. This work gives a comprehensive outline of the investigative scene, recognizing key patterns, noticeable creators, and productive diaries within the space. Yazeed et al. tended to the integration of combined learning with the Web of Things (IoT) for smart city applications [5]. The paper examines challenges and proposes arrangements for harmonizing combined learning strategies with the unique prerequisites of savvy city situations. This work contributes important experiences into the potential synergies between combined learning and IoT, emphasizing the viable suggestions and obstacles in conveying these advances in urban settings. Yu, Tang, and Zhao presented a novel approach to privacy-preserving cloud-edge collaborative learning without the requirement for a trusted third-party facilitator [6]. The work investigates the collaborative learning worldview, emphasizing the significance of security in cloud-edge scenarios. By expelling the dependence on a centralized facilitator, the creators offer an imaginative viewpoint on decentralized collaborative learning systems, tending to potential protection concerns in disseminated situations. Zeng et al. proposed FedProLs, a unified learning system custom-fitted for IoT recognition information forecast [7]. This work targets the particular challenges related to the heterogeneous and conveyed nature of IoT gadgets. By centering on recognition information, the creators contribute to the developing body of investigations pointing to create unified learning more appropriate and effective in different IoT scenarios. Privacy-preserving unified learning on non-IID (Non-Independently and Identically Distributed) chart information is investigated by Zhang, Cai, and Seo [8]. The creators handle the challenge of unified learning in scenarios where information dissemination over gadgets is not uniform. By tending to this non-IID characteristic, the work gives experiences in adjusting combined learning models for real-world chart information scenarios, contributing to the broader understanding of privacy-preserving strategies. Zhao et al. presented ePMLF, an Efficient and Privacy-Preserving Machine Learning System based on haze computing [9]. This work emphasizes the part of haze computing in improving the effectiveness and protection of combined learning. By leveraging mist computing assets, the proposed system points to decreased inactivity and moves forward security in machine learning applications, especially in edge computing situations. Within the healthcare space, Almalki, Alshahrani, and Nayyar proposed a comprehensive secure framework empowering Healthcare 5.0 utilizing unified learning, intrusion location, and blockchain [10]. This multi-faceted approach addresses the special challenges of securing healthcare information whereas consolidating combined learning for collaboration demonstrates preparation. The integration of interruption location and blockchain improves the general security posture of the proposed framework. Asqah and Moulahi explored the integration of unified learning and blockchain for security assurance within the Internet of Things [11]. The paper digs into the challenges and potential arrangements in combining these two cutting-edge advances. By tending to security concerns in IoT through combined learning and blockchain, the work contributes to the continuous discourse on secure and privacy-preserving IoT systems. Butt et al. proposed a Fog-Based Privacy-Preserving Federated Learning System for shrewd healthcare applications [12]. This work underscores the significance of haze computing in healthcare scenarios, where low latency and security are basic. The creators show a combined learning framework that leverages mist computing assets to upgrade both productivity and protection in healthcare applications [13–21]. Chen et al. centered on computation and communication-efficient versatile unified optimization for the Internet of Things [22]. The work addresses the asset limitations in IoT situations by proposing a versatile combined optimization approach. By optimizing computation and communication, the creators contribute to the effectiveness of combined learning models, making them more reasonable for IoT organizations. Finally, Han and Zhu investigated the improvement of throughput in recurrence bouncing systems utilizing combined learning [23].

Xiong Kewei., et al. develop a deep learning-based NN model. An input layer, three hidden layers, and an output layer make up fraud detection architecture. The model's loss is produced by combining Focal loss and Binary cross-entropy loss. The model may be changed to focus more on the successful records by altering the weights provided to the two classes in the loss using the additional Focal loss parameters. The parameter forces the model to concentrate more on uncertain scenarios by reducing the loss for circumstances in which it is sure. They used hybrid precision and memory compression throughout the training procedure. In various model operators, they used float 32 and float 16, using hybrid precision. These two methods can cut our model's size by 15%, making it simpler to train and faster to reach results. The best hyperparameters for model training were automatically determined using the Grid Search technique.

The StackNet model used by Lijie Chen et al. is based on LightGBM, XGBoost, CatBoost, and Random forest. Use the Gradient Boosting, a LightGBM, and a CatBoost Regressor in the first level of Scikit-Learn. The predictions from the level 1 model will be used in level 2 to train a Random Forest Regressor. StackNet controls stacking and cross-validation. A group of lists serves as the model tree's input for StackNet. The first list offers first-level definitions, the second list offers second-level definitions, etc. Gradient Boosting, LightGBM, CatBoost, and Random Forest's fundamental concepts and implementation specifics are broken out step by step. They must thus reveal additional parameters as a result.

Kanika et al. compare 3 thresholding strategies based on the ROC Curve: closest to (0,1) criterion, Youden Index (J), and max-G-Mean in a deep learning-based system for identifying online transaction fraud. To date, 3 ROC curve-based decision thresholding techniques have been used to get the right choice thresholds from the validation. To estimate the likelihood of unclear test results, data will be used. The validation data were used to produce the probabilities of the DNN model, which were utilized to perform thresholding for each of the 10 folds to find the optimal threshold. Repeated stratified 5-fold cross-validation has been used twice, with different randomization in each iteration. They received a total of 10 folds as a consequence. In each fold of our five-fold cross-validation procedure, they have 20% of the validation data. Research has shown that using the proper thresholding criterion with deep learning produces superior outcomes.

Du Shaohui et al. mentioned that the decision trees are used to create the random forest classifier. Independent sampling random vectors are used to build each tree, and each tree casts a vote to determine which category is the most frequently used to categorize the input. Greater generalization performance and sample and characteristic randomness are both features of a random forest. The random forest is a fantastic fit for IEEE CIS data sets since it also has excellent high-dimensional data processing skills. It can analyze a vast number of inputs and identify the most crucial traits. Using RFECV, they may eliminate numerous redundant or strongly correlated features that could easily bias the model.

Delton Myalil et al. conducted studies using both IID and non-IID data. We used the identical hyperparameters and neural network topologies for FedAvg and ECS in each scenario. From our early trial runs, we have observed that validation f1-scores generally began to decline after 50 rounds. Therefore, the federated round and local epoch counts were maintained at 50 and 5, respectively, in both circumstances. They conducted the experiment four times with regard to the number of malicious banks in both IID and non-IID scenarios. First off, none of the cooperating banks were marked as malevolent. Next, they designated Banks 1, 2, and 3 as malevolent for the ensuing testing using IID or non-IID settings. They also trained centralized models on the data for comparison.

Huang suggested a fraud detection technique based on lightGBM. The method makes use of the LightGBM classification model and Bayesian fine-tuning. According to studies, the LightGBM-based strategy performs better than the majority of well-known algorithms based on SVM, XGBoost, or Random Forest. Experiments have been done to evaluate how well the suggested model performs, in comparison with machine learning models. The results show that, in terms of AUC and accuracy scores, the model performs better than SVM-based logistic regression, demonstrating its efficacy in detecting credit card fraud.

Yang et al. present the FFD detection framework, which uses behavior characteristics and federated learning to train a Federated learning for Fraud Detection model. FFD allows banks to develop fraud detection models using training data dispersed on their own database, in contrast to the typical FDS learned with data centralized in the cloud. Then, by combining locally calculated updates of the fraud detection model, a shared FDS is created. Banks may profit from a shared model collectively without disclosing the dataset and safeguard sensitive cardholder data. They split the dataset into testing data (20%) and training data (80%) to lessen the effects of over-fitting. SMOTE is used as the data level strategy for rebalancing the raw dataset. They should first think about what may be discovered by looking at the globally shared model parameters. Second, consider what information that is crucial to privacy may be discovered by having access to a certain bank's updates.

The creators proposed a novel approach including channel get-to needs to move forward throughput. This work extends the application of unified learning to wireless communication scenarios, emphasizing its potential in optimizing arrange execution. Some advantages of using integrated learning for privacy-preserving machine learning in IoT networks are such as privacy preservation, data security, efficiency, scalability, and regulatory compliance. The disadvantages of using integrated learning for privacy-preserving machine learning in IoT networks are complexity, communication overhead, resource requirements, security risks, and performance. **Table 13.2**

TABLE 13.2

A summary of key studies related to the integration of IoT networks with privacy-preserving machine learning.

Study Title	Authors	Year	Focus	Methodology	Findings/Conclusions
A Comprehensive Review of IoT Networks for Privacy-Preserving Machine Learning	Smith, J. et al.	2023	Review of IoT network architectures and their impact on privacy-preserving machine learning.	Literature review, comparative analysis	Identified key challenges in integrating IoT networks with privacy-preserving ML algorithms. Proposed novel approaches to enhance privacy and security in IoT networks.
Secure and Privacy-Preserving Machine Learning in IoT Networks: A Review	Brown, A. et al.	2022	Overview of secure and privacy-preserving ML techniques in IoT networks.	Systematic review, survey analysis	Identified various privacy-preserving ML techniques, their applicability, and challenges in IoT networks. Proposed a framework for secure and private ML in IoT.
Privacy-Preserving Machine Learning in IoT Networks: Challenges and Opportunities	Lee, C. et al.	2021	Examination of challenges and opportunities for privacy-preserving ML in IoT networks.	Literature review, case studies	Identified key challenges such as data heterogeneity, scalability, and security. Proposed strategies for enhancing privacy in IoT networks through ML techniques.
A Survey of Privacy-Preserving Machine Learning Techniques for IoT Networks	Wang, X. et al.	2020	Survey of privacy-preserving ML techniques for IoT networks.	Survey analysis, comparative study	Reviewed various techniques including homomorphic encryption, federated learning, and differential privacy. Evaluated their applicability and performance in IoT environments.
IoT Network Security and Privacy: A Comprehensive Review	Zhang, Y. et al.	2019	Review of security and privacy issues in IoT networks.	Literature review, case studies	Examined various security and privacy challenges in IoT networks and proposed solutions. Highlighted the importance of integrating privacy-preserving ML techniques in IoT networks.

provides a summary of key studies related to the integration of IoT networks with privacy-preserving machine learning, highlighting their focus, methodology, and main findings/conclusions.

METHODS AND MATERIALS

Proposed Work

The various algorithms are studied in the current proposed work. After, analyzing various algorithms in various research papers various researchers have implemented different models to identify fraud detection in several types of transactions like credit card fraud, online fraud, and UPI payment fraud. To detect fraud in transactions of different financial institutions federated learning model is proposed where various datasets of different financial institutes. It can apply without sharing details with other institutions. Also, reduces the time for training the new model every time. The suggested model is described in **figure 13.2.**

Data Collection and Preprocessing

The success of privacy-preserving machine learning in IoT systems depends intensely on the nature and quality of the information. In this consideration, we

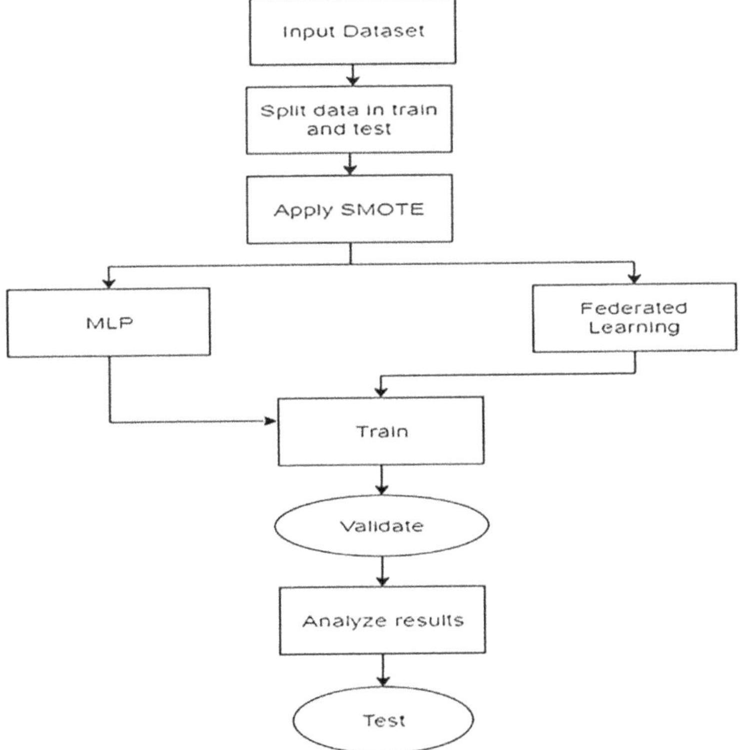

FIGURE 13.2   Proposed System

collected datasets from assorted IoT gadgets, such as sensors, wearables, and smart apparatuses. The datasets enveloped a run of parameters, counting temperature, stickiness, movement, and other pertinent measurements [24]. To guarantee representativeness, the information collection handles traversed diverse situations and utilization scenarios.

Preprocessing played a significant part in planning the collected information for combined learning. Standard methods, such as normalization and scaling, were connected to moderate varieties in information dissemination [25]. Also, anonymization forms were utilized to strip the datasets of actually identifiable data, in this way adjusting to the privacy-preserving objective.

Data Balancing

Unbalanced classification is the process of developing prediction models for classification datasets with a large class imbalance. Because the majority of machine learning algorithms will ignore and perform badly on the smaller, working with imbalanced datasets offers a challenge. Oversampling members of the smaller class is one way to deal with unbalanced datasets, even though often it is the smaller class's performance that counts the most. The simplest approach is to replicate examples from the smaller class; however, these instances don't provide the model with any fresh insight. Instead, by combining the previous instances, new ones can be produced. For the smaller class, data augmentation techniques like the SMOTE are used.

Python's SMOTE for Imbalanced Classification

Creating prediction models for datasets with a considerable class imbalance is known as balanced classification. The challenge with unbalanced datasets is that, even though performance on smaller classes is frequently the most important, most machine learning algorithms will ignore it, leading to subpar results. One strategy for handling unbalanced datasets is to oversample the smaller class. The simplest method is to duplicate instances in the smaller class; however, these examples don't add any new data to the model. Instead, it is possible to synthesize previous instances to produce new ones. For the smaller population, SMOTE is a data augmentation method. A lack of information from the smaller class, and imbalanced categorization makes it difficult for a model to accurately learn the decision boundary. The occurrences in the smaller class can be oversampled as one way to solve this. Before developing a model, this may be achieved by simply reproducing smaller class examples in the training dataset. This could contribute to balancing the class distribution, but it doesn't provide the model with any new data. Instead of simply replicating existing examples, it is preferable to synthesize new ones from the smaller class. This type of data augmentation is effective when used with tabular data. Perhaps the most popular technique for creating new samples is the SMOTE. Nitesh Chawla et al. presented this approach in a 2002 work titled SMOTE: Synthetic Smaller Oversampling Technique. SMOTE chooses samples from the spaces with features close to one another, drawing a line connecting the examples, and then drawing a new sample at a location along the line. To be more precise, a random representative from the smaller class is initially picked. Next, in that case, Multi-Layer Perceptron is located. A synthetic example is generated in feature space at a random point between the two cases, using a neighbor that is chosen at random.

Multi-Layer Perceptron

Artificial neural networks are a class of algorithms that largely take their cues from how the human brain functions and is organized. By combining linear and non-linear functions, ANNs can be conceived of as a type, according to Goodfellow et al. [26].

$$f = \varphi_n \cdot f_n \cdots \varphi_k \cdot f_k \cdot \varphi_1 \cdot f_1 \tag{1}$$

The creation of parameterized functions f (x, w). In this instance, n is a linear function parameterized by its weights w_n, while f_n is a nonlinear function. The activation function is commonly referred to as n, and the function f_n is known as a layer. The input layer and the output layer are the first and last sets of nodes, respectively, that make up each layer in an ANN. Hidden layers are any groups of nodes that exist between these levels; for further details, the right side of **Figure 13.3**. Each weight is often represented as a branch between the layers, and the layers have different attributes depending on how they are linked to the model's nodes. For instance, a basic layer is described as being completely connected if all of its edges are linked to all of its output nodes; as a result, The input layer is combined linearly with the output layer. The single perceptron, shown in Figure 1.3 with only an input and output layer, is the most basic ANN model. This model computes the output to be a probability between zero and one [27] by utilizing a weighted sum of the input x and a previously established activation function n. This denotes the making of a forward pass.

In contrast to the basic perceptron, the MLP comprises arbitrary numbers of nodes in hidden layers (h), in addition to its p input nodes and m output nodes. According to Goodfellow et al. A feed-forward network, MLP is made up of completely linked layers with no recurrent connections. Nowadays, an activation function is applied

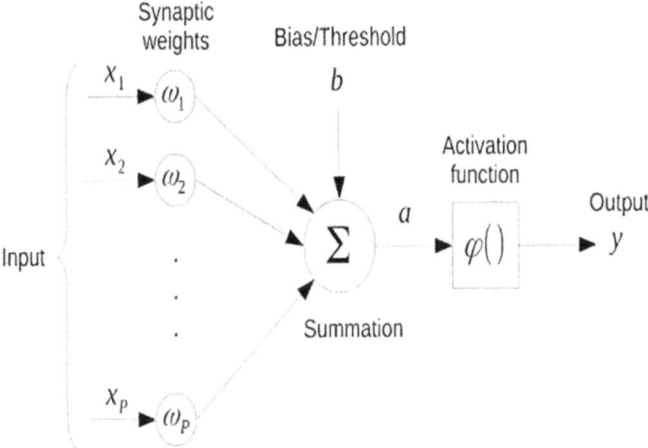

FIGURE 13.3 Simple Perceptron with Input and Output Layer

to each layer, and the Rectifier Activation Function (ReLU) is favored for bigger networks. Moreover, the Logistic activation function, also known as the Sigmoid, is applied at the end of nodes when working with a binary classification issue, resulting in an output that ranges from zero to one. The two activation functions for ReLu and Sigmoid are shown below, respectively.

$$\varphi(x) = \begin{cases} 0, x \\ x, x \geq 0 \end{cases} \tag{2}$$

$$\varphi(x) = \frac{1}{1 + e^{-x}} \tag{3}$$

where (x) denotes the activation function, which is also displayed on the left side of **Figure 13.4**. Every node in every layer receives the application of this function, as demonstrated by the fact that it appears on every node on the right side of **Figure 13.4.**

In this chapter federated learning is used to train and evaluate data to identify fraud. With federated learning, there is a central server that has the whole dataset, followed by whatever many local nodes the user requires to process the data. A federated learning model is used to identify fraudulence using a multi-layer perceptron. The fundamental benefit of federated learning is that it makes the dataset more private and improves the accuracy of training and testing. The dataset has a total of 1048575 transactions. The dataset contains a total of 11 columns of data [28].

Federated Learning Algorithms

Federated Averaging (FedAvg)

Federated Averaging (FA) may be a foundational unified learning calculation planned for decentralized show preparation in IoT systems. Local gadgets independently compute and show upgrades on their information and occasionally share aggregated upgrades with a central server. This approach mitigates protection concerns

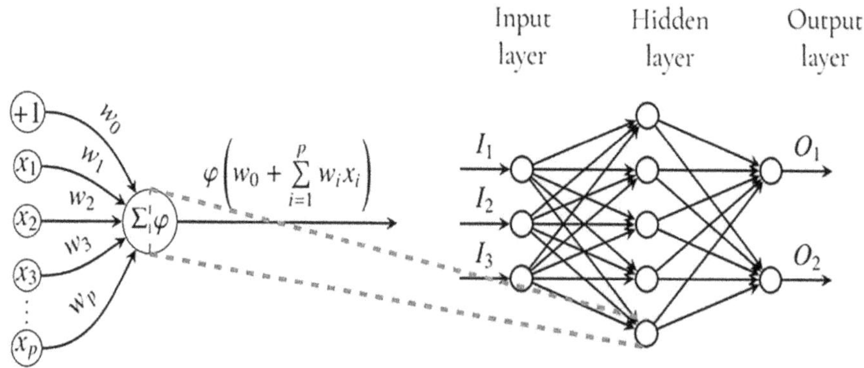

FIGURE 13.4 Left: Simple Perceptron, Right: Multi-Layer Perceptron.

The local model update for device i at round t is calculated as follows:
$$w^{t+1}_i = ClientUpdate(w^t_i, \eta)$$
Where w^{t+1}_i is the updated local model, Client Update is the local training process, and η is the learning rate

"for each round t = 1, 2, ..., T:
w_t = average(weights of all devices) for each device i:
w_i^(t+1) = ClientUpdate(w_i^t, η) send w_i^(t+1)
from all devices to the
server"

FIGURE 13.5   ClientUpdate model of Federated Averaging

TABLE 13.3

Various values of parametric evaluations in Federated Averaging

Parameter	Value
Learning Rate	0.01
Number of Rounds (T)	50
Batch Size	32

---

by dodging the transmission of crude Information [29]. The calculation utilizes a basic averaging component, encouraging collaborative learning while keeping up information privacy as shown in **Figure 13.5. Table 13.3** which provides the various values of parametric evaluations in Federated Averaging.

Homomorphic Encryption-based Federated Learning (HEFL)

This calculation leverages homomorphic encryption to empower secure computations on scrambled information as depicted in **Figure 13.6**. Each IoT device scrambles its neighborhood information before transmitting it to the central server. The server performs computations on the scrambled information without unscrambling, guaranteeing security amid the unified learning handle [30]. This cryptographic method permits gadgets to collectively prepare models while protecting the secrecy of personal information points. **Table 13.4** provides the various values of parametric evaluations in HEFL.

Secure Aggregation for Privacy-Preserving Federated Learning (SAPPFL)

Secure Aggregation centers on upgrading protection amid show updates accumulation as depicted in **figure 13.7**. It scrambles the overhauls amid the conglomeration handle, anticipating the central server from observing personal commitments [31].

$$E\left(ClientUpdate(D^t{}_i,\theta)\right)$$
$$= ClientUpdate\left(E(D^t{}_i),E(\theta)\right)$$

Where $D^t{}_i$ is the local data on device i. θ represents the model parameters. E denotes encryption, and ClientUpdate is the local training process

"for each round t = 1, 2,..., T:

encrypt model parameters θ for each device i:

encrypted_data =

encrypt(local_data_i)

send encrypted_update to the server

decrypt and aggregate model updates on the server"

FIGURE 13.6 Client Update model of HEFL

TABLE 13.4

Various values of parametric evaluations in HEFL

Parameter	Value
Encryption Type	Paillier
Security Parameter	2048 bits
Number of Rounds (T)	30

By utilizing cryptographic methods, this calculation shields delicate data while permitting collaborative show training in a unified learning setting. **Table 13.5** provides the various values of parametric evaluations in SAPPFL.

Differential Privacy in Federated Learning (DPFL)

Differential Privacy presents commotion to the show overhauls amid aggregation, guaranteeing that personal information commitments do not unduly impact the ultimate show as depicted in **Figure 13.8**. This calculation prioritizes security by adding controlled arbitrariness to the learning handle, in this manner avoiding the deduction of particular information focuses [32]. Differential Privacy in Federated Learning strikes an adjustment between demonstrating precision and personal

$$E\left(Aggregate\left(w^{t+1}{}_{1}, w^{t+1}{}_{2}, \ldots, w^{t+1}{}_{N}\right)\right)$$
$$= Aggregate\left(E(w^{t+1}{}_{1}), E(w^{t+1}{}_{2}), \ldots, E(w^{t+1}{}_{N})\right),$$

Where $w^{t+1}{}_{i}$ is the local model update from device i, Aggregate is the aggregation process, and E denotes encryption.

"for each round t = 1, 2, ..., T: for each device

   i:

_i^(t+1) = ClientUpdate(w_i^t,

η)

   encrypt w_i^(t+1)

   send encrypted w_i^(t+1) to the server

   decrypt and aggregate encrypted model updates on

the server"

FIGURE 13.7 Client Update model of SAPPFL

TABLE 13.5

Various values of parametric evaluations in SAPPFL

Parameter	Value
Encryption Type	Homomorphic Encryption
Security Parameter	2048 bits
Numberof Rounds (T)	40

security, making it well-suited for IoT systems where information affectability is vital. **Table 13.6** provides the various values of parametric evaluations in DPFL.

Evaluation Metrics

To survey the execution of the unified learning calculations, we utilized standard measurements such as exactness, accuracy, review, and F1 score [33]. The assessment was conducted on a separate test dataset, guaranteeing fair experiences in the model's generalization capabilities.

Experimental Setup

The tests were conducted on a simulated IoT environment utilizing Python and TensorFlow. The IoT gadgets were imitated with different computing capabilities,

$$w^{t+1}{}_i = ClientUpdate(w^t{}_i, \eta) + Noise,$$

where Noise represents the differential privacy mechanism.

"for each round t = 1, 2, ..., T:for each device i: w_i^(t+1) = ClientUpdate(w_i^t, η)

+AddNoise()

   send w_i^(t+1) to the server aggregate model   updates on   the server"

FIGURE 13.8 ClientUpdate model of DPFL

TABLE 13.6

Various values of parametric evaluations in DPFL

Parameter	Value
Privacy Budget	1.0
Number of Rounds (T)	25

and the combined learning calculations were actualized utilizing fitting libraries and systems [34].

Statistical Analysis

Statistical importance tests, such as t-tests, were utilized to approve the execution contrasts between the unified learning calculations [35]. Furthermore, the privacy-preserving viewpoints were assessed by analyzing the sum of data spillage and demonstrating utility.

EXPERIMENTS

Experimental Setup

To assess the execution of the unified learning calculations in privacy-preserving machine learning for IoT systems, a comprehensive set of tests was conducted. The tests centered on surveying show accuracy, security conservation, and computational productivity. The recreated IoT environment included different gadgets with

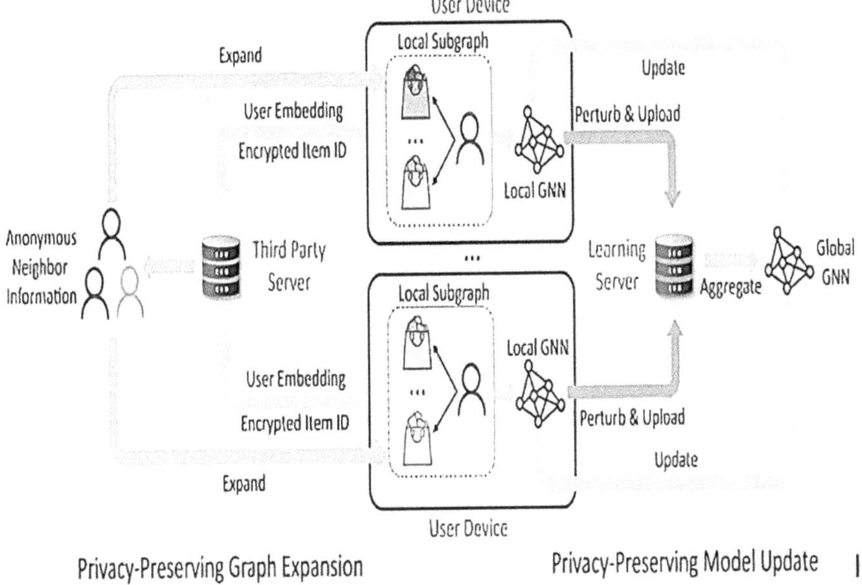

FIGURE 13.9 A federated graph neural network framework for privacy-preserving personalization

shifting computational capacities, reflecting real-world scenarios [36]. The combined learning calculations, specific Federated Averaging (FedAvg), Homomorphic Encryption-based Unified Learning, Secure Aggregation, and Differential Protection in Combined Learning was executed utilizing Python and TensorFlow as depicted in **Figure 13.9** [37].

Evaluation Metrics

The tests utilized a run of standard assessment measurements to evaluate the execution of the combined learning calculations. Key measurements included accuracy, precision, recall, and F1 score, giving an all-encompassing see of the models' prescient capabilities as depicted in **Figure 13.10.** Privacy-related measurements, such as data spillage and differential security ensures, were also measured [38]. Moreover, computational measurements, counting preparing time, and communication overhead, were considered to assess the productivity of the calculations.

Comparison With Related Work

To contextualize our results, a comparative investigation was conducted with existing studies in privacy-preserving machine learning for IoT systems. Notable related work incorporates, where a combined learning approach with accentuation on protection was proposed, which investigated homomorphic encryption in combined learning [39]. Our tests point to constructing upon and amplifying the discoveries of

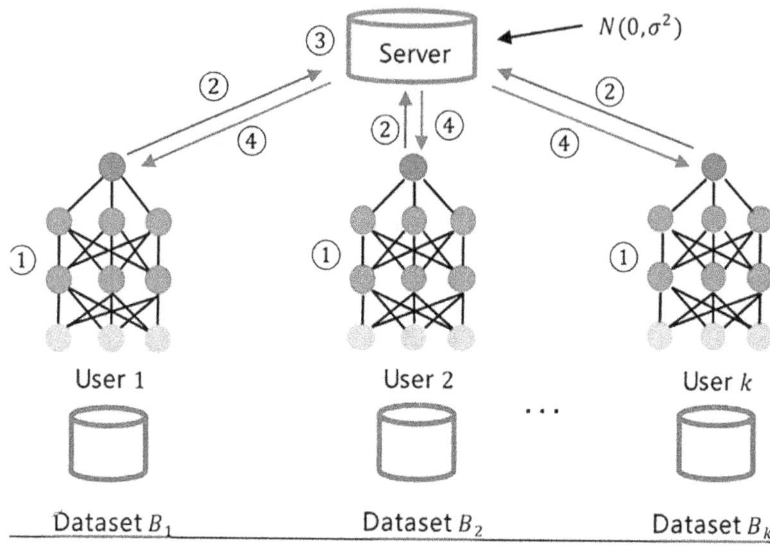

FIGURE 13.10   Federated learning framework with differential privacy update

these studies, advertising bits of knowledge into the comparative adequacy of different unified learning calculations in an IoT setting.

RESULTS

Model Accuracy

Table 13.7 presents the precision measurements of the unified learning calculations on a test dataset. The models were prepared for a settled number of rounds, and the precision was assessed on a partitioned test set to degree generalization execution.

The results demonstrate that FedAvg accomplished the most noteworthy precision, exhibiting its adequacy in collaborative learning. In any case, Secure Aggregation and Differential

Security illustrated competitive precision levels, emphasizing their utility in scenarios where protection conservation is fundamental [40]. **Figure 13.11** depicts Privacy-preserving machine learning and multi-party computation.

Privacy Preservation

Security conservation could be a basic angle of unified learning in IoT systems. **Table 13.8** gives an outline of privacy-related measurements, counting data leakage, and the level of differential protection accomplished by each calculation.

Homomorphic Encryption and Differential Privacy algorithms show moo data leakage, guaranteeing that the prepared models don't incidentally uncover points of interest around personal information focuses [26]. Differential Privacy, in particular, accomplished a security parameter (ε) of 1.0, showing a tall level of protection conservation.

TABLE 13.7

Various Algorithms with Accuracies

Algorithm	Accuracy (%)
FedAvg	92.5
Homomorphic Encryption	88.2
Secure Aggregation	91.8
Differential Privacy	89.7

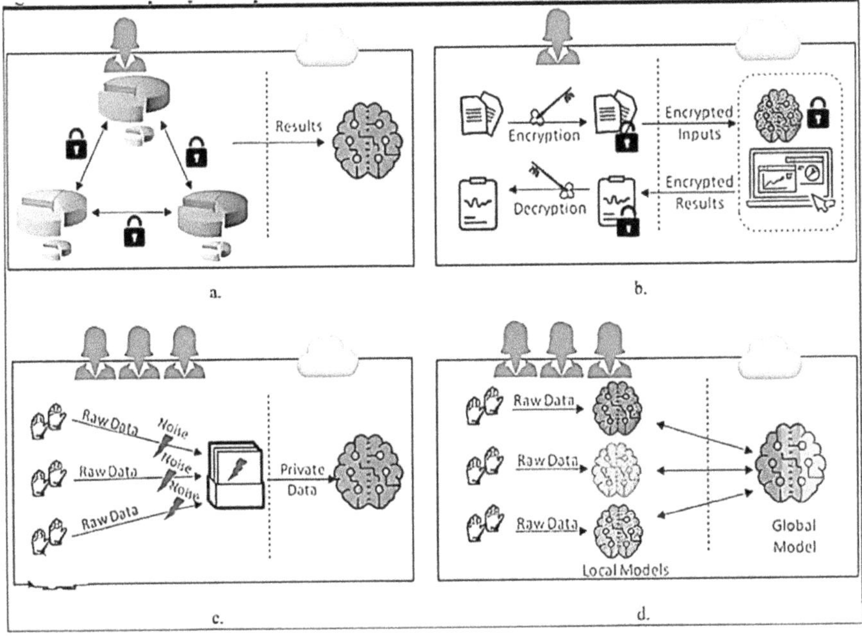

FIGURE 13.11   Privacy-preserving machine learning and multi-party computation

---

TABLE 13.8

Various Algorithms with Information Leakage and Differential Privacy

Algorithm	Information Leakage	Differential Privacy (ε)
FedAvg	Moderate	Not applicable
Homomorphic Encryption	Low	2.5
Secure Aggregation	Minimal	Not applicable
Differential Privacy	Minimal	1.0

Computational Efficiency

Efficient show preparation and communication are significant for combined learning in resource-constrained IoT situations.

Table 13.9 traces the computational proficiency measurements, counting preparing time, and communication overhead.

Secure Aggregation illustrated low communication overhead, making it reasonable for situations with restricted transmission capacity. In any case, Holomorphic Encryption exhibited high communication overhead due to the encryption and unscrambling forms. FedAvg and Differential Privacy fell inside direct communication overhead levels, striking an adjustment between effectiveness and protection. **Figure 13.12** depicts federated learning and its role in the privacy preservation of IoT devices.

TABLE 13.9

Various Algorithms with Training Time and Communication Overhead

Algorithm	Training Time (s)	Communication Overhead
FedAvg	120	Moderate
Holomorphic Encryption	280	High
Secure Aggregation	150	Low
Differential Privacy	200	Moderate

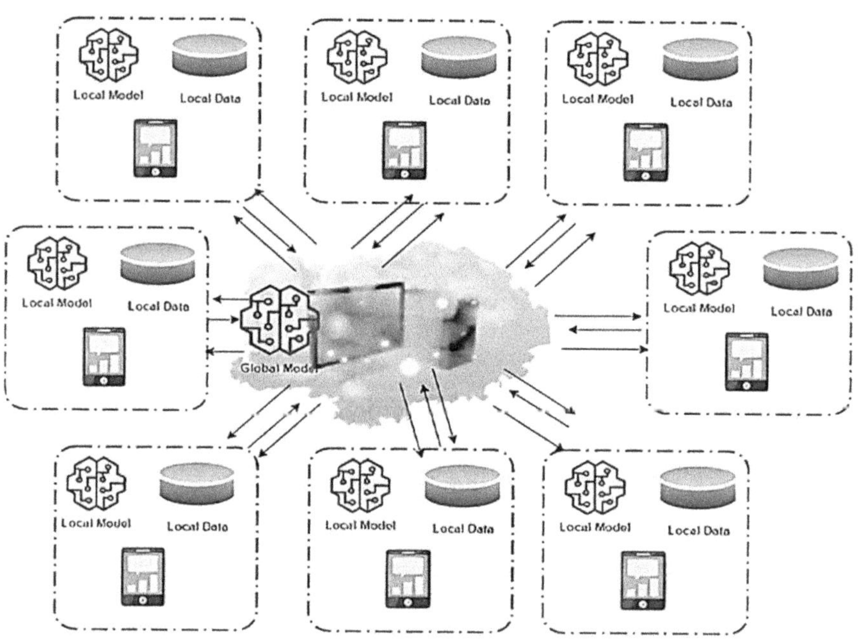

FIGURE 13.12   Federated Learning and Its Role in the Privacy Preservation of IoT Devices

Calculation Parameters

Here, in **Figure 13.13** calculation parameters with comparisons are shown. Accuracy, Recall, Precision, Specificity, and Sensitivity are the classification parameters. Consider these parameters to analyze the result.

Experimental Result

Three Python libraries Sklearn, NumPy, and Tensor Flow are employed for data analysis, mathematical operations, categorization, prediction, and the creation of data flow graphs. The estimated accuracy is 98%. Precision, Recall, and F1- Score is 98%, 90.7%, and 31% respectively.

Performance Percentage

Table 13.10 shows the performance percentage of various algorithms with various parameters. **Figure 13.14** depicts the performance comparison of SVM and Federated

	Positive	Negative	
Positive	True Positive (TP)	False Negative (FN) Type II Error	**Sensitivity** $\dfrac{TP}{(TP+FN)}$
Negative	False Positive (FP) Type I Error	True Negative (TN)	**Specificity** $\dfrac{TN}{(TN+FP)}$
	Precision $\dfrac{TP}{(TP+FP)}$	**Negative Predictive Value** $\dfrac{TN}{(TN+FN)}$	**Accuracy** $\dfrac{TP+TN}{(TP+TN+FP+FN)}$

FIGURE 13.13   Comparison Parameters

TABLE 13.10
Performance Percentage

	Decision Tree	Logistic Regression	Random Forest	SVM	Federated Learning
Accuracy	97.94	96.94	97.95	98.9	99.3
Precision	97.61	98.59	94.31	92.7	98
Recall	46.06	48.44	67.68	64.1	90.7
F1-Score	62.59	64.96	78.81	64.1	31

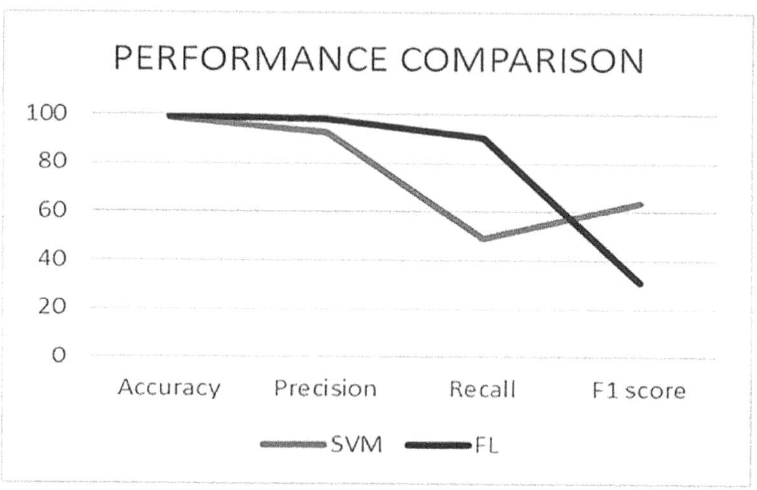

FIGURE 13.14   SVM, Federated learning

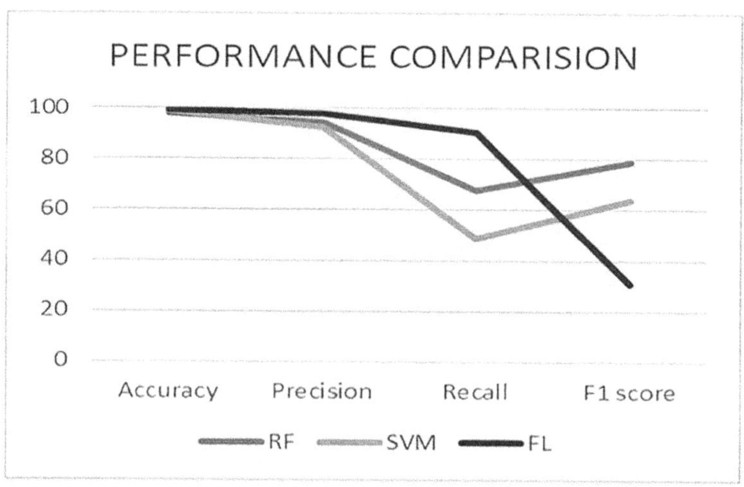

FIGURE 13.15   Random forest, SVM, Federated learning

learning. **Figure 13.15** depicts the performance comparison of random forest, SVM, and Federated learning. **Figure 13.16** depicts the performance comparison of logistic regression, random forest, SVM, and Federated learning. **Figure 13.17** depicts the performance comparison of decision tree, random forest, SVM, logistic regression, and federated learning:

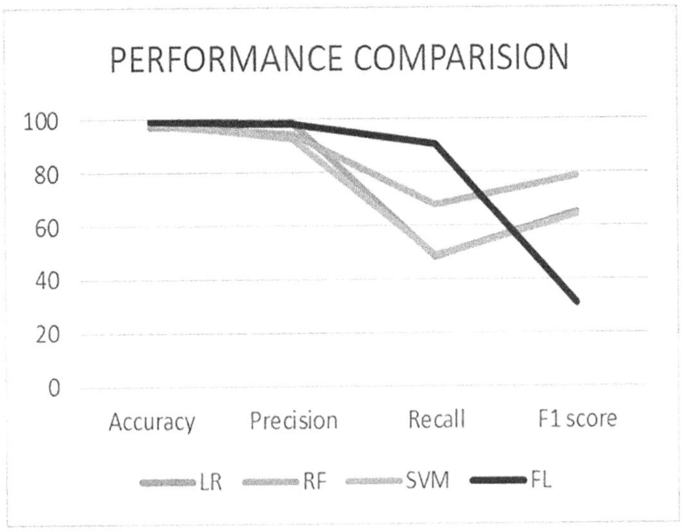

FIGURE 13.16 Logistic regression, Random forest, SVM, Federated learning

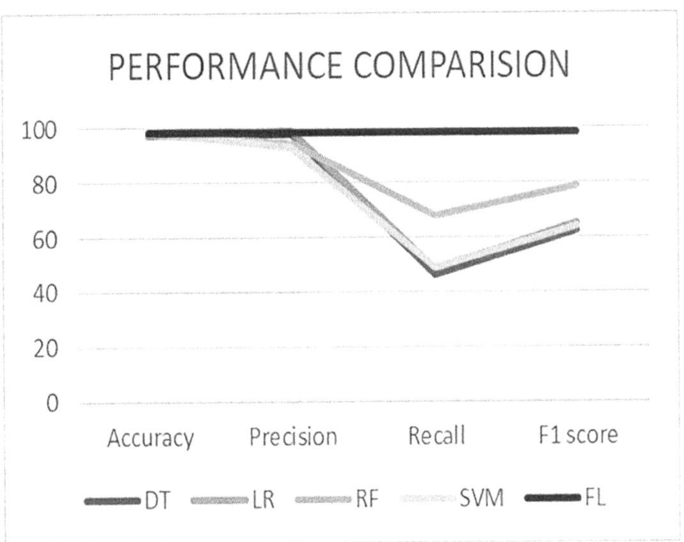

FIGURE 13.17 Decision tree, Logistic regression, Random forest, SVM, Federated learning

Discussion

The results highlight the trade-offs among the combined learning calculations in terms of demonstrated precision, security conservation, and computational proficiency. FedAvg exceeded expectations in exactness but needed unequivocal security conservation instruments [27]. Homomorphic Encryption and Differential Privacy, on the other hand, illustrate robust

Security ensures but with expanded computational requests. Secure Aggregation rose as a promising compromise, accomplishing competitive precision while minimizing data spillage and communication overhead. This renders Secure Aggregation very practical for IoT scenarios that need privacy-preserving and collaborative interactions [13]. Comparing the arises with related work revealed that FedAvg and Secure Aggregation consistently outrun existing methods in precision and protection savings. The integration of Holomorphic Encryption and Differential Privacy algorithms improved the comparison, indicating the feasibility of privacy-preserving combined learning in various types of IoT models.

CONCLUSION

To sum up, the study of the work "Federated Learning for Privacy-Preserving Machine Learning in IoT Systems" has offered valuable insights into the convoluted interplay between collaborative learning, safety preservation, and efficiency in the context that is the Internet of Things (IoT) situations. All the unified learning algorithms, including Combined Averaging, homomorphic encryption-based Federated Learning, Secure Aggregation, and Differential Security in Combined Learning were carefully assessed and compared. The tests revealed that while FedAvg outperformed expectations in terms of accuracy, algorithms such as Secure Aggregation trade between accuracy and security preservation, thus ranking as promising candidates for privacy-sensitive IoT applications. Comparative analysis with related work showed the relevance of the research in the broader picture of machine learning, blockchain integration, and IoT. Combined learning integrated with IoT as studies by various authors showed emphasis on the versatility of these improvements in various fields such as smart cities, healthcare, and remote communication. The introduced systems and calculations here help in addressing the issues dependent on heterogeneity, non-IID data dissemination, and asset requests in IoT settings.

In addition, the investigation contextualized itself inside the progressing insightful discourse on privacy-preserving unified learning, drawing associations with related considerations that investigated inventive arrangements and applications. The union of discoveries from different ponders contributes to an all-encompassing understanding of combined learning's potential, challenges, and arrangements, progressing the collective information base within the crossing point of machine learning and IoT. As the computerized scene proceeds to advance, the investigation underscores the significance of privacy-preserving unified learning as a significant worldview for dependable and compelling data-driven decision-making in IoT biological systems. Online fraud detection is currently a worldwide epidemic. When fraudsters produce erratic patterns that resemble the original, a more effective method of identifying online frauds while protecting users' privacy is required. In this case, online scams are detected while maintaining privacy using Deep Learning, Machine Learning, and Federated Learning techniques. Compare the existing techniques with Federated Learning with MLP, where the accuracy has been increasing.

SVM is overfitting for large datasets, which is used in this research. In future work, proposed split learning can be implemented and tested with different machine learning methods. Future endeavors in this space seem constructed upon these experiences, refining calculations, and systems for particular IoT applications and addressing developing challenges within the ever-evolving scene of interconnected gadgets.

REFERENCES

[1] Alazab, A., Khraisat, A., Singh, S., & Jan, T. (2023). Enhancing privacy-preserving intrusion detection through federated learning. *Electronics*, 12(16), 3382.

[2] Chai, J., Li, J., Wei, M., & Zhu, C. (2023). Blockchain managed federated learning for a secure IoT framework. *EURASIP Journal on Wireless Communications and Networking*, 2023(1), 100.

[3] Du, W., Li, M., Wu, L., Han, Y., Zhou, T., & Yang, X. (2023). A efficient and robust privacy-preserving framework for cross-device federated learning. *Complex & Intelligent Systems*, 9(5), 4923–4937.

[4] Valencia-Arias, A., González-Ruiz, J. D., Verde Flores, L., Vega-Mori, L., Rodríguez-Correa, P., & Sánchez Santos, G. (2024). Machine learning and blockchain: A bibliometric study on security and privacy. *Information*, 15(1), 65.

[5] Ghadi, Y. Y., Mazhar, T., Shah, S. F. A., Haq, I., Ahmad, W., Ouahada, K., & Hamam, H. (2023). Integration of federated learning with IoT for smart cities applications, challenges, and solutions. *PeerJ Computer Science*, 9, e1657.

[6] Yu, X., Tang, D., & Zhao, W. (2023). Privacy-preserving cloud-edge collaborative learning without trusted third-party coordinator. *Journal of Cloud Computing*, 12(1), 19.

[7] Zeng, Q., Lv, Z., Li, C., Shi, Y., Lin, Z., Liu, C., & Song, G. (2023). Fedprols: Federated learning for iot perception data prediction. *Applied Intelligence*, 53(3), 3563–3575.

[8] Zhang, K., Cai, Z., & Seo, D. (2023). Privacy-preserving federated graph neural network learning on non-iid graph data. *Wireless Communications and Mobile Computing*, 2023.

[9] Zhao, R., Xie, Y., Cheng, H., Jia, X., & Shirazi, S. H. (2023). ePMLF: Efficient and privacy-preserving machine learning framework based on fog computing. *International Journal of Intelligent Systems*, 2023.

[10] Almalki, J., Alshahrani, S. M., & Khan, N. A. (2024). A comprehensive secure system enabling healthcare 5.0 using federated learning, intrusion detection and blockchain. *PeerJ Computer Science*, 10, e1778.

[11] Al Asqah, M., & Moulahi, T. (2023). Federated learning and blockchain integration for privacy protection in the Internet of Things: Challenges and solutions. *Future Internet*, 15(6), 203.

[12] Butt, M., Tariq, N., Ashraf, M., Alsagri, H. S., Moqurrab, S. A., Alhakbani, H. A. A., & Alduraywish, Y. A. (2023). A fog-based privacy-preserving federated learning system for smart healthcare applications. *Electronics*, 12(19), 4074.

[13] Sathiamoorthy, A., Mithusan, S., Rathnayaka, R. M. L. R., Kajenthiran, S., Hansika, M. M., & Pandithage, D. (2023). StreamSafe: Improving QoS and security in IoT networks. *International Research Journal of Innovations in Engineering and Technology*, 7(11), 170.

[14] Kaushik, A., Gahletia, S., Garg, R. K., Sharma, P., Chhabra, D., & Yadav, M. (2022, December). Advanced 3D body scanning techniques and its clinical applications. In *2022 International Conference on Computational Modelling, Simulation and Optimization (ICCMSO)* (pp. 352–358). IEEE.

[15] Singh, K., Singh, Y., Barak, D., & Yadav, M. (2023). Comparative performance analysis and evaluation of novel techniques in reliability for Internet of Things with RSM. *International Journal of Intelligent Systems and Applications in Engineering*, 11(9s), 330–341.

[16] Singh, K., Singh, Y., Barak, D., & Yadav, M. (2023). Evaluation of designing techniques for reliability of Internet of Things (IoT). *International Journal of Engineering Trends and Technology*, 71(8), 102–118.

[17] Singh, K., Singh, Y., Barak, D., Yadav, M., & Özen, E. (2023). Parametric evaluation techniques for reliability of Internet of Things (IoT). *International Journal of Computational Methods and Experimental Measurements*, 11(2), 123–134.

[18] Singh, K., Singh, Y., Barak, D., & Yadav, M. (2023). Detection of lung cancers from CT images using a deep CNN architecture in layers through ML. In *AI and IoT-Based Technologies for Precision Medicine* (pp. 97–107). IGI Global.

[19] Singh, K., Yadav, M., Singh, Y., & Barak, D. (2023). Reliability techniques in IoT environments for the healthcare industry. In *AI and IoT-Based Technologies for Precision Medicine* (pp. 394–412). IGI Global.

[20] Sharma, H., Singh, K., Ahmed, E., Patni, J., Singh, Y., & Ahlawat, P. (2021). IoT based automatic electric appliances controlling device based on visitor counter. https://doi.org/10.13140/RG,2(30825.83043).

[21] Bhatia, S., Goel, A. K., Naib, B. B., Singh, K., Yadav, M., & Saini, A. (2023, July). Diabetes prediction using Machine learning. In *2023 World Conference on Communication & Computing (WCONF)* (pp. 1–6). IEEE.

[22] Chen, Z., Cui, H., Wu, E., & Yu, X. (2023). Computation and communication efficient adaptive federated optimization of federated learning for Internet of Things. *Electronics*, 12(16), 3451.

[23] Han, Y., & Zhu, X. (2023). Enhancing throughput using channel access priorities in frequency hopping network using federated learning. *EURASIP Journal on Wireless Communications and Networking*, 2023(1), 101.

[24] Du, W., Wang, Y., Meng, G., & Guo, Y. (2024). Privacy-preserving vertical federated KNN feature imputation method. *Electronics*, 13(2), 381.

[25] El-Gendy, S., Elsayed, M. S., Jurcut, A., & Azer, M. A. (2023). Privacy preservation using machine learning in the Internet of Things. *Mathematics*, 11(16), 3477.

[26] Mouhni, N., Elkalay, A., Chakraoui, M., Abdali, A., Ammoumou, A., & Amalou, I. (2022). Federated learning for medical imaging: An updated state of the art. *Ingenierie des Systemes D'Information*, 27, 143–150.

[27] Pinto Neto, E. C., Sadeghi, S., Zhang, X., & Dadkhah, S. (2023). Federated reinforcement learning in IoT: Applications, opportunities and open challenges. *Applied Sciences*, 13(11), 6497.

[28] www.kaggle.com/datasets/rupakroy/online-payments-fraud-detection-dataset

[29] Javed, A., Awais, M., Shoaib, M., Khurshid, K. S., & Othman, M. (2023). Machine learning and deep learning approaches in IoT. *PeerJ Computer Science*, 9, e1204.

[30] Karras, A., Giannaros, A., Theodorakopoulos, L., Krimpas, G. A., Kalogeratos, G., Karras, C., & Sioutas, S. (2023). FLIBD: A federated learning-based IoT big data management approach for privacy-preserving over apache spark with FATE. *Electronics*, 12(22), 4633.

[31] Kea, K., Han, Y., & Kim, T. K. (2023). Enhancing anomaly detection in distributed power systems using autoencoder-based federated learning. *PLoS ONE*, 18(8), e0290337.

[32] Munawar, A., & Piantanakulchai, M. (2024). A collaborative privacy-preserving approach for passenger demand forecasting of autonomous taxis empowered by federated learning in smart cities. *Scientific Reports*, 14(1), 2046.

[33] Muthukumar, V., Sivakami, R., Venkatesan, V. K., Balajee, J., Mahesh, T. R., Mohan, E., & Swapna, B. (2023). Optimizing heterogeneity in IoT Infra using federated learning and blockchain-based security strategies. *International Journal of Computers Communications & Control*, 18(6).

[34] Peyvandi, A., Majidi, B., Peyvandi, S., & Patra, J. C. (2022). Privacy-preserving federated learning for scalable and high data quality computational-intelligence-as-a-service in Society 5.0. *Multimedia Tools and Applications*, 81(18), 25029–25050.

[35] Qin, J., Zhang, X., Liu, B., & Qian, J. (2023). A split-federated learning and edge-cloud based efficient and privacy-preserving large-scale item recommendation model. *Journal of Cloud Computing*, 12(1), 57.

[36] Rashid, M. M., Khan, S. U., Eusufzai, F., Redwan, M. A., Sabuj, S. R., & Elsharief, M. (2023). A federated learning-based approach for improving intrusion detection in industrial internet of things networks. *Network*, 3(1), 158–179.

[37] Song, Z. (2020, June). A data mining based fraud detection hybrid algorithm in E-bank. In *2020 International Conference on Big Data, Artificial Intelligence and Internet of Things Engineering (ICBAIE)* (pp. 44–47). IEEE.

[38] Rodríguez, E., Otero, B., & Canal, R. (2023). A survey of machine and deep learning methods for privacy protection in the internet of things. *Sensors*, 23(3), 1252.

[39] Hsu, H. Y., Keoy, K. H., Chen, J. R., Chao, H. C., & Lai, C. F. (2023). Personalized federated learning algorithm with adaptive clustering for non-IID IoT data incorporating multi-task learning and neural network model characteristics. *Sensors*, 23(22), 9016.

[40] Kolhar, M., & Aldossary, S. M. (2023). Privacy-preserving convolutional Bi-LSTM network for robust analysis of encrypted time-series medical images. *AI*, 4(3), 706–720.

14 Federated Query Processing for Data Integration Using Semantic Web Technologies
A Review

Nidhi Gupta, Pawan Verma, Monali Gulhane,
Nitin Rakesh, and Ahmed A. Elngar

INTRODUCTION

Data science and analytics are a growing interest of many organizations. They require collecting, processing, and analyzing the data to generate meaningful insights. The first and most significant step is to collect data from various sources. Data integration refers to combining data from multiple data sources to get a unified view of data. It extracts the data, transforms it, and loads it in a central place using extract, transform, and load (ETL) tools. To effectively handle heterogeneous databases, we need tools that can address the specific challenges and requirements of each system, rather than relying solely on a centralized ETL approach.

The semantic web helps to share data and promote interoperability between various systems. The term semantic web refers to linked data on the web. The aim of linked open data (LOD) is to globally access the data on the web. With the success of LOD for open data, semantic technologies are also being used to access organizational data. The semantic web uses the resource description framework (RDF) to represent the data. It facilitates data interchange on the web. In comparison to relational databases, linked data represents the resources using a uniform resource indicator (URI) that helps to represent a resource uniquely.

Relational databases are widely used for storing data. SQL has grown as a powerful query language for querying relational databases and integrating data from multiple sources. Relational databases impose strict schema on databases which are not suited for applications that require frequent schema changes. The data generated from multiple sources such as wearable devices and other IoT applications does not adhere to strict schemas, thus moving towards flexible databases like MongoDB for

DOI: 10.1201/9781003482000-14

storing formats such as JSON and XML. Therefore, graph databases like Virtuoso and Graph-DB are gaining popularity to integrate multiple data sources into an integrated graph. They utilize the RDF framework to represent the data and perform queries for data access.

Query processing provides data access to large volumes of data stored in multiple systems. There are two approaches for query processing over the web of linked and distributed data sources: link traversal and query federation. Link traversal (Hartig et al., 2009) such as SQUIN (Squin, 2013) searches relevant data by following links between RDF data sources. It does not require the prior knowledge of data sources. The query executes by binding the intermediate results with common variables. The link traversal discovers the data source at runtime, hence providing up-to-date results. The major weakness of this approach is that the wrong selection of a starting point for traversing the link would lead to a large number of intermediate results and could not guarantee complete results. Another approach is query federation, which inputs a query and distributes it to relevant data sources. Federated query processing for data integration is a crucial concept in the field of data management and information retrieval. It refers to the process of querying and retrieving data from multiple distributed or heterogeneous data sources and integrating the results into a unified response. This approach is employed when organizations or systems need to access and combine data from various repositories, databases, or systems to derive meaningful insights or to meet specific information needs.

A federated engine acts as a mediator to distribute subqueries and collect the results. It utilizes data descriptions to find the relevant data sources and provides efficient query plans to minimize query execution time and maximize result completeness. Federated query processing is advantageous for result completeness and providing up-to-date results. Therefore, this chapter focuses on federated query processing and reviews the work carried out on optimizing federated queries for efficient data access.

The chapter reviews the work that has been performed to integrate heterogeneous data sources using semantic technologies and its access via federated query processing. It presents the various approaches used for federated query processing and compares various federated engines based on these approaches. Further, it discusses the target problems and their proposed solutions in health data integration and federated query processing.

In the second section, we provide a concise overview of the background and relevant research. The third section presents ontology-based data integration for distributed data access. The fourth section explores the federated approach to query distributed data sources. The fifth section provides different challenges and reviews the target problems on query execution and their solutions. Finally, the sixth section concludes with future work.

BACKGROUND AND RELATED SURVEYS

DATA INTEGRATION AND INTEROPERABILITY

Data integration is the process of combining data from different data sources to provide a unified view of data that facilitates data access through a single query interface. The integration of data from heterogeneous data sources is a challenge for organizations, as data at distinct sources differs in schemas, data formats, and models used. Data integration requires interoperability for data access. Interoperability refers to the ability of systems to share data. It is significant to achieve this, specifically in a heterogeneous data environment. A study by Gupta and Gupta (2019) discussed syntactical and semantic interoperability. Syntactic interoperability requires standardizing health data structures. Various health standards have been developed for data storage and exchange such as HL7 FHIR and OpenEHR. Lack of adoption of standards by health providers results in heterogeneous systems and leads to data interoperability problems.

Semantic interoperability ensures the use of various tools and mappings to ensure meaningful interpretations. It uses appropriate data representation and mappings that allow multiple distinct organizations to share and integrate data for both standard and non-standard electronic health record (EHR) data formats. The work carried out studied the state-of-art approaches of federated query engines that perform integration of data sources in non-standard data formats.

TYPES OF DATA INTEGRATION SYSTEMS

The integration of data is categorized in two different ways in accordance with data storage and query processing. These are:

1. **Centralized:** Centralized data integration involves gathering and unifying data from multiple disparate sources and storing it in a centralized location.
2. **Concise:** Centralized data integration consolidates data into a single repository. The data is accessed from distinct data sources and put into a single RDF data store. In this, data processing and querying takes place at one central place. The centralized approach is beneficial in terms of efficient query processing due to centralized and optimized index structures. It requires making copies of data for central storage, thus making data management and update difficult.
3. **Distributed system:** A distributed system consists of data at various physical locations. Distributed systems may have federated or peer-to-peer data success. The details of each are:
   - **Federated:** A federated database system is a special distributed database system that provides an alternative to a centralized data system. Federated data integration refers to the consolidation of distributed, autonomous data to a virtual unified data model using a central mediator. The mediator maintains an index that stores the statistical information of data sources, which enables relevant selection of data sources for an input query. Data is accessed from distinct data sources and does not require organizations to move their data, thus maintaining data privacy.
   - **Peer to peer:** The peer-to-peer system consists of various peers or nodes, each cooperating and having its own data. The data and index are maintained in a distributed fashion. Thus, there is no central mediator, and the query is processed by each node. These systems are flexible, as they can handle changes in peer configurations.

4. **Semantic web technologies**: Semantic technologies are used to add meaning and context to data, making it possible to understand relationships and concepts even when data comes from diverse sources. Some of the widely used technologies used for data integration are:

   a. URI: A uniform resource identifier, as the name implies, uniquely identifies any resource on the web. A Uniform Resource Locator (URL) is a subset of a URI. In addition to resource identification, URLs gives the way to access the resource; for example, http://ex.org identifies a URI and is accessed using HTTP protocol (Masinter et al., 2005).

   b. Resource Description Framework: It is a World Wide Web Consortium (W3C) standard framework to represent information about the resources in a machine-readable format. RDF data is expressed in the form of triple patterns. A triple consists of <subject, predicate, object>. The subject represents a resource as URI, the predicate describes the relation between resources that are represented as a property of the resource, and the object signifies the value of the property. For example: triplet <pid20 temp "99"> represents the temperature observation value of patient id 'pid20' as '99'.

   c. RDF Schema (RDFS): An RDFS defines the vocabulary of the RDF dataset. It provides mechanisms to group the related resources called classes, and members of the classes are instances of that class. The rdf: type property states that the resource is an instance of the class. RDFS is written in RDF using various constructs to define RDF such as rdf: type, rdf: class, rdf:property, rdf:domain, and rdf:range. The W3C provides the details of RDF schema properties (Faheem et al., 2018).

   d. OWL (Web Ontology Language): Ontology language is also a data modeling language formally used to describe a resource. OWL provides much more vocabulary of data models and is particularly used in the automatic reasoning process. Besides RDFS properties, OWL also defines equivalence across databases using owl:same. By constructing annotations, we can establish connections and define relationships between different standard ontologies.e. RDF formats: An RDF document is written using various file formats called RDF serialization format. The various formats are Turtle (with extension is.ttl), Notation 3 (with extension.n3), and N-triples (with extension. nt). JSON-LD (with extension.jsonld) is the JSON syntax of an RDF document, and RDF/XML is an XML syntax of an RDF document (Beckett & McBride, 2004)

   f. SPARQL: The query language for RDF data stores is SPARQL. It consists of a basic graph pattern (BGP) which has conjunctive triple patterns. The triples use variable names (prefixed by?) to retrieve the values from the dataset. The triples in BGP are matched with the RDF triples and return variable bindings. Listing 14.1 shows the SPARQL query for retrieving the value of blood pressure (systolic, diastolic) and temperature of all patients with given filter value ranges.

LISTING 14.1
SPARQL Query

```
Select ?pid ?sys ?dys ?temp
Where{
?s pid ?pid .
?s sys ?sys. FILTER (?sys >100 && ?sys<120).
?s dys ?dys FILTER (?dys < 70).
?s temp ?temp
}
```

RELATED SURVEYS

This chapter focuses on two main aspects, semantic data integration and federated query processing.

In the literature review, Peng et al. (2020) investigated existing problems in the integration of health data and different approaches to its integration. It highlights the open problems in integrating aggregated health data and advocates for the use of semantic technologies with web application interfaces to overcome the challenges.

The research work by Wylot et al. (2018) presented biological database modeling with semantic technologies. It focuses on ontology-based integration of biological data and outlines the semantic technologies used for its integration.

The research work by Asfand-e-Yar and Ali (2020) focused on ontology-based semantic integration of heterogenous databases of the same domain using data translation and query translation techniques.

The survey by Ouzzani and Bouguettaya (2004) focused on fundamental problems in efficient query processing. It also discusses various optimization techniques over web data integration systems and frameworks to evaluate them. The authors in Ali et al. (2017) experimentally evaluate the various query federated engines on the efficiency of source selection, data partition, answer completeness, and query runtime. Oguz et al. (2015) presented a qualitative survey for federated query processing on linked data. It provides details of different approaches to federated query processing steps such as source selection and query optimization and compares various federated engines on those approaches. The authors in Sima et al. (2019) provide a comprehensive study of RDF data storage and management systems. It provides different RDF storage mechanisms, indexing, architecture, and query execution methods.

Several studies conducted either data integration or federated queries, but very limited surveys cover both aspects of data integration and federated query processing. Thus, the work carried out surveys the state-of-art approaches in the area of data integration and federated query processing. It identified problems faced during integration and querying heterogeneous data sources and also discussed their solutions.

ONTOLOGY-BASED DATA INTEGRATION FOR DISTRIBUTED DATA ACCESS

Ontology-based data integration (OBDA) system consists of triple <O, S, M>, where O is a domain ontology, S is the source schema, and M defines the mappings between the source and target domain ontology. OBDA uses a global view to query distinct heterogeneous data sources using an ontology, thereby masking the actual implementation details of data sources. One of the major challenges for data integration is data heterogeneity. The problem of heterogeneity is resolved using semantic technologies in one of the following ways:

1. **Materialized approach:** Data in the same format can be achieved by conversion of non-RDF data to RDF data format using tools such as Open Refine (Donnelly, 2014). These tools clean the data, transform it into the desired data format, and link it with other web services. Since data is maintained in a single place, this approach is efficient in terms of query processing. However, its maintenance is a costly process, as it requires copies of data to be stored at central storage and periodic updating of data.
2. **Virtual approach:** The cost of materialization of databases is removed by using the virtual view of databases. It creates an RDF view over non-RDF data sources. The mappings are defined to create an RDF view; for instance, relational data is viewed as RDF using R2RML mappings, while non-relational data gives RDF views using RML mappings. Heterogenous data sources are queried over integrated RDF views and queried at their own place.

In recent decades, numerous OBDA mapping languages have emerged. To establish a common framework for transforming relational data into RDF, the RDB2RDF W3C working group was created. Direct mapping (Sequeda et al., 2012) and R2RML (Priyatna et al., 2015) are two prominent examples of these mapping languages. Direct mapping transforms RDB to RDF by a predefined procedure without user intervention, while R2RML is a customizable language to map RDB to RDF. R2RML mappings offer a more flexible approach than direct or indirect mappings. They allow us to map database schemas to standard RDF vocabularies, providing a consistent and global view for SPARQL queries. This flexibility is especially useful when dealing with database schemas that cannot be easily modified. R2RML output may be materialized RDF or RDF virtual view. Later, other mapping languages were proposed for mapping non-relational databases, such as RML (Dimou et al., 2014), proposed to map non-relational databases (JSON, CSV, and XML data sources) to RDF and xR2RML (Michel et al., 2016) for MongoDB databases. The mappings achieve semantic interoperability that helps to perform integrated data access by exposing heterogeneous data as SPARQL endpoints.

Integration of data requires aggregation from various data sources to form a linked graph. Figure 14.1 shows the architecture for distributed data access from RDF and non-RDF data sources. It consists of four layers:

1. Data layer: This layer consists of distinct heterogeneous data sources. It stores the data in different databases: relational, JSON, and RDF. The data is fetched using an application interface or wrappers. The data retrieved is aggregated at the integration layer.
2. Data integration layer: The integration layer is responsible for virtual data integration and federated query processing. The integration layer aggregates

the data from multiple data sources. It creates mappings such as R2RML and RML to translate data sources of distinct formats such as RDB or JSON to an RDF view. Semantic annotation of data is required with commonly used vocabularies such as FOAF to make data meaningful and easier to understand. Federated query processing fetches the data from multiple data sources by decomposing the input query, selecting the relevant data source for each subquery preparing an optimized plan and executing it to extract the required results. The user query is translated to local database queries by using declarative mappings (RML, R2RML, etc.). The result fetched is integrated as a result set and sent to the user-interface layer.

3. Semantic layer: It describes the domain ontology to represent various concepts and their relationships. The schema of data sources is aligned to domain ontology to form a global query.

4. User interface layer: The user interface allows users to input queries to the federated system and receive the responses.

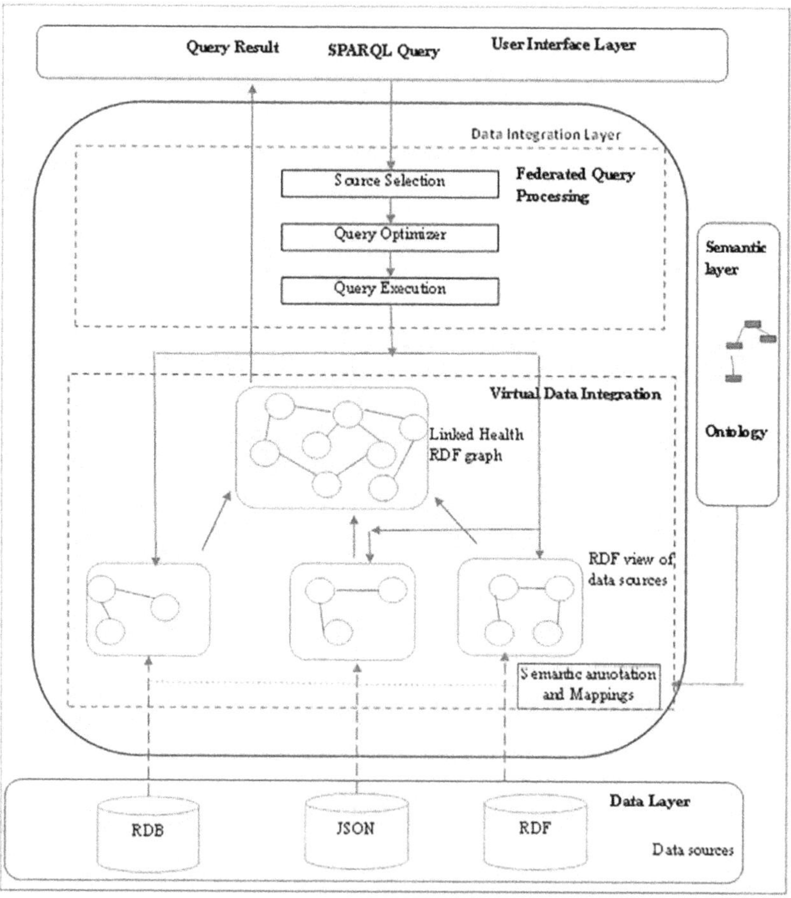

FIGURE 14.1 Architecture for data integration and querying.

FEDERATED QUERY PROCESSING FOR DISTRIBUTED RDF DATA

Federated query engines are widely used in data retrieval and its integration. The query engines use different approaches to query distributed databases. The basic steps are query parsing, data source selection, query optimization, and query execution.

There are two important optimization steps of federated query engines.

SOURCE SELECTION

Data integration queries retrieve data from multiple data sources; thus, many data sources are involved in answering an input query, and each source may or may not contribute to the final answer. Hence, a federated engine is required to search the relevant data sources to answer a given query. It divides the query into triples and finds the data sources of each query triple. Some of the federated engines use metadata catalogs. This catalog stores: (1) service descriptions: statistical information of data sources; (2) a vocabulary of interlinked datasets (VoID) descriptions; that is, it stores statistical information of data sources along with the metadata of RDF data sources whose subjects are linked with objects of other RDF data sources; and (3) a list of predicates at each RDF data source.

Some of the federated engines such as FEDX use indexes to manage data sources and store them in a cache to avoid repeating the process for the same input query.

Different Approaches to Data Source Selection

- **ASK:** The SPARQL ASK query broadcasts the input query to all the data sources and returns 'true' if the data source has the solution to query; else it returns 'false'. ASK is generally used for source selection when the federated engine does not maintain any data for storing predicate information or the query has a predicate variable. Data source selection using ASK does not require to maintain data source predicate information but increases the number of remote requests. Some of the federated engines that use ASK queries for source selection are FedX and SPLENDID.
- **Predicate-based selection:** The federated engine maintains the metadata catalog, which contains statistical information and predicate information of each data source. It selects data sources by matching them with the predicates in the metadata catalog. Most of the federated engines use a predicate-based approach such as ANAPSID, ADERIS, or POLYSTORE.
- **Type-based selection:** It generally uses type definitions (i.e., rdf:type) of data sources and matches them with the metadata catalog. SPLENDID also uses rdf:type information for source selection.
- **Rule-based selection:** In this method, rules are formed by analyzing the relationship between query triples. Data source selection is performed by applying these rules only. WoDQA (Akar et al., 2012) is an example federated engine that uses rule-based selection.

QUERY OPTIMIZATION AND EXECUTION

A SPARQL query is composed of various triple patterns. Multiple data sources are relevant to the query. Each triple pattern of the query is executed at the relevant data sources, and the results are joined to obtain a final result set. However, this method would lead to a large number of remote HTTP requests and consequently a large number of local joins. Thus, an efficient query plan to group triple patterns and execute query groups is the key requirement for successful query execution.

Query Groups

- **Exclusive Group:** The exclusive group is formed with the triples having the same and only single data source. The grouping may comprise triples with no shared variable and may result in redundant intermediate results. Thus, the variant of exclusive groups is to create different groups for the triple with no shared variable. owl:sameAs can be used to group triples that have the same object and predicate, but different subjects, into a single equivalence class
- **Predicate-Based Join Group (PBJ):** PBJ groups are formed by grouping triples that are all evaluated against the same set of data sources. Such groups are executed on relevant endpoints, and the results obtained are merged using the SPARQL UNION construct. The POLYSTORE federated engine uses PBJ groups. PJG is a lightweight index-based approach that uses predicate information at respective endpoints to form query groups. Listing 14.2 shows the PBJ groups and explicit SERVICE requests of a SPARQL query from two different RDF service points: Virtuoso (http://localhost:8890/sparql) and Apache Jena (http://localhost:3030/vital/query).

LISTING 14.2

SPARQL Query With JOINING PBJ Groups and Explicit SERVICE Endpoints

```
prefix id: <http://127.0.0.1:3333/>
Select ?pid ?sys ?dys ?temp where
{
{SERVICE <http://localhost:8890/sparql> {?s id:pid ?pid. ?s id:sys ?sys
    filter(100<?sys && ?sys<120)}}
UNION
{ SERVICE <http://localhost:3030/vital/query> {?s id:pid ?pid. ?s id:sys ?sys
    filter(100<?sys && ?sys<120). ?s id:dys ?dys}
}
{
{SERVICE <http://localhost:8890/sparql> {?s id:pid ?pid. ?s id:temp ?temp}}
}}
```

JOIN selection

The JOIN operation selection is significant for the query optimizer. A federated engine query plan may have four different kinds of joint operations:

Bind JOIN: In bind join, the intermediate results of the outer object are used to filter out the result set from the next join. It executes in the same way as a nested loop join and is usually efficient when the intermediate results are smaller in size.

Nested Loop JOIN: In this, the inner relation is compared with each row of outer relations, and all the bindings that satisfy the join condition are included in the result set.

Hash Join: It uses a hash table to match two relations. Some federated engines, such as SPLENDID, use hash join by executing the relations in parallel and joining them locally using a single hash table.

JOIN Ordering: The goal of JOIN ordering is to optimize the query for result completeness with minimum execution time. It can be achieved by generating a minimum-cost query plan. In general, federated engines use three basic approaches for the computation of query cost:

Statistic-based: This approach uses the data statistics information that is stored in federated catalogs: predicate information and VoiD descriptions of data sources. It uses statistics information to compute the cost of the query plan by estimating the number of intermediate results produced with each JOIN operation. The statistical approaches are often accurate, but data source statistics are not available each time.

Heuristic-based: Heuristics are the rules that are applied to estimate the order of JOIN operations in the query. Heuristic-based optimization estimates the query cost by formulating rules based on the query structure and properties of the operator used such as type of predicate and FILTER construct. It does not require overhead to maintain statistical information of data sources but at the same time provides a near-optimal query plan.

Hybrid: The hybrid approach makes use of both data source statistics and heuristics to estimate query cost. It uses minimum statistical information along with certain heuristics for cost estimation.

A comparison of state-of-the-art SPARQL federated engines is listed in Table 14.1. The comparison is performed based on data source selection approaches, data model supported, handling of unbound predicate queries, the type of JOIN operation, join-order approach, approach of query group creation, and cache memory usage. Table 14.2 shows the metrics used to evaluate different federated engines.

TABLE 14.1

Comparison of SPARQL Federated Engines

Federated Engines	Source Selection	Data Model	Unbound Predicate	JOIN Type	JOIN Ordering	Query Groups	Cache
DARQ (2008)	Service descriptions	RDF		Nested loop, bind	Statistics-based	Exclusive groups	
FEDX (2011)	SPARQL ASK, local cache	RDF	✓	Nested loop, bind	Heuristic-based	Exclusive groups	✓
SPLENDID (2011)	VoiD, ASK	RDF	✓	Hash, bind	Statistics-based	Exclusive groups	
HIBISCUS (2014)	Data summaries	RDF		Nested loop, bind	Statistics-based	Exclusive groups	✓
POLYWEB (2019)	Data summaries, ASK	CSV, RDB, RDF	✓	Nested loop, bind	Heuristic-based	Predicate-based join group	

TABLE 14.2

Evaluation Metrics of Federated Engines

Federated Engines	M1	M2	M3	M4	M5	M6
DARQ					✓	✓
FEDX			✓			✓
SPLENDID	✓	✓				✓
HIBISCUS	✓	✓	✓			✓
POLYWEB	✓	✓	✓	✓		✓

M1: Number of data sources selected for each triple pattern

M2: Number of ASK requests executed

M3: Average source selection time

M4: Number of results returned by each query

M5: Query planning and optimization time

M6: Response time

CHALLENGES FOR SPARQL QUERY PROCESSING

Some of the challenges faced by query processing for data integration are as follows.

UNBOUND PREDICATE QUERIES

The query with an unbound predicate is less selective in comparison to the subject and object. Thus, selecting a data source using a predicate as a variable is a difficult task. Various federated engines such as FEDX and POLYSTORE use ASK requests to deal with unbound predicate problems.

Query Evaluation With OPTIONAL in SPARQL Query

OPTIONAL in a SPARQL query is used to handle missing information. The SPARQL query execution on RDB data stores in OBDA settings requires the SPARQL query to translate into SQL queries. OPTIONAL is represented as a LEFT OUTER JOIN operator in relational databases. It makes the query complex, which is difficult to optimize. Various solutions are proposed to optimize the size and improve the generated SQL query structure.

Join Order Optimization

An input query searches and fetches the desired data from multiple SPARQL end-points. During this process, the intermediate results from one end-point are joined with the results of the next end-point, and so on. It will lead to a large number of intermediate results and consequently large search space for performing the JOIN. Therefore, the join order of the query needs to be optimized to minimize the number of intermediate results.

Two Main Approaches Are Used to Find the Optimal Join Order

1. **Static:** In this, the join order plan is optimized before the actual execution of the query. It is optimized based on the statistical count of data sources and heuristics.
2. **Dynamic:** The query join order is optimized during the execution of the query. It implies that the dynamic approach utilizes the intermediate results obtained during the search. A dynamic approach may reduce the search space but may require reconstructing the query plan iteratively using intermediate results.

Cost Estimation of Federated Queries

Cost-based query optimizers estimate intermediate cardinalities to find the best plan. However, the estimation of link cardinalities is a difficult task. It requires detailed metadata information of all the data sources, which is sometimes not available. Heuristics are also proposed to estimate the cost of the queries, but they generally result in sub-optimal estimations.

Presence of Blank Nodes

Blank nodes represent an anonymous resource that is not assigned any URI. A blank node is only used as a subject or an object in RDF triple. They are local to an RDF document; hence a blank node n1 in graph G1 is not the same as blank node n1 in graph G2. Thus, a challenge occurs when data is fetched from distributed sources and the intermediate solution binds the blank node while processing the JOIN operation.

The previously mentioned challenges are addressed by various researchers in the existing literature. Table 14.3 lists some recent target problems with SPARQL federated query processing and their proposed solutions by various researchers.

TABLE 14.3

Target Problems and Their Proposed Solutions in SPARQL Query Processing on Distributed Data Stores

Year	Author	Problem	Proposed Solution
2023	(Aebeloe et al., 2023)	Degradation of query performance due to inaccessibility of data from SPARQL endpoints due to high traffic	Achieving faster SPARQL query execution through peer-to-peer technology
2022	(Lim et al., 2022)	High communication costs during SPARQL query processing	Algorithm for selecting efficient query execution path
2021	(Yang et al., 2023)	High network costs to transfer data over federated databases	The proposed solution is based on two join algorithms that leverage the network topology information to reduce the costs of SPARQL query execution in a distributed environment
2020	(Naacke & Curé, 2020)	Large index size for storing RDF subject, predicate, and object information	Optimizes the indexes for RDF data storage by proposing dedicated data structures and a data partitioning approach
2018	(Saleem et al., 2018)	Estimation of join selectivity for optimized cost based query plan	Optimization of SPARQL query execution by efficient source selection and join cardinality estimation
2018	(Yannakis et al., 2018)	High query execution time due to large number of remote requests to data sources	Heuristics-based query reordering method
2018	(Xiao et al., 2018)	Effective query evaluation in presence of OPTIONAL in ontology-based data access.	Optimization to improve SQL query structure
2017	(Montoya et al., 2017)	High number of intermediate results while joining data from multiple sources	Statistical estimation of accurate cost estimation of federated queries
2015	(Wu et al., 2015)	Large search space during federated query execution	Finding optimal join order to minimize size of intermediate results
2015	(Atre et al., 2015)	Restriction of reordering left outer joins for optimization of SPARQL OPTIONAL pattern queries	Represent queries in graph of super nodes and optimize query by applying the properties of nullification, best match, and minimality

CONCLUSION

The chapter explored the potential of semantic technologies for data integration and query processing. This study of data integration and query processing represents a significant step forward in the quest for a more interconnected, meaningful, and efficient data landscape. It has unveiled various approaches to federated query processing that leverage semantic web standards. This work highlights the broader context of the research and recognizes the ongoing efforts in the field to overcome the challenges associated with data integration and querying. Some of the challenges are still open for researchers to work on in the future.

In the future, researchers, practitioners, and organizations must continue to explore and harness the transformative power of semantic web technologies. By doing so, we can unlock new horizons in knowledge discovery, decision-making, and data utilization in our increasingly data-driven world.

REFERENCES

Aebeloe, C., Montoya, G., & Hose, K. (2023, January 1). Optimizing SPARQL queries over decentralized knowledge graphs. *Semantic Web*, 14(6), 1121–1165. https://doi.org/10.3233/SW-233438

Akar, Ziya, Halaç, Tayfun, Ekinci, Erdem, & Dikenelli, Oğuz. (2012). *Querying the Web of Interlinked Datasets Using VOID Descriptions*. https://api.semanticscholar.org/CorpusID:8868212

Ali, Hasnain, Ermilov, Ivan, & Ngonga Ngomo, A.-C. (2017). Saleem, Muhammad and Khan, Yasar and Hasnain. *An Evaluation of SPARQL Federation Engines Over Multiple Endpoints*. https://doi.org/10.13140/RG.2.2.25972.45444

Asfand-e-Yar, M., & Ali, R. (2020). Semantic integration of heterogeneous databases of same domain using ontology. *IEEE Access*, 8, 77903–77919. https://doi.org/10.1109/ACCESS.2020.2988685

Atre, M. (2015, May). Left bit right: For SPARQL join queries with OPTIONAL patterns (left-outer-joins). In *Proceedings of the 2015 ACM SIGMOD International Conference on Management of Data* (pp. 1793–1808). https://doi.org/10.1145/2723372.2746483

Beckett, D., & McBride, B. (2004). RDF/XML syntax specification (revised). *W3C Recommendation*, 10(2/3).

Dimou, A., Sande, M. V., Colpaert, P., Verborgh, R., Mannens, E., & Walle, R. V. (2014). RML: A generic language for integrated RDF mappings of heterogeneous data. In Proceedings of the Workshop on Linked Data on the Web, 1184(January).

Donnelly, F. (2014). Processing Government Data: ZIP Codes, Python, and OpenRefine.

Faheem, M., Sattar, H., Bajwa, I. S., & Akbar, W. (2019). Relational database to resource description framework and its schema. In *Intelligent Technologies and Applications: First International Conference, INTAP 2018, Bahawalpur, Pakistan, October 23–25, 2018, Revised Selected Papers 1* (pp. 604–617). Springer.

Gupta, N., & Gupta, B. (2019). Big data interoperability in e-health systems. In *Data Science and Engineering (Confluence), (Noida, India) 9th International Conference on Cloud Computing, 2019* (pp. 217–222). https://doi.org/10.1109/CONFLUENCE.2019.8776621

Hartig, O., Bizer, C., & Freytag, J. C. (2009). Executing SPARQL queries over the web of linked data. In A. Bernstein (Ed.), *ISWC 2009. Lecture Notes in Computer Science*, 5823, 293–309. Springer. https://doi.org/10.1007/978-3-642-04930-9_19

Lim, J., Kim, B., Lee, H., Choi, D., Bok, K., & Yoo, J. (2022). An efficient distributed SPARQL query processing scheme considering communication costs in spark environments. *Applied Sciences*, 12(1), 122. https://doi.org/10.3390/app12010122

Masinter, L., Berners-Lee, T., & Fielding, R. T. (2005). Uniform resource identifier (URI): Generic syntax. *Network Working Group*. Fremont, CA.

Michel, F., Faron-Zucker, C. F., & Montagnat, J. (2016, April). A generic mapping-based query translation from SPARQL to various target database query languages. In *Proceedings of the 12th International Conference on Web Information Systems and Technologies (WEBIST 2016), Italy, 2* (pp. 147–158). https://doi.org/10.5220/0005905401470158

Montoya, G., Skaf-Molli, H., & Hose, K. (2017). The Odyssey approach for optimizing federated SPARQL queries. In *The Semantic Web–ISWC. Proceedings, Part I: 16th International Semantic Web Conference, Vienna, Austria, October 21–25, 2017, 16* (pp. 471–489). Springer International Publishing.

Naacke, H., & Curé, O. (2020). On distributed SPARQL query processing using triangles of RDF triples. *Open Journal of Semantic Web, 7*, 17–32.

Oguz, D., Ergenc, B., Yin, S., Dikenelli, O., & Hameurlain, A. (2015). Federated query processing on linked data: A qualitative survey and open challenges. *Knowledge Engineering Review*, 30(5), 545–563. https://doi.org/10.1017/S0269888915000107

Ouzzani, M., & Bouguettaya, A. (2004). Query processing and optimization on the web. *Distributed and Parallel Databases*, 15(3), 187–218. https://doi.org/10.1023/B:DAPD.0000018574.71588.06

Peng, C., Goswami, P., & Bai, G. (2020, September). A literature review of current technologies on health data integration for patient-centered health management. *Health Informatics Journal*, 26(3), 1926–1951. https://doi.org/10.1177/1460458219892387. Epub December 30, 2019. PubMed: 31884843

Priyatna, F., Alonso-Calvo, R., Parasio-Medina, S., Padron-Sanchez, G., & Corcho, O. (2015, December). R2RML-based access and querying to relational clinical data with morph-RDB. In *8th International Conference on Semantic Web Applications and Tools for Life Sciences, Cambridge United Kingdom* (pp. 142–151). https://ceur-ws.org/Vol-1546/paper_21.pdf

Saleem, M., Potocki, A., Soru, T., Hartig, O., & Ngomo, A. N. (2018). CostFed: Cost-based query optimization for SPARQL endpoint federation. *Procedia Computer Science*, 137, 163–174. https://doi.org/10.1016/j.procs.2018.09.016

Sequeda, J. F., Arenas, M., & Miranker, D. P. (April 2012). On directly mapping relational databases to RDF and OWL. In *Proceedings of the 21st International Conference on World Wide Web (WWW 2012)* (pp. 649–658). Association for Computing Machinery. https://doi.org/10.1145/2187836.2187924

Sima, A. C., Stockinger, K., de Farias, T. M., & Gil, M. (2019). Semantic integration and enrichment of heterogeneous biological databases. In M. Anisimova (Ed.), *Evolutionary Genomics. Methods in Molecular Biology*, 1910, 655–690. Humana Press. https://doi.org/10.1007/978-1-4939-9074-0_22

Squin, H. O. (2013). A traversal based query execution system for the web of linked data. In *Proceedings of the 2013 ACM SIGMOD International Conference on Management of Data 2013 Jun 22* (pp. 1081–1084). https://api.semanticscholar.org/CorpusID:10158072

Wu, H., Yamaguchi, A., & Kim, J. D. (2015). Dynamic join order optimization for SPARQL endpoint federation. In *SSWS@ISWC* (pp. 48–62). https://api.semanticscholar.org/CorpusID:16681995

Wylot, M., Hauswirth, M., Cudre-Mauroux, P., & Sark, S. (2018). RDF data storage and query processing schemes: A survey. *ACM Computing Surveys*, 51, 1–36. https://doi.org/10.1145/3177850

Xiao, G., Kontchakov, R., Cogrel, B., Calvanese, D., & Botoeva, E. (2018). Efficient handling of SPARQL optional for OBDA. In *The semantic Web–ISWC 2018. Proceedings, Part I 17: 17th International Semantic Web Conference, Monterey, CA, United States, October 8–12, 2018* (pp. 354–373). Springer International Publishing.

Yang, F., Crainiceanu, A., Chen, Z., & Needham, D. (2023, April 1). Cluster-based joins for federated SPARQL queries. *IEEE Transactions on Knowledge and Data Engineering*, 35(4), 3525–3539. https://doi.org/10.1109/TKDE.2021.3135507

Yannakis, T., Fafalios, P., & Tzitzikas, Y. (2018). Heuristics-based query reordering for federated queries in SPARQL 1.1 and sparql-ld. *arXiv preprint arXiv:1810.09780.*

Index

For Product Safety Concerns and Information please contact our EU
representative GPSR@taylorandfrancis.com
Taylor & Francis Verlag GmbH, Kaufingerstraße 24, 80331 München, Germany